# 3D Printing in Chemical Sciences
## Applications Across Chemistry

# *3D Printing in Chemical Sciences*

## *Applications Across Chemistry*

By

**Vipul Gupta**
*University of Tasmania, Australia*
*Email: Vipul.Gupta@utas.edu.au*

**Pavel Nesterenko**
*Lomonosov Moscow State University*
*Email: P.Nesterenko@phys.chem.msu.ru*

and

**Brett Paull**
*University of Tasmania, Australia*
*Email: Brett.Paull@utas.edu.au*

ROYAL SOCIETY
OF CHEMISTRY

Print ISBN: 978-1-78801-440-3
PDF ISBN: 978-1-78801-574-5
EPUB ISBN: 978-1-78801-766-4

A catalogue record for this book is available from the British Library



The Royal Society of Chemistry is a charity, registered in England and Wales, Number 207890, and a company incorporated in England by Royal Charter (Registered No. RC000524), registered office: Burlington House, Piccadilly, London W1J 0BA, UK, Telephone: +44 (0) 20 7437 8656.

Visit our website at www.rsc.org/books

Printed in the United Kingdom by CPI Group (UK) Ltd, Croydon, CR0 4YY, UK

# Preface

The impact of 3D printing on our everyday lives shows no signs of slowing, and indeed is likely to gain pace as the various printer technologies continue to develop and mature. From the manufacture of everyday plastic objects, components and consumables, to the printing of advanced integrated systems and devices, the time will soon be upon us (if we are not unknowingly already there), when a quick glance around the room will reveal that we are surrounded by 3D-printed objects. In fact, it's fair to say that in less than a decade, the evolution of 3D printing has rapidly transformed to a revolution, with uptake of this technology now so widespread, that it is already considered a routine manufacturing tool across many commercial sectors. Few design and manufacturing facilities across the globe would not have already invested in additive manufacturing, or else plan to do so very soon.

Both the evolution and the current revolution of 3D printing go hand-in-hand with advances in chemical science. At the very heart of 3D printing is the print material, the development of which sits upon the successes and developments of the past and current polymer and materials research communities. The printer technology itself will only go as far as printable materials permit, and parameters such as print speed and print resolution will in the end demand improved material properties and compositions. The properties and performance of the printed objects are of course also dependent upon the nature of the materials used, and significant interest in new composite and functional printable materials will remain for the foreseeable future.

This successful techno-chemical partnership has delivered new possibilities in how we design, manufacture, and indeed simply think about objects, and in particular has enabled us to learn more about structure–function relationships in 3D. Within the chemical sciences we stubbornly continue to sub-categorise our field into traditional disciplines of analytical, organic,

3D Printing in Chemical Sciences: Applications Across Chemistry
By Vipul Gupta, Pavel Nesterenko and Brett Paull

Published by the Royal Society of Chemistry, www.rsc.org

inorganic, and physical chemistry, together with the intersection with biochemistry, and indeed chemical education. To that end this book has been structured in a similar way, to review and consider the developments to-date under each of these classifications. The majority of the contents discussed within this book are relatively new developments, mostly dating between 2012 and 2017. The impact of 3D printing within the chemical sciences is moving at such speed that at the time of writing new publications are emerging at a rate of more than a dozen a week, so our commentary is aligned to a period in time, with a view to the future, which is already upon us.

Within this book we have attempted to cover as much of the research to-date as possible in a limited window. There is little doubt that some of the activity from the very recent past has been missed given the rate and scope of ongoing research. However, what is provided is a demonstration of the breadth and impact of this new exciting and powerful technology, which we hope will give some inspiration to the chemical science community, together with other scientists and engineers, educators, and all those working within related fields with an eye to the future.

Vipul Gupta, Pavel Nesterenko, and Brett Paull

This book is dedicated to Arushi, Ekaterina and Dmitry, and Eliza and Fleur.

# Contents



3D Printing in Chemical Sciences: Applications Across Chemistry
By Vipul Gupta, Pavel Nesterenko and Brett Paull

Published by the Royal Society of Chemistry, www.rsc.org

CHAPTER 1

# *An Introduction to 3D Printing*

## 1.1 History

It is certainly true to say that the history of additive fabrication technology, now commonly referred to as 3D printing, is rather complex to unravel. The precise definition of what constitutes 3D printing has much to do with this confusion, and so to begin this book we must present our concise definition, upon which we make all our observations and comments from this point forward. Techniques which fall under our definition of 3D printing include all production technologies that are based upon the automated and computer-controlled formation of three-dimensional objects from one or more starting materials, either liquid or solid, through an additive process based upon solidification. If we apply this definition, we can take the origin of 3D printing back beyond the common claims of an early 1980s origin, and step back to 1977, when Swainson and Kremer patented their proposed scheme for the creation of 'Three dimensional systems'. This was based upon the exposure of reactive monomer systems to the intersection of radiation beams, which following a direct or indirect polymerisation process, produced a sensible 3D object or solid structure.[1] A few years later in 1980, Hideo Kodama filed a patent application for a 3D printing process (for rapid prototyping), which was followed by a research article in 1981, detailing the work, entitled 'Automatic method for fabricating a three-dimensional plastic model with photo hardening polymer'.[2] The approach utilised a commercial liquid photo-hardening polymer known at the time as 'Tevista' (from Teijin Ltd.), which was a mixture of unsaturated polyester, acrylic ester, and styrene monomer cross-linkers, a polymerisation initiator, and a UV sensitiser. A xenon lamp and optical fibre coupled to a moving *x*/*y* plotter were the other essential elements to these very early

3D Printing in Chemical Sciences: Applications Across Chemistry
By Vipul Gupta, Pavel Nesterenko and Brett Paull

Published by the Royal Society of Chemistry, www.rsc.org

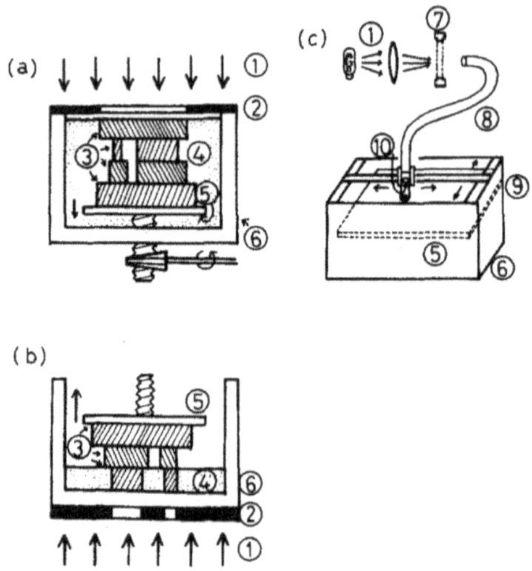


**Figure 1.1** Schematics of the 3D printers that were presented by Hideo Kodama: (a) build plate is moving into the liquid resin, (b) build plate is moving away from the liquid resin, and (c) raster with an optical fibre. 1 – ultraviolet rays, 2 – mask, 3 – solidified layers, 4 – liquid photo-hardening polymer, 5 – movable plate, 6 – receptacle, 7 – shutter, 8 – optical fibre, 9 – *XY* plotter, and 10 – optical lens. Reproduced from ref. 2 with the permission of AIP Publishing.

3D printers (Figure 1.1). However, the patent application by Kodama did not progress to a full patent, as the author failed to file the full patent application before the one-year deadline. Despite this, the published paper confirms this historical contribution and marks the beginning of an exponential growth in the field.

Only a few years after Kodama's work, in 1983, Charles (Chuck) Hull, who is widely regarded (particularly in North America) as the father of 3D printing, invented a stereolithographic apparatus (SLA), and subsequently secured a patent on the technology in 1986.[3] The development of this first stereolithography based printer marked the beginning of commercial 3D printing technology, and rightly earned Hull a place in the National Inventors Hall of Fame on May 21st, 2014.

The company '3D Systems' introduced the first commercial SLA (SLA-1) 3D printer in 1987, which began selling in 1988 after rigorous testing. This was the spark that ignited the fire, which throughout the 1990s saw several pioneers and their new alternative additive manufacturing processes developed, patented, and commercialised. These included Scott Crump, who patented fused deposition modelling (FDM) in 1992,[4] Deckard, Beaman, and Darrah, who together patented their technology for selective laser sintering (SLS) in 1992,[5] Sachs, Haggerty, Cima, and Williams, whose 1993 patent 'three-dimensional printing' formed the basis of what is now commonly referred to as '3D inkjet printing',[6] and finally, Feygin, Shkolnik, Diamond, and Dvorskiy, who patented laminated object manufacturing (LOM) in 1998.[7] SLA-1

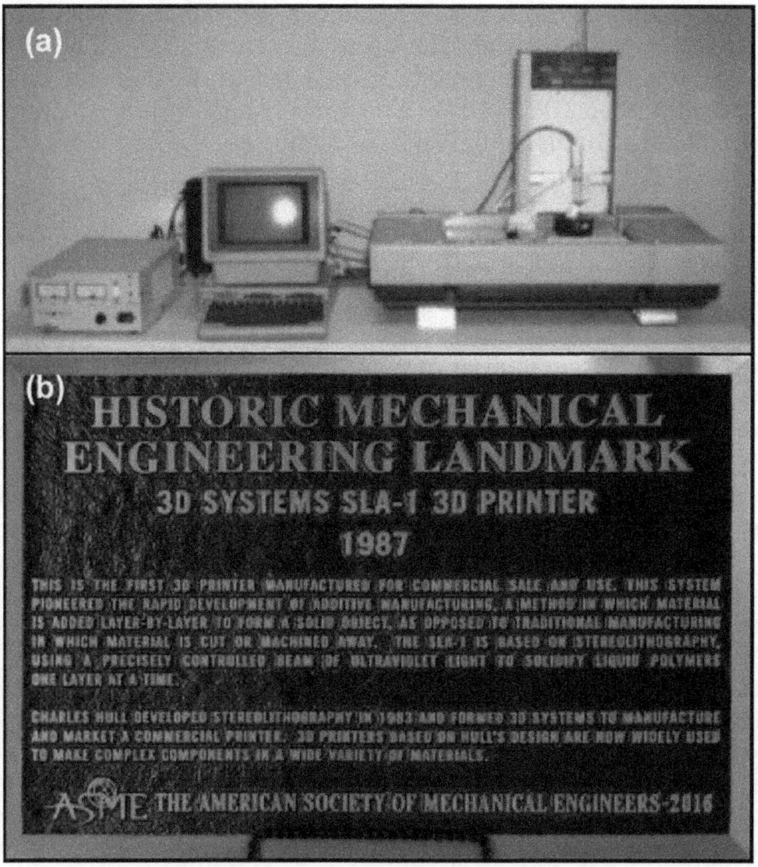


**Figure 1.2** (a) SLA-1 printer and (b) its enrolment as a historic mechanical engineering landmark.

earned recognition as an American Society of Mechanical Engineers Historic Mechanical Engineering Landmark (Figure 1.2), and the award citation read; "*This is the first 3D printer manufactured for commercial sale and use. This system pioneered the rapid development of additive manufacturing. A method in which material is added layer-by-layer to form a solid object, as opposed to traditional manufacturing in which material is cut or machined away. The SLA-1 is based on stereolithography, using a precisely controlled beam of ultraviolet light to solidify liquid polymers one layer at a time. Charles Hull developed stereolithography in 1983 and formed 3D Systems to manufacture and market a commercial printer. 3D printers based on Hull's design are now widely used to make complex components in a wide variety of materials.*"

Throughout the 1990s and 2000s, additive manufacturing gained its foothold as an attractive alternative to traditional manufacturing processes, and so quickly began to impact a wide spectrum of industrial, scientific, educational, and social activities. This gave birth to various new terms for much the same thing, such as rapid prototyping, rapid tooling, rapid casting, and rapid manufacturing. However, in 1993, Sachs *et al.* at the Massachusetts Institute of Technology, coined the term 'three-dimensional printing' and trademarked 3DP™, which was readily embraced globally as an umbrella term for all similar additive manufacturing processes, regardless of the field.

The full range of commercial 3D printers can be roughly divided into two major groupings. The first group, typically high-end and advanced systems, is geared towards high-quality industrial applications, such as aerospace and automotive parts, medical implants, and jewellery. The devices in the second class are typically of much lower cost and sophistication, and thus geared towards desktop-based applications, such as working prototypes, models, and small components and objects. The printers in the second category are sometimes referred to as 'concept modellers' and were developed as user-friendly and cost-effective systems. The range and diversity of these lower cost printers is increasing annually, with a great number of smaller specialist manufacturers emerging to meet the growing demand of small industry for 3D printers, together with the increasing masses of 'social' users, *e.g.* hobby and educational applications.

Growth in the user base has been facilitated by a number of innovative supporting activities and organisations. Internet-based community funding (so-called crowdsourcing) has been behind much of this activity, spawning a staggering diversity of free or low-cost hardware and software to new users. For example, the open-source RepRap project, developed by Adrian Bowyer, helped to increase worldwide access to 3D printers through the provision of plans and manuals for users to build their own printers. The RepRap project was launched in 2005 to create a 3D printer that could, for the most part, 3D print itself. In 2008, the first generation self-replicating printer was released, named Darwin. Since that time a large number of similar systems have emerged of increasing complexity and capability.

The historical journey mentioned above, plus other major commercial milestones that have contributed to the current state-of-the-art of 3D printing, since the origin of commercial 3D printing in 1986, are outlined in Figure 1.3. How these commercial products and technologies have been applied within, and impacted upon, the chemical sciences over this period is detailed throughout this book.

## 1.2 3D Print Files

The 3D printing process starts with the generation of a virtual 3D model of the desired object (Figure 1.4), which can be generated using any one of a number of free or commercial computer-aided design (CAD) programs, or *via* scanning data,[8] or from mathematical equations.[9] Currently, CAD programs are the predominant means to generate 3D models, and free libraries of designs and popular models can now be found readily throughout the internet. Designs can be easily extended, developed and customised by the

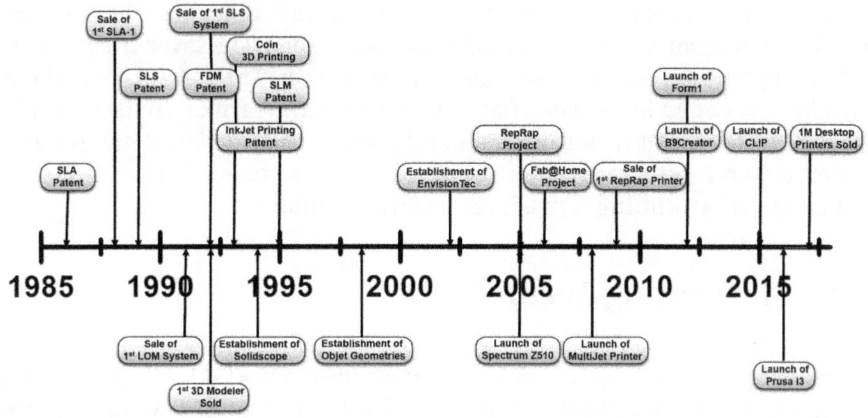


**Figure 1.3** Historical commercial milestones in the field of 3D printing from 1985 to 2017.

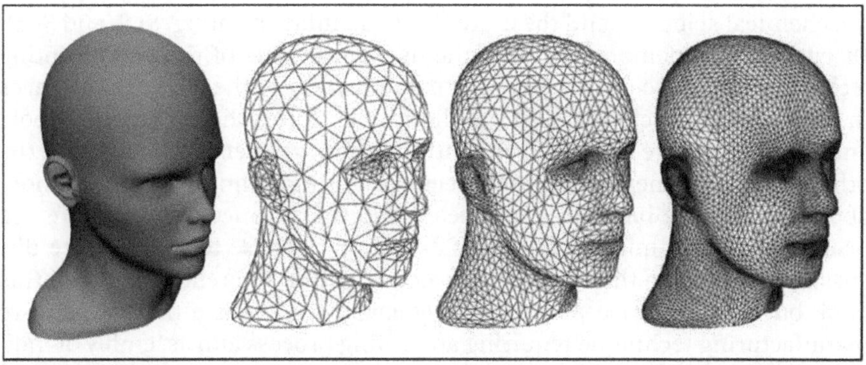

**Figure 1.4** Computer-aided design and tessellation (with triangles) of a human face.

user, without a great deal of technical expertise. CAD programs range in complexity, from those for the social or recreational user, to those designed for professionals. 3D models generated with CAD programs hold information about the shape and size of an object, however, these models cannot be interpreted directly by a 3D printer. Currently, the most commonly used and *de facto* method for preparing CAD models for 3D printing is by converting them to a standard tessellation language (STL) file.[10] The STL file is created by tessellating a 3D model with small triangles (facets) (Figure 1.4) and recording the vertices and components of the unit normal vector of the triangles in ASCII or binary formats. STL files store information about the object's geometry, however, they lack any information on the object's colour. Hence, STL files can only be used for monochromatic printing. Multi-colour printing requires virtual reality modelling language (VRML) files, which store both structure and colour information.

Each 3D printer is accompanied by slicing software that slices the 3D models obtained from STL or VRML files into 2D layers. The layer thickness is defined by the step size (z-resolution) of the printer. The information about each layer is coded in a G-code file, which directs the various components of the 3D printer (print heads, lasers, build platforms, and extruders) to move in defined and synchronised paths. Thus, 3D printers build a three-dimensional object by printing it in a layer-by-layer fashion.

## 1.3 3D Printing Techniques

In the three decades since the origin of 3D printing, the following 3D printing techniques have been developed: stereolithography (SL), inkjet printing, fused deposition modelling (FDM), selective laser sintering (SLS), selective laser melting (SLM), direct-ink writing, two-photon polymerisation (2PP), laminated object manufacturing (LOM), solid ground curing (SGC), and ballistic particle manufacturing (BPM). For the sake of brevity here we will focus more attention upon 3D printing techniques that have been widely applied in the chemical sciences, and these are FDM, SL, inkjet printing, SLS, and SLM. Discussion and demonstration of the use and impact of these 3D printing technologies on the advancement and application of the chemical sciences in all its various disciplines and sub-disciplines, forms the basis of this book, and as such they are described and introduced in greater detail below. Of the other techniques mentioned, the direct-ink writing technique has seen more application in the biological sciences,[11] and the advanced 2PP systems are primarily used in micro- and nanofabrication.[12] These techniques are discussed as and when they occasionally occur within different chapters in this book but are otherwise not extensively covered. LOM is a pseudo-additive manufacturing technique requiring an etching process and assembly of individual layers,[13] and techniques such as BPM and SGC are now rarely used and will not be discussed further.

### 1.3.1 Stereolithography (SL)

The SL technique is based on the spatially controlled photopolymerisation of liquid resins using a laser. There are two major forms: direct laser writing and mask-based laser writing. Exact classification is based on the method used for laser exposure and the orientation of the laser to the resin reservoir.

In direct laser writing, the laser directly rasters on the surface of a resin reservoir (Figure 1.5(a)). The resin exposed to the laser is polymerised to form the solid components of the layer, and the unexposed resin acts as a support material for the consecutive layers. Completion of each layer is followed by the lowering of the build platform by a specified step size to replenish the top layer with fresh resin. The process is repeated for all the consecutive layers

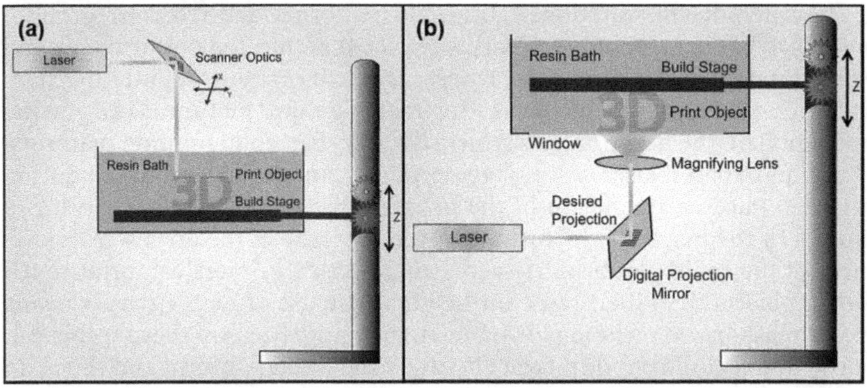


**Figure 1.5** Schematic of the stereolithography printer: (a) direct laser writing printer, and (b) mask-based laser writing printer.

to build a 3D object. The common name for direct laser writing is SLA printing (from *stereolithographic apparatus*), although it is also known as free surface printing.[14–17]

For mask-based laser writing, a digital light projection (DLP) and a digital projection mirror casts an image on the surface of the resin reservoir (Figure 1.5(b)). The resin exposed to the pixels of light is polymerised to form the solid components of the layer, and the resin exposed to the dark pixels remains unpolymerised. In mask-based laser writing printers, the completion of each layer is followed by the raising of the build platform (as opposed to the lowering in the direct laser writing). The raising of the platform replenishes the bottom of the resin reservoir with fresh resin for consecutive layers. The common name for mask-based laser writing is DLP printing, although it is also known as constrained surface printing[18] or 'bat' configuration printing.[19] In both direct and mask-based laser writing printers, the printed objects are detached from the build platform, and any non-polymerised resin remaining in the object is recognised to be removed using a suitable solvent.

The *z*-resolution of both types of SL printers is dependent on the step size of the build platform, which can be less than 10 µm.[20] The resolution of direct laser writing in the *x*- and *y*-direction is limited by the laser spot size, which is typically between 80 µm to 250 µm.[14,20–22] In mask-based laser writing, the *x*- and *y*-resolution is dependent on the pixel size, which can be much smaller than the laser spot size. Direct laser writing suffers from longer print times because the laser rastering process is much slower than the image casting process. The resolution of both types of SL-printed objects is also dependent on the resin chemistry and the characteristics of the light source. The thickness of the cured resin layer ($C_D$) can be determined by the equation: $C_D = D_P \ln(E/E_C)$,[23] where, $D_P$ is the light penetration depth, $E$ is the intensity of the light source, and $E_C$ is the critical energy of the resin.[23]

The different configurations of the build platforms for direct laser (ground) and mask-based laser writing (bat) is reflected within the performance and capabilities of each printer type. Direct laser writing exerts a lower mechanical force on the print object by resting it on a ground platform. The gravitational pull on the heavy objects attached to the 'bat' configuration platforms of the mask-based laser writing type printers can cause them to fall during printing. Further, the height of the printed object in direct laser writing is limited by the height of the resin reservoir, because of the downward movement of the build platform. However, mask-based printers can print much longer objects than their reservoir heights. The use of high viscosity resins in the mask-based technique is often challenging, because these resins can fail to flow as uniform thin layers between the build platform and the resin reservoir.[24]

Mask-based laser writing is often preferred over direct laser writing, as it offers higher print speeds and greater resolution (by which we mean smaller feature production capability). Conventional mask-based printers cure the entire layer of resin between the build platform and the reservoir floor, which creates a challenge when separating the printed object from the floor.[13] This challenge is overcome with the relatively recent advent of continuous liquid interface production (CLIP) technology, which uses an oxygen permeable membrane below the resin to form a 'dead zone' at the reservoir floor. The presence of oxygen at the window and reservoir interface prevents photopolymerisation and allows easy separation of the cured layers from the reservoir floor. CLIP printers have up to 100 times the print speeds of other 3D printers.[25]

The high resolution of SL printing in general ensures accuracy and a smooth surface finish on the printed objects. SL printing also does not require additional support material, because the uncured resin acts as a support material for each subsequent layer. However, most SL print resins are acrylic and epoxide monomers, which become brittle and can shrink after polymerisation.[21,26] Multi-material SL printers are now being developed which presents an exciting new capability,[27–29] however their practical use is presently limited as they require either repeated swapping or repeated cleaning of the resin reservoirs, which is time consuming and costly compared to alternative multi-material printers. Finally, SL-printed objects often require a post-curing step to ensure complete polymerisation of the resin.

### 1.3.2 Selective Laser Sintering and Melting (SLS/M)

SLS and SLM printers are based on the process of the selective fusion of powder particles by laser sintering and melting, respectively (Figure 1.6). The printing process starts by exposing a layer of powder particles to a high-energy laser raster. The laser increases the local bed temperature to either slightly below the melting point (for sintering) or above the melting point (for melting) of the powder particles. The sintered or melted powder particles fuse together to form the solid components of the layer, and the unfused

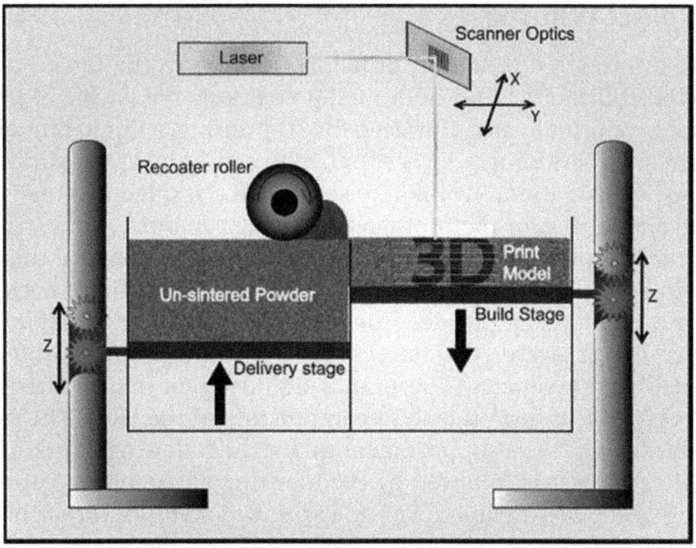


**Figure 1.6** Schematic of the selective laser sintering/melting printer.

particles act as a support material for the following layers. The completion of each layer is followed by the lowering of the build platform and the raising of the delivery platform by a specified step size, which defines the layer thickness. This allows the next layer of powder particles to be rolled down to repeat the same process for the consecutive layers. The printed object is obtained from the powder bed, and it is post-processed to remove the unbound particles by dusting off with compressed air, blasting with sand particles or glass beads, or by taping and sonication.

SLS is typically used with polymers because they do not have a particularly sharp melting point, whereas SLM is applied to metals as they have a much more defined melting point. SLS-printed objects are typically more porous than those printed by SLM. The resolution of both SLS and SLM printers is dependent both on particle and laser spot size.

As mentioned above, SLS and SLM printers can print with a variety of materials, such as polymers, metals, ceramics, and glass. The cost of SLS and SLM printers and print materials is the highest of all 3D printers and print materials, and so is rarely seen outside of industrial or research environments. Individual layers in SLS and SLM-printed objects are thermally fused together, rendering the objects high purity (based upon the starting materials) and of high mechanical strength.[30] However, on the negative side, the high thermal stress endured during the fusion process can result in the deformation and shrinkage of the printed objects, particularly for larger objects of greater thermal mass.[31] In addition, SLS and SLM printing processes are relatively slow, and the printed objects can require tedious post-processing steps to remove unbound powder particles from inaccessible internal zones.

### 1.3.3 Inkjet Printing

3D inkjet printers are based on principles similar to the most commonly used 2D inkjet printers. They both modulate the spatial jetting of inks from their respective print heads. However, for 3D printing, a liquid binder, a photopolymer, or a thermomaterial is supplied in a layer-by-layer fashion. Based on the type of inks used, 3D inkjet printers are classified as binder jetting, photopolymer jetting, or thermomaterial jetting printers.

Binder jetting printers are powder-based systems; and hence their operation is similar to the SLS/M printers (Figure 1.7(a)). The print process begins by rolling a fine layer of powder from the powder bed onto the build platform. The powder layer is then selectively exposed to a binder liquid as per the desired layer geometry. The binder liquid binds the required powder particles together to form the solid components of the layer. The unbound powder particles act as a support material for the following layers. The completion of each layer is followed by the lowering of the build platform and the raising of the delivery platform by a specified step size (layer thickness). This allows the next layer of powder particles to be rolled down to repeat the same process for the following layers. Like the SLS/M printers, the printed object is obtained from the powder bed, and the unbound powder particles are removed.

Binder jetting printers offer an advantage in that they can print with a variety of materials (as listed in Table 1.1), using different binders for different materials. The resolution of binder jetting printers is primarily dependant on the size of the powder particles (usually a few hundred micrometres). The typical resolution of a binder jetting printer is 50–100 μm in the $x$- and $y$-direction and 200 μm in the $z$-direction.[32] The material properties of the binder dominate the physical and chemical properties of the printed objects. It is not unusual that binder jetting printed objects contain incompletely bound powder particles,[33] which results in loss of material homogeneity, can

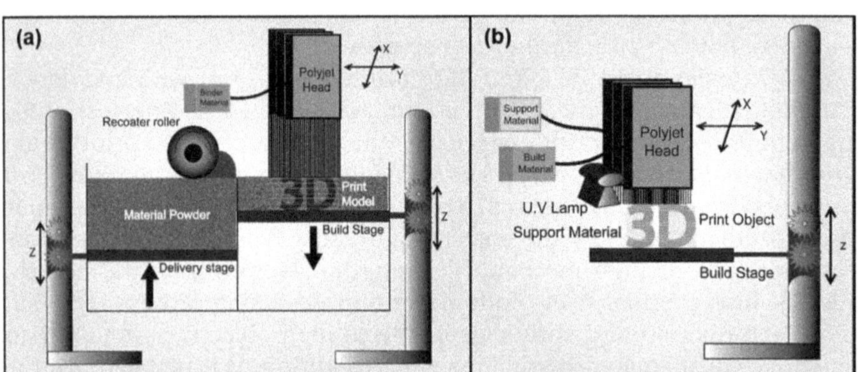


**Figure 1.7** Schematic of the inkjet printer: (a) binder jetting printer, and (b) photopolymer jetting printer.

**Table 1.1** Properties of the stereolithography, inkjet printing, selective laser sintering, selective laser melting, and fused deposition modelling 3D printing techniques.

| | Stereolithography (SL) | Inkjet printing | Selective laser sintering and melting (SLS/M) | Fused deposition modelling (FDM) |
|---|---|---|---|---|
| Synonyms | Digital light projection (DLP), stereolithographic apparatus (SLA) | PolyJet printing/modelling, multiJet printing/modelling (MJM), drop on demand (DOD), thermo jetting, binder jetting | Direct metal laser sintering (DMLS), laser curing | Fused filament fabrication (FFF), thermoplastic extrusion, fused layer modelling/manufacturing (FLM) |
| Principle | Photopolymerisation of a liquid resin by direct laser writing or mask-based laser writing to form each individual layer. | Spatially controlled jetting of a binder, a photopolymer, or a thermomaterial, which bonds powder particles together, photopolymerises, or solidifies, respectively, to form each individual layer. | Laser selectively sinters or melts the powder particles, which fuses them together to form each individual layer. | Thermal extrusion of thermoplastics, which deposits and solidifies forming each individual layer. |
| Types | (1) Direct laser writing (free surface printing)<br>(2) Mask-based laser writing (constrained surface printing) | (1) Binder jetting<br>(2) Photopolymer jetting<br>(3) Thermomaterial jetting | (1) Laser sintering<br>(2) Laser melting | |
| Commercially available materials | Polyacrylate composites and acrylonitrile-butadiene-styrene (ABS) | **Binder jetting**: Ceramic, silica sand, stainless steel alloys, nickel alloys, iron alloys, chromium alloys, cobalt alloys, zircon, soda lime glass, and tungsten.<br>**Photopolymer jetting**: Polyacrylate composites and acrylonitrile-butadiene-styrene (ABS)<br>**Thermomaterial jetting**: Wax and poly(methyl methacrylate) (PMMA). | **SLS**: Nylon, polyamide, polyether ether ketone (PEEK), polyetherimide (PEI), polycarbonate, polystyrene, acrylonitrile-butadiene-styrene (ABS), polyester, polypropylene, silica sand, thermoplastic polyurethane (TPU), and polyamide and nylon composites.<br>**SLM**: Aluminium, cobalt–chromium alloys, copper alloys, gold, stainless steel, nickel alloys, silver, and titanium alloys. | Polylactic acid (PLA), polycarbonate, polystyrene, acrylonitrile-butadiene-styrene (ABS), nylon, polyphenylsulfone (PPSF), polyetherimide (PEI), polyamide, kevlar, acrylonitrile styrene acrylate, polyether ether ketone (PEEK), poly(methyl methacrylate) (PMMA), polyoxymethylene, polypropylene, poly(vinyl alcohol) (PVA), polyethylene terephthalate (PET), thermoplastic elastomers, polyurethanes, wax, and PLA composites (with carbon fibres, metals, magnetic irons, *etc.*) |

(*continued*)

**Table 1.1** (*continued*)

| | Stereolithography (SL) | Inkjet printing | Selective laser sintering and melting (SLS/M) | Fused deposition modelling (FDM) |
|---|---|---|---|---|
| Advantages | **Direct laser writing**: Exerts a low mechanical force on the building parts<br>**Mask-based laser writing**: Enables high throughput printing and the print height is independent of the reservoir height.<br>**Common**: Offers high-resolution printing, the process is free from solid support materials, and printed parts can have high optical transparency. | **Binder jetting**: Offers a wide variety of print materials.<br>**Photopolymer jetting**: Enables high throughput printing, offer multi-material and multi-colour printing capabilities, and allow fabrication of complex structures.<br>**Thermomaterial jetting**: Easy operation. | Availability of a wide range of polymeric and metallic print materials, high purity and mechanical strength of the printed objects, and additional support material is not required. | Offers multi-material and multi-colour printing capabilities, low-cost printers and materials, availability of a wide variety of print materials, minimal post-processing requirements, and the availability of miniaturised hand-held printing pens. |
| Limitations | **Direct laser writing**: Requires long print times and the print height is dependent on the reservoir height.<br>**Mask-based laser writing**: Exerts a high mechanical force on the building parts, limits the flow of high viscosity resins, and the cured layer often sticks to the reservoir's bottom surface.<br>**Common**: Suffers from the limited availability of print materials, multi-material printing capabilities are limited, and printed objects often require a post-photocuring step. | **Binder jetting**: Offers a low-resolution print capability, physical and chemical properties of the printed parts are dictated by the binder, and printed parts suffer from material heterogeneities and bulk defects.<br>**Photopolymer jetting**: Suffers from limited availability of print materials, limited material customisation opportunities, and printed objects often require tedious post-processing steps.<br>**Thermomaterial jetting**: Printed objects are mechanically weak and have poor surface finish. | The high cost of printers and materials, shrinkage and deformation of the printed objects, and a low throughput printing. | Printed objects suffer from material heterogeneities, bulk defects, low optical transparency, and staircase and chordal surface defects. It limits the fabrication of complex structures. |

cause bulk defects, reduce mechanical strength, and reduced optical clarity. Objects produced using this approach also require extensive post-processing steps, such as thermal curing, infiltration with hardeners/sealants, and/or coating.

Photopolymer jetting printers jet liquid build and support resins from individual print heads, which are then immediately exposed to UV light for photopolymerisation to generate the solid and hollow sections, respectively, of the individual layers (Figure 1.7(b)). The build resin is polymerised into a solid material, and the support resin into a gel-like material. The completion of each layer is followed by the lowering of the build platform to repeat the same process for the following layers. The printed object is detached from the platform, and the support material is removed by washing with a suitable solvent. The photopolymer jetting technique is commonly known as PolyJet printing.

The typical resolution of a PolyJet printer is 40 μm in the *x*- and *y*-direction and 16 μm in the *z*-direction,[34] giving higher resolution than binder jetting printers. The PolyJet print speed is also much higher. PolyJet printers can easily employ multiple print heads for multi-material and multi-colour 3D printing. The use of dense support structures in PolyJet printers allows the successful fabrication of highly complex architectures. PolyJet-printed objects are mechanically more robust and optically more transparent compared to the binder jetting printed objects. However, PolyJet print materials are mostly limited to polyacrylate composites. Moreover, the print materials cannot be easily customised, as use of any uncertified print material generally voids the printer warranty and poses a risk of blocking the print heads. PolyJet-printed objects usually require long and tedious post-processing steps to remove the dense support material, particularly when used to print objects of complex geometries and small internal channels and voids.[35]

Thermomaterial jetting printers jet a molten wax-like material, which solidifies on the build platform to produce each layer. These printers use the same material to generate both the build and support structures. The use of identical support and build materials makes the support removal process difficult and results in a poor surface finish. Moreover, thermomaterial jetting printed objects are mechanically weak, because of the use of soft wax-like print materials. These limitations have restricted the use of thermomaterial jetting printers in the chemical sciences.

### 1.3.4 Fused Deposition Modelling (FDM)

The FDM technique was introduced in 1992, and has now become the most widely used 3D printing technique.[33] It is based on the extrusion and layer-by-layer deposition of thermoplastics (Figure 1.8). Thermoplastic filaments are fed through rollers in the print heads followed by thermal extrusion through fine nozzles and deposition from the print heads onto the build platform. Each layer is composed of build and support components.

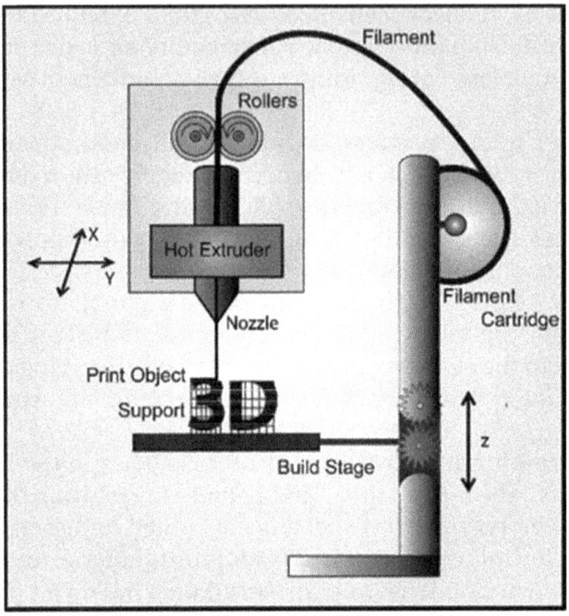


**Figure 1.8** Schematic of the fused deposition modelling printer.

Multiple extruders can be used to deposit one or more materials for each build and support component. The completion of each layer is followed by the lowering of the build platform to repeat the same process for all the consecutive layers to build a three-dimensional object. The printed object is mechanically detached from the platform, and any support structure is either broken off, dissolved, or washed off with a suitable solvent.

FDM printers and print materials are usually the cheapest 3D printing option available, and as such have become the printer of choice for the social user.[36] A wide variety of FDM filaments are now produced, and the filament material and its properties can be easily customised. Recently, there has been a marked increase in the availability of functional print materials for FDM printers, which has impacted upon the types of applications for this type of printer. For printed objects, the wall thickness and infill density can be independently controlled to modulate the appearance and mechanical properties of the object. The ability to use multiple extruders allows multi-material and multi-colour 3D printing. FDM-printed objects usually do not require rigorous post-processing steps, because the support structures, if any, can be easily broken off. FDM printers are based on a simple filament extrusion set-up, which allows system miniaturisation; with miniaturised FDM pens already available in the market.[37]

However, there are currently a few limitations to FDM printing. The $x$- and $y$-resolution of FDM printers is dependent on the extrusion diameter, which is usually between 200 and 800 μm. This low-resolution results in staircase and chordal surface defects, and high surface roughness.[38] The low-resolution and the absence of dense support materials mean that complex structures cannot be fabricated.[39] FDM-printed objects usually have material heterogeneities, bulk defects, and low-pressure resistant seals because of poor bonding between adjacent layers,[13] and they also have low optical transparency because of the high surface roughness and heterogeneities in the layers.[40]

Table 1.1 summarises the key characteristics and the commercially available materials for the SL, inkjet printing, SLS/M, and FDM 3D printing techniques.

## 1.4 The 3D Revolution

As mentioned above, 3D printing was introduced in the early 1980s. Since then it has seen enormous growth and impact, the like of which is seen with relatively few new technologies. The technology itself is of course very much at its core a development based upon materials science. Photocurable polymers, thermoplastics, and metallic micro-particles, are all from the chemistry and materials laboratories of the past (and present), with new functional 'printable' materials emerging from chemistry laboratories today to continue this developing capability.

However, this revolutionary technology only really began to impact upon the chemical sciences relatively recently, perhaps just within the past decade. To demonstrate this we can review the scientific literature (although this is not the only measure we could use). Figure 1.9 classifies the reports highlighted in this book according to their year of publication, and clearly shows an exponential increase in the use of 3D printing in the chemical sciences since 2011. The 3D printing market observed a steep growth from 2012 ($288 million) to 2013 ($2.5 billion).[41] The projected value for the 3D printing market in 2022 is $33.6 billion with a compound annual growth rate of 27.3% from 2017 to 2022.[42] In the 2013 State of the Union address, President Barack Obama highlighted the importance of 3D printing by stating, "*3D printing has the potential to revolutionize the way we make almost everything.*"

In just three decades since its introduction, 3D printing has been extensively used in various fields such as biotechnology and life sciences,[43,44] engineering,[45] robotics,[46] arts,[47] and food sciences.[48] The use of 3D printing is now rapidly expanding to all areas of research and manufacturing. This book discusses the use of 3D printing in all major branches of the chemical sciences: analytical, pharmaceutical, physical, and synthetic chemistry, biochemistry, and chemical education. The relative popularity of different 3D printing techniques in different branches of chemistry is indicated in Figure 1.10, with the inexpensive and easy to use FDM being the most popular.

The popularity of 3D printing can be attributed to its many advantages over conventional fabrication techniques. Conventional lithographic techniques require expensive facilities,[38] whereas the low cost of 3D printers and the minimum wastage of the print materials make 3D printing a highly cost-effective technique. Conventional techniques require extensive user

**Table 1.2** Commonly used 3D printers in chemical sciences and their printing technique, resolution, and approximate cost.

| 3DP technique | Printer | | Highest resolution | Approximate cost (USD) |
|---|---|---|---|---|
| | Manufacturer | Model | | |
| FDM | Stratasys | Dimension 1200es printer | 254 μm layer thickness | 19 000 |
| | | Dimension Elite | 178 μm layer thickness | 32 000 |
| | | Dimensions SST 768 | 254 μm layer thickness | 25 000 |
| | | uPrint SE Plus | 254 μm layer thickness | 20 000 |
| | 3D Printing Systems | UP Plus | 150 μm layer thickness | 1000 to 2000 |
| | | UP Plus 2 | 150 μm layer thickness | 1000 to 2000 |
| | 3Dfactories | EASY3DMAKER | 80 μm layer thickness | 3200 |
| | | Profi3Dmaker | 80 μm layer thickness | 6800 |
| | Makerfarm | Prusa i3 | 50 μm layer thickness | 270 |
| | | RepRap | 100 μm layer thickness | 700 |
| | Felix | 1.5 | 50 μm layer thickness | 1000 to 2000 |
| | | 3 | 50 μm layer thickness | 1000 to 2000 |
| | | 3.1 | 50 μm layer thickness | 1000 to 2000 |
| | MakerBot | Replicator 2X | 100 μm layer thickness | 2500 |
| | 3D Systems | 3D Touch | 125 μm layer thickness | 3000 |
| | Baoyan Automation | HOFI-X1 | 200 μm layer thickness | 350 |
| | Zmorph | Zmorph2.0SX | 50 μm layer thickness | 2700 |
| | Deltaprintr | Delta | 100 μm layer thickness | 500 |
| | Ultimaker | 2 | 20 μm layer thickness | 2500 |
| | ROVA | 3D | 10 μm layer thickness | 7000 |
| | Cliever | CL1 Black | 100 μm layer thickness | 1500 |
| | GTMax | Graber i3 | 100 μm layer thickness | 700 |
| | Aurora | Z-603S | 100 μm layer thickness | 800 |
| | MakerMex | MM1 Beta | 20 μm layer thickness | 1000 to 2000 |
| | Makers Tool Works | MendelMax 2 | 100 μm layer thickness | 1600 |
| | Fusion3 | F306 | 20 μm layer thickness | 4600 |
| | Flashforge | Creator Pro | 100 μm layer thickness | 900 |
| | Velleman | K8200 | 200 μm layer thickness | 450 |

| | | | | |
|---|---|---|---|---|
| SL | 3D Systems | Projet 1200 | 43 × 43 × 30 μm (*X*,*Y*,*Z*) | 4900 |
| | | Projet 7000 HD | 30 × 30 × 50 μm (*X*,*Y*,*Z*) | >250 000 |
| | | Projet 3510 HD | 34 × 34 × 29 μm (*X*,*Y*,*Z*) | 50 000 to 100 000 |
| | envisionTEC | Prefactory Desktop Aureus | 18 × 24 × 25 μm (*X*,*Y*,*Z*) | 10 000 to 50 000 |
| | | Micro Plus HD | 60 × 60 × 25 μm (*X*,*Y*,*Z*) | 10 000 to 50 000 |
| | | Ultra 3SP | 50 × 50 × 50 μm (*X*,*Y*,*Z*) | 10 000 to 50 000 |
| | Formlabs | Form 1+ | 25 μm layer thickness | 3000 |
| | | Form 2 | 25 μm layer thickness | 4000 |
| | Asiga | PicoPlus 27 | 27 × 27 × 1 μm (*X*,*Y*,*Z*) | 7000 |
| | B9 Creator | V1.2HD | 30 × 30 μm (*X*,*Y*) | 4600 |
| | Miicraft | | 56 × 56 × 56 μm (*X*,*Y*,*Z*) | 5000 to 10 000 |
| PolyJet printing | Stratasys | Objet Eden 350V | 42 × 42 × 16 μm (*X*,*Y*,*Z*) | 100 000 to 250 000 |
| | | Objet Eden 260VS | 42 × 84 × 16 μm (*X*,*Y*,*Z*) | 50 000 to 100 000 |
| | | Objet260 Connex | 42 × 42 × 16 μm (*X*,*Y*,*Z*) | 120 000 |
| | | Objet500 Connex | 42 × 42 × 16 μm (*X*,*Y*,*Z*) | 330 000 |
| | | Objet30 | 42 × 42 × 28 μm (*X*,*Y*,*Z*) | 10 000 to 50 000 |
| | | Objet350 Connex | 42 × 42 × 16 μm (*X*,*Y*,*Z*) | 100 000 to 250 000 |
| | 3D Systems | ProJet 3000 HD | 39 × 39 × 32 μm (*X*,*Y*,*Z*) | 90 000 |
| | | ProJet 3500 HDPlus | 34 × 34 × 16 μm (*X*,*Y*,*Z*) | 70 000 |
| SLM | Realizer | SLM 50 | 75 × 75 × 25 μm (*X*,*Y*,*Z*) | > 250 000 |
| SLS | EOS | Formiga P100 | 100 μm layer thickness | 175 000 |
| | Sintratec | Sintratec Kit | 100 μm layer thickness | 5800 |

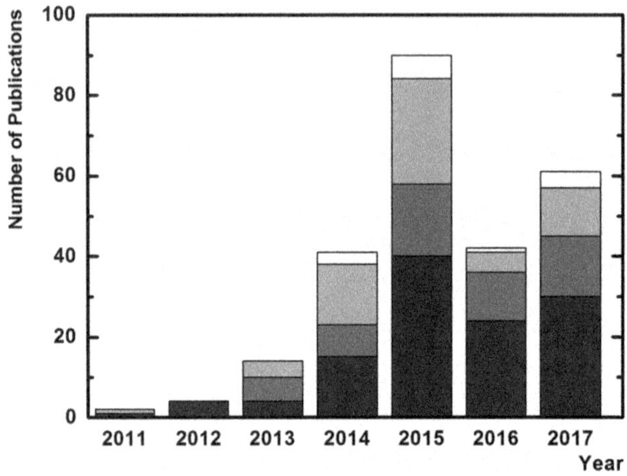


**Figure 1.9** Classification of the reports highlighted in this book as per their year of publication (2011 to 2017) and the printing technique that was used. Dark grey – FDM, grey – SL, light grey – inkjet printing, and white – SLS/M.

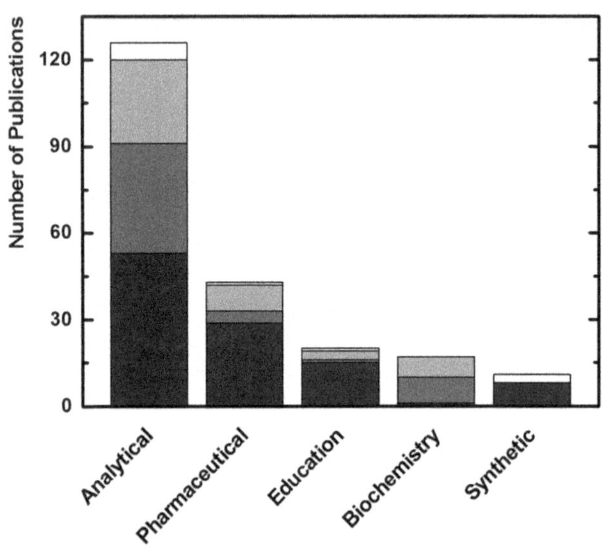


**Figure 1.10** Classification of the reports highlighted in the book as per their application in different branches of chemistry (analytical chemistry, pharmaceutical chemistry, chemical education, biochemistry, and synthetic chemistry) and the printing technique that was used. Dark grey – FDM, grey – SL, light grey – inkjet printing, and white – SLS/M.

training programs,[38] whereas most 3D printers are straightforward to operate and require minimal user training. Conventional fabrication techniques are limited to two-dimensions, with the fabrication of three-dimensionally complex objects by conventional techniques often impractical. 3D printing techniques allow easy and rapid development of three-dimensionally complex objects in a single-bodied configuration. Hence, 3D printing is the preferred method for fabricating three-dimensionally complex designs. Moreover, 3D printing allows rapid design customisation and offers a 'fail fast and often' approach,[49] which is an invaluable asset for research and development. 3D printers offer unique capabilities for on-site and remote production. The low-cost and desktop size of most of printers allow their installation at local sites. The autonomous behaviour of 3D printers also makes them ideal candidates for operation in remote settings – NASA is even developing printers for use in space.[50] 3D print files that can be customised or directly used for printing can be easily shared through online sharing platforms.[51]

Today, a variety of 3D printers are available on the market, offering different capabilities and suiting different applications. A list of 3D printers that are commonly used in chemical sciences has been provided in Table 1.2 with their employed 3D printing technique, highest available resolution, and approximate cost.

## References

1. W. K. Swainson, *US Pat.*, 4041476, 1977.
2. H. Kodama, *Rev. Sci. Instrum.*, 1981, **52**, 1770–1773.
3. C. W. Hull, *US Pat.*, 4575330, 1986.
4. S. S. Crump, *US Pat.*, 5121329, 1992.
5. C. R. Deckard, J. J. Beaman and J. F. Darrah, *US Pat.*, 5155324, 1992.
6. E. M. Sachs, J. S. Haggerty, M. J. Cima and P. A. Williams, *US Pat.*, 5204055, 1993.
7. M. Feygin, A. Shkolnik, M. N. Diamond and E. Dvorskiy, *US Pat.*, 5730817, 1998.
8. N. J. Mankovich, D. Samson, W. Pratt, D. Lew and J. Beumer 3rd, *Otolaryngol. Clin. North Am.*, 1994, **27**, 875–889.
9. K. M. Griffith, R. de Cataldo and K. H. Fogarty, *J. Chem. Educ.*, 2016, **93**, 1586–1590.
10. A. Dolenc and I. Mäkelä, *Comput.-Aided Des.*, 1994, **26**, 119–126.
11. J. A. Lewis, *Adv. Funct. Mater.*, 2006, **16**, 2193–2204.
12. S. Wu, J. Serbin and M. Gu, *J. Photochem. Photobiol.*, 2006, **181**, 1–11.
13. A. K. Au, W. Huynh, L. F. Horowitz and A. Folch, *Angew. Chem., Int. Ed.*, 2016, **55**, 3862–3881.
14. M. N. Cooke, J. P. Fisher, D. Dean, C. Rimnac and A. G. Mikos, *J. Biomed. Mater. Res., Part B*, 2003, **64**, 65–69.

15. K.-W. Lee, S. Wang, B. C. Fox, E. L. Ritman, M. J. Yaszemski and L. Lu, *Biomacromolecules*, 2007, **8**, 1077–1084.
16. I. K. Kwon and T. Matsuda, *Biomaterials*, 2005, **26**, 1675–1684.
17. J. W. Lee, P. X. Lan, B. Kim, G. Lim and D.-W. Cho, *Microelectron. Eng.*, 2007, **84**, 1702–1705.
18. Y.-M. Huang, S. Kuriyama and C.-P. Jiang, *Int. J. Adv. Manuf. Technol.*, 2004, **24**, 361–369.
19. E. Rune, *US Pat.*, 2566443, 1951.
20. P. J. Bártolo, *Stereolithography: Materials, Processes and Applications*, Springer Science & Business Media, New York, 2011.
21. R. A. Harris, R. J. Hague and P. M. Dickens, *Int. J. Mach. Tools Manuf.*, 2004, **44**, 59–64.
22. D. W. Hutmacher, M. Sittinger and M. V. Risbud, *Trends Biotechnol.*, 2004, **22**, 354–362.
23. P. Jacobs, *Rapid Prototyping and Manufacturing: Fundamental of Stereolithography*, Society of Manufacturing Engineers, Dearborn, Michigan, 1st edn, 1992.
24. Y.-M. Huang, S. Kuriyama and C.-P. Jiang, *Int. J. Adv. Manuf. Technol.*, 2004, **24**, 361–369.
25. J. R. Tumbleston, D. Shirvanyants, N. Ermoshkin, R. Janusziewicz, A. R. Johnson, D. Kelly, K. Chen, R. Pinschmidt, J. P. Rolland and A. Ermoshkin, *Science*, 2015, **347**, 1349–1352.
26. F. P. Melchels, J. Feijen and D. W. Grijpma, *Biomaterials*, 2010, **31**, 6121–6130.
27. R. Holtrup, *Design and Construction of a Multi-material 3D DLP Printer*, I. design, University of Twente, Twente, 2015.
28. C. Zhou, Y. Chen, Z. Yang and B. Khoshnevis, presented in part at the Annual solid freeform fabrication symposium, Austin, TX, 2011.
29. R. Wicker, F. Medina and C. Elkins, *US Pat.*, 7556490B2, 2009.
30. V. Gupta, M. Talebi, J. Deverell, S. Sandron, P. N. Nesterenko, B. Heery, F. Thompson, S. Beirne, G. G. Wallace and B. Paull, *Anal. Chim. Acta*, 2016, **910**, 84–94.
31. R.-J. Wang, L. Wang, L. Zhao and Z. Liu, *Int. J. Adv. Manuf. Technol.*, 2007, **33**, 498–504.
32. A. Pfister, R. Landers, A. Laib, U. Hübner, R. Schmelzeisen and R. Mülhaupt, *J. Polym. Sci., Part A: Polym. Chem.*, 2004, **42**, 624–638.
33. B. C. Gross, J. L. Erkal, S. Y. Lockwood, C. Chen and D. M. Spence, *Anal. Chem.*, 2014, **86**, 3240–3253.
34. B. J. de Gans, P. C. Duineveld and U. S. Schubert, *Adv. Mater.*, 2004, **16**, 203–213.
35. K. B. Spilstead, J. J. Learey, E. H. Doeven, G. J. Barbante, S. Mohr, N. W. Barnett, J. M. Terry, R. M. Hall and P. S. Francis, *Talanta*, 2014, **126**, 110–115.
36. G. W. Bishop, J. E. Satterwhite, S. Bhakta, K. Kadimisetty, K. M. Gillette, E. Chen and J. F. Rusling, *Anal. Chem.*, 2015, **87**, 5437–5443.
37. 3Doodler, https://intl.the3doodler.com/, (accessed 29 August, 2018).

38. J. L. Moore, A. McCuiston, I. Mittendorf, R. Ottway and R. D. Johnson, *Microfluid. Nanofluid.*, 2011, **10**, 877–888.
39. V. Gupta, P. Mahbub, P. N. Nesterenko and B. Paull, *Anal. Chim. Acta*, 2018, **1005**, 81–92.
40. G. W. Bishop, J. E. Satterwhite-Warden, I. Bist, E. Chen and J. F. Rusling, *ACS Sens.*, 2015, **1**, 197–202.
41. S. M. Oskui, G. Diamante, C. Liao, W. Shi, J. Gan, D. Schlenk and W. H. Grover, *Environ. Sci. Technol. Lett.*, 2015, **3**, 1–6.
42. O. Research, Global 3D Printing Market By Type, Technology Used, Process, Industry, Geography, Trends and Forecast to 2022, https://www.reuters.com/brandfeatures/venture-capital/article?id=10871, (accessed 27 August, 2018).
43. S. M. Peltola, F. P. Melchels, D. W. Grijpma and M. Kellomäki, *Ann. Med.*, 2008, **40**, 268–280.
44. S. V. Murphy and A. Atala, *Nat. Biotechnol.*, 2014, **32**, 773–785.
45. P. Marks, *New Sci.*, 2011, **211**, 17–18.
46. N. W. Bartlett, M. T. Tolley, J. T. Overvelde, J. C. Weaver, B. Mosadegh, K. Bertoldi, G. M. Whitesides and R. J. Wood, *Science*, 2015, **349**, 161–165.
47. T. Allard, M. Sitchon, R. Sawatzky and R. Hoppa, 6th International Symposium on Virtual Reality, Archaelogy and Cultural Heritage, Pisa, Italy, 2005.
48. J. Sun, Z. Peng, W. Zhou, J. Y. Fuh, G. S. Hong and A. Chiu, *Procedia Manuf.*, 2015, **1**, 308–319.
49. C. I. Rogers, K. Qaderi, A. T. Woolley and G. P. Nordin, *Biomicrofluidics*, 2015, **9**, 016501.
50. K. Cooper and M. Griffin, *Microgravity manufacturing via fused deposition*, S. T. A. SAFETY Report NASA/TM-2003-212636, NASA, Huntsville, AL, 2003.
51. Thingverse, https://www.thingiverse.com/, (accessed 24 September, 2016).

CHAPTER 2

# 3D Printing of Micro- and Macro-fluidic Devices

## 2.1 Introduction

Within the chemical sciences over the past two decades or more, the study of microfluidics and related technologies has gained considerable momentum, as the scientific community has realised the significant potential benefits of working at the micro-scale. Over this period, there has been a wealth of impressive applications reported, covering a multitude of application areas, including for example, point-of-care diagnostics,[1–4] chip-based electrophoresis,[5,6] microreactions,[7–9] immunoassays,[10–13] particle synthesis,[14–17] tissue engineering,[18,19] drug discovery,[20] cellular analysis,[21–24] DNA analysis,[25–28] micro-RNA analysis,[29,30] and protein analysis.[31–34] Microfluidics (also somewhat erroneously (*or often optimistically*) commonly referred to as 'lab-on-a-chip') offers various well known advantages to the wider science community, such as the ability to work with small volumes (femtolitre to nanolitre), the ability to generate portable and miniaturised systems, and the capacity to integrate multiple processes and capabilities upon or within a single platform (*towards the elusive 'lab-on-chip'*).[35] However, fabrication of microfluidic devices, which are reliable, functional and reproducible, is by no means trivial, and has long restricted a mass movement towards this technology. For greater utilisation of the unique advantages and potential of microfluidic devices, user-friendly and low-cost (in both time and resources) fabrication tools are required.

Up until recently, microfluidic devices and platforms have been typically produced *via* techniques such as direct micro-milling, micro-moulding (*e.g.* injection moulding and hot embossing) or lithographic processing of soft polymers, such as polydimethylsiloxane. These traditional fabrication

3D Printing in Chemical Sciences: Applications Across Chemistry
By Vipul Gupta, Pavel Nesterenko and Brett Paull

Published by the Royal Society of Chemistry, www.rsc.org

techniques typically require long preparation times, specialised skills, involve manual processes, and in some instances (*e.g.* micro-milling) involve expensive facilities. Moreover, these processes are typically restricted (or at least more suited) to certain materials, for example, soft-lithography with polydimethylsiloxane, which does not offer the mechanical strength necessary to allow high-pressure applications[36] and are typically not compatible with robust 'Plug-and-Play' assembly.[37] The typical soft-lithographic process for the production of polydimethylsiloxane microchannels within a microfluidic chip is shown in Figure 2.1.[38] It takes approximately 24 hours to complete this fabrication process.

The advent of 3D printing has had a significant impact upon the microfluidics community. Those involved in the design and development of microfluidic devices and platforms, across all fields of application, now have the ability to rapidly prototype and iterate their devices and simplify fabrication processes considerably. This so-called 'fail fast and often'[39] approach to

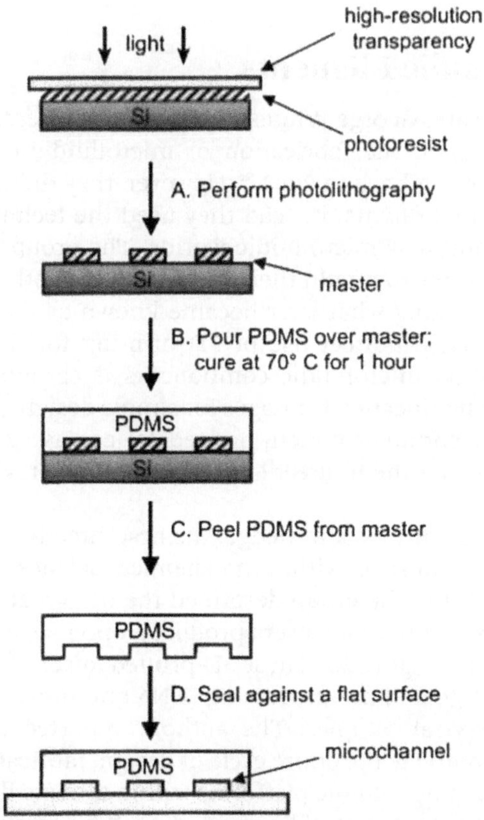


**Figure 2.1** Schematic representation of the soft lithographic processing of polydimethylsiloxane (PDMS) for fabricating micro-channels. Reproduced from ref. 38 with permission from American Chemical Society, Copyright 2002.

fabrication, by eliminating time-consuming and labour-intensive multi-step production processes, has opened up research and development in the field of microfluidics to a whole new generation of young and eager engineers and scientists.[40–42] Perhaps of even more impact, however, has been the significant increase in the range of materials available (and growing) that can be utilised in new microfluidic devices and products delivered using 3D printing technologies. The days where practitioners of the microfluidic arts were frustrated by the many limitations of the polydimethylsiloxane microfluidic chip seem to be behind us.

Various fluid manipulators and controllers, such as microfluidic elements (mixers, droplet generators, and particle generators), valves, and pumps have all been developed and produced using 3D printing techniques within the past decade. These devices have been used individually, and in conjunction with, or within, more complex systems, to serve various physical and analytical purposes. The development and application of these various 3D-printed designs and devices is described, reviewed, and discussed throughout the following chapter.

## 2.2 Microfluidic Elements

As far as we are aware, George Whitesides' group represents one of the first groups to report upon the fabrication of microfluidic components with the help of 3D printing back in 2002.[43] However, they did not directly print the microfluidic components, instead they used the technique as a bridge between the old and new microfluidic worlds. The group used 3D-printed structures as templates to mould their desired polydimethylsiloxane microfluidic devices, inventing what later became known as the 'print, cast and peel' approach.[44] This indirect use of 3D printing for the production of templates for casting microfluidic components is certainly a time-saving approach and has significant advantages for simple designs, although herein we will focus our attention on directly printed devices, as a greater discussion on achievements using the indirect approach has been reviewed previously elsewhere.[40,42,45]

To the best of the authors' knowledge, the first directly 3D-printed microfluidic device for application within the chemical sciences was reported by Bonyár *et al.*[46] in 2010. The group described the design and functioning of 3D-printed mixers and homogenisers produced upon a simple microfluidic platform, as shown in Figure 2.2. These 3D-printed mixers and homogenisers were used for storage and pre-processing (DNA and protein separations) of gynaecological cervical samples. The authors' reported that 3D printing allowed them to complete the entire cycle of design, fabrication, testing, and modification of the microfluidic platforms within a single day.

At a similar time, Moore *et al.*[47] reported on their work using 3D printing to produce capillary valves for centrifugal microfluidic devices. To produce their devices, Bonyár *et al.* used the PolyJet technique and at the time were only able to produce relatively large channels of 1 mm height and 2 mm

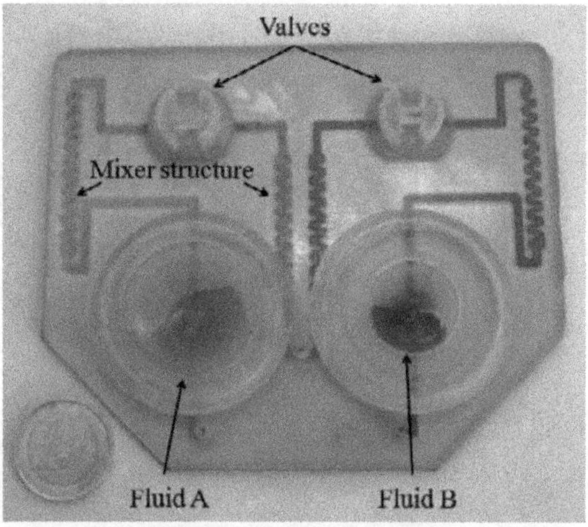


**Figure 2.2** PolyJet-printed polyacrylate composite microfluidic platform with a mixer and a homogeniser. Reproduced from ref. 46 with permission from Elsevier, Copyright 2010.

width, whereas Moore *et al.* used an FDM printer that could deliver channels of considerably smaller dimensions, namely 168.7 ± 76.0 μm (peak-to-peak) and 328.2 ± 92.1 μm (groove-to-groove), from designed dimensions of 127 μm. These differences in design and delivery, plus the channel variation of close to 50%, reflects what was achievable at the time with the first generation of low-cost desktop FDM printers.

In these early studies, both groups printed platforms with open channels, which were later sealed with polyester foil or a PMMA cover. The need to include this post-print manual procedure was obviously a significant disadvantage at the time. However, in 2014, Shallan *et al.*[48] investigated and demonstrated the ability to 3D-print enclosed microfluidic channels, together with sub-mm scale features (*e.g.* pillars). Examples of the 3D-printed enclosed microfluidic channels and equivalent sized pillars produced in this study can be seen in Figure 2.3. The group obtained these results using an SL-based printer and could achieve open (unblocked) channels as small as 250 μm high and 250 μm wide.

More recently, a number of studies into 3D print capabilities have emerged, providing comparative evaluations of the performance and resolution of a number of well-known commercial 3D printers, as applied to the production of simple microfluidic platforms.[49–51] These studies generally conclude that FDM and binder jetting printers result in higher surface roughness and lower print accuracy, as compared to the PolyJet and SL printers when applied to the printing of micro- and macro-fluidic channels. The absolute values of surface roughness and print accuracy are highly dependent on the channel architecture, such as its length, tortuosity, and diameter. The minimum closed channel diameter that can be printed with

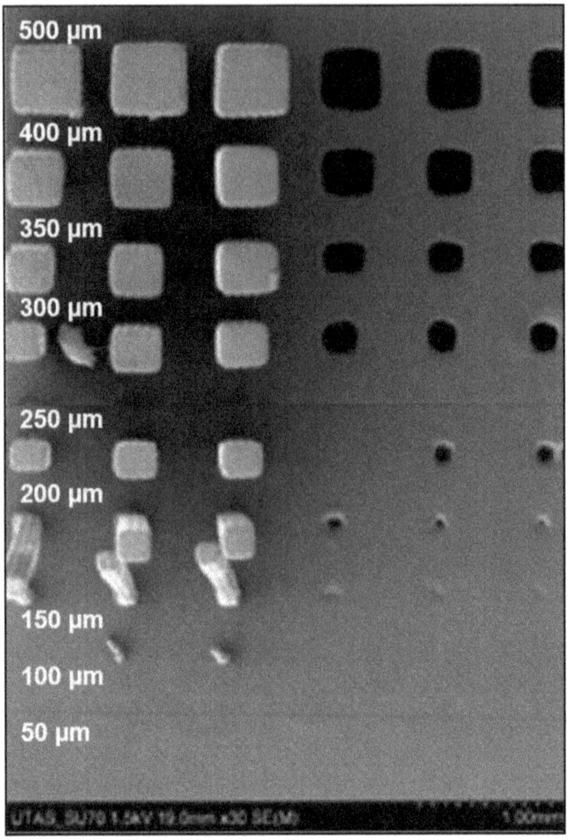


**Figure 2.3** SEM micrographs of the SL-printed enclosed square channels and pillars with the height and width as indicated. The length of each channel and pillar was 2000 µm each. Reproduced from ref. 48 with permission from American Chemical Society, Copyright 2014.

a typical FDM, PolyJet, or SL printer has been reported as 500 µm, 205 µm, and 154 µm, respectively.[51] The absolute surface roughness values observed for such sized channels was given as 11 µm, 1 µm, and <0.5 µm, respectively. For SL-printed laminar flow microfluidic chips, the lower channel wall roughness contributed significantly towards their better performance compared to similar FDM and PolyJet-printed chips (Figure 2.4).[51]

Macdonald *et al.*, in their comparative study, reported that FDM and PolyJet-printed closed channels are usually smaller than their designed dimensions, whereas SL-printed closed channels are usually larger.[51] Other reported observations included that FDM-printed channels suffer from deformation primarily due to the thick and less precise flow of the extruded fibres, and that narrow and complex PolyJet, SLS, and SLM-printed closed channels often suffer from the incomplete removal of the support materials, and hence channel blockage. It was also noted how SL

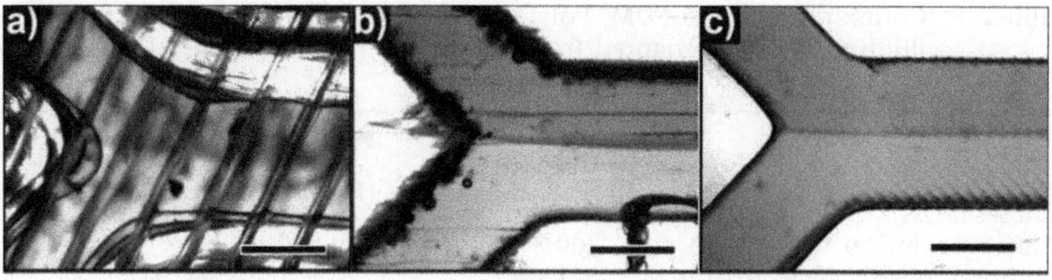


**Figure 2.4** 3D-printed laminar flow microfluidic chips: (a) micrograph of the FDM-printed chip, (b) micrograph of the PolyJet-printed chip, and (c) micrograph of the SL-printed chip. Scale bar is 500 μm. Reproduced from ref. 51 with permission from American Chemical Society, Copyright 2017.

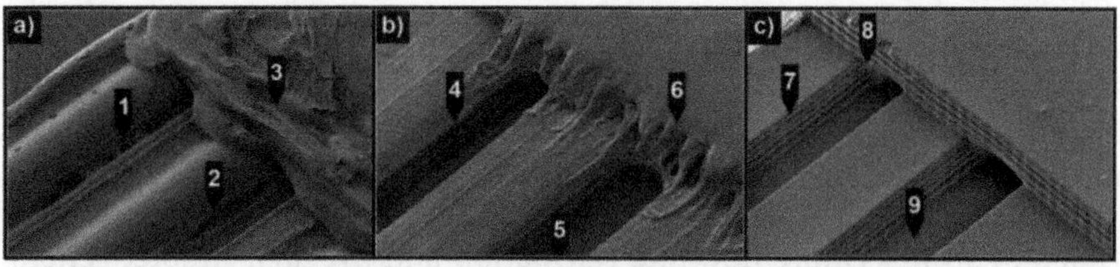


**Figure 2.5** SEM micrographs of the 3D-printed microfluidic channels. (a) FDM-printed channels: (1) 250 μm channels, (2) 350 μm channels, and (3) surface deformation. (b) PolyJet-printed channels: (4) 16 μm layer thickness, (5) stitch marks, (6) surface deformation. (c) SL-printed channels: (7) 32 μm layer thickness, (8) collapsed channel, (9) scanline pattern. Reproduced from ref. 51 with permission from American Chemical Society, Copyright 2017.

printing can result in the collapse of the narrow and complex channels, due to under or over exposure of the printed layers. Figure 2.5 shows scanning electron micrographs of the typical FDM, PolyJet, and SL-printed microfluidic channels compared and contrasted by Macdonald *et al.* in their recent study. The data on the resultant micro- and macro-fluidic channels produced with typical FDM, PolyJet, and SL printers is provided in Table 2.1.

## 2.2.1 Droplet-based Microfluidics

Droplet-based microfluidic devices have been developed to allow convenient handling and automated control of small liquid volumes, and have been used for many varied applications, such as drug delivery, food processing, and development of cosmetics and pharmaceutics.[52–55] As with other microfluidic platforms, droplet generators have been conventionally fabricated using hot embossing,[56] soft lithography,[57] and glass capillary assemblies.[58] Once again, all of these techniques require multiple manual procedures, expensive equipment, and skilled practitioners. Moreover, these techniques rely on the use of elastic polydimethylsiloxane material, which deforms at

**Table 2.1** Comparison of the FDM, PolyJet, and SL printers for fabricating microfluidic channels. Adapted from ref. 51 with permission from American Chemical Society, Copyright 2017.

| | FDM (Felix 3.0) | PolyJet (Eden 260VS) | DLP-SLA (Miicraft+) |
|---|---|---|---|
| Printer resolution $X,Y,Z$ (μm) | 300,300,50 | 250,250,16 | 56,56,50 |
| Minimum positive feature (μm) | 321 ± 5 | 270 ± 70 | 94 ± 7 |
| Minimum negative feature (μm) | 370 ± 7 | 205 ± 13 | 154 ± 10 |
| Build area $X,Y,Z$ (mm) | 240,205,235 | 255,252,200 | 43,27,180 |
| Surface roughness, $R_a$ (μm) | 10.97 | 0.99 | 0.35 |
| Mixing – 5 mm at 100 μL $min^{-1}$ (%) | 71 ± 12 | 27 ± 10 | 3 ± 1 |
| Post-processing time | <1 hour | >1 day | <15 min |
| Printing time per chip (min) | 6 | 30 | 12 |
| Chips printed per hour | 10 | 33 | 4 |
| Cost per chip (USD) | 0.1 | 4 | 2 |
| Printer cost (USD) | 1.3 k | 85 k | 3 k |

pressures higher than $10^5$ Pa[59,60] and swells with many organic solvents, such as decane.[61,62] The deformation of polydimethylsiloxane channels often results in an inaccurate flow-rate calculation and irregular droplet generation and trajectories. Hence, 3D printing has been recently investigated for its potential to overcome these limitations and develop customised and low-cost droplet generators. As described in Chapter 1, a wide variety of print materials are available for 3D printing that can be used as more stable alternatives to polydimethylsiloxane. For example, PolyJet printing has been used to develop droplet generators using both acrylonitrile butadiene styrene[63] and polyacrylate composites.[64]

One of the most commonly used methods to generate droplets is through T-junctions. These T-junctions are used to pinch a droplet of the dispersed phase, which is injected from the terminating arm into the stream of a continuous phase, which flows through the continuous arm. Donvito *et al.*[63] have 3D-printed a droplet generator with six independent and identical T-junctions within a single 3 × 8 × 0.5 cm chip (Figure 2.6). Channel widths of 200 μm and 400 μm were used for the dispersed phase and the continuous phase, respectively. The 3D-printed droplet generator was shown to deliver similar droplet sizes and polydispersity index values to conventionally fabricated droplet generators and could be applied to produce a stable water-in-oil emulsion.

Planar T-junctions (as shown in Figure 2.6) usually require wall surface treatment to modify their wetting properties for droplet formation. The availability of favourable wetting properties is even more challenging in multiple emulsion generators, as these require a switch between hydrophobic and hydrophilic surfaces within the space of a few micrometres.[65] In order to overcome this limitation, a non-planar T-junction with graduated thickness was proposed by Rotem *et al.*, which could generate droplets even in

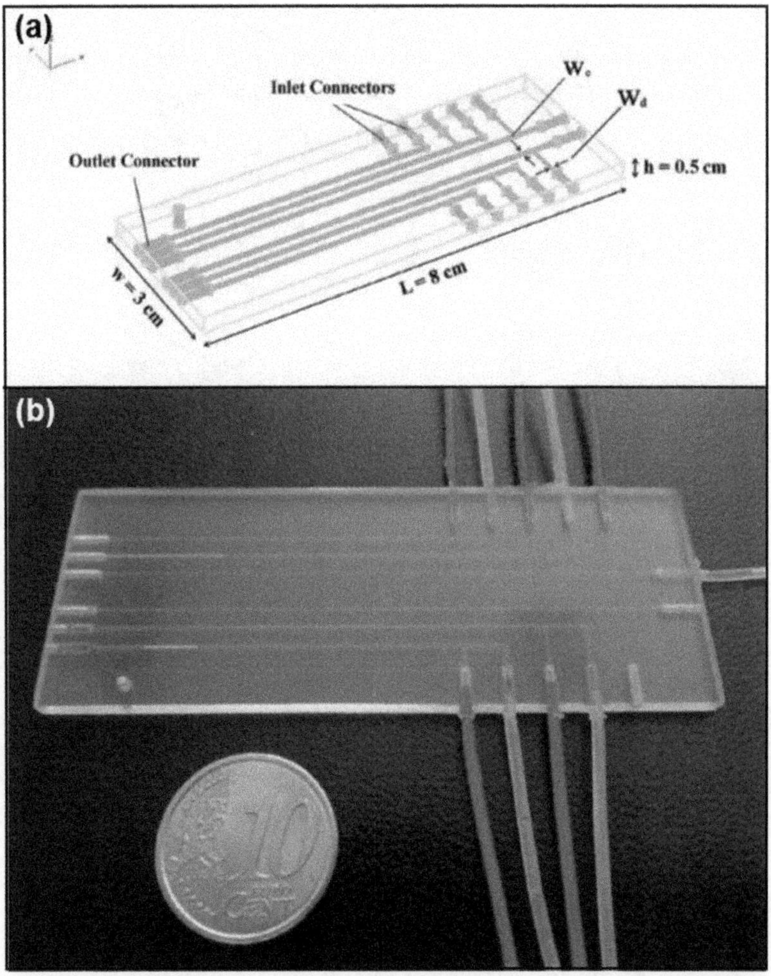


**Figure 2.6** PolyJet-printed chip with six T-junction based droplet generators. (a) Render: $W_c$ represents continuous phase and $W_d$ represents dispersed phase and (b) photograph. Reproduced from ref. 63 with permission from Institute of Physics Publishing, © IOP Publishing.

the absence of favourable wetting surfaces (Figure 2.7).[65] However, actually producing the microfluidic device proved problematic, due to the complex fabrication requirements when applying conventional soft lithographic techniques. Zhang *et al.*[64] recently applied 3D printing to a similar design to simplify the fabrication steps, and thus further exploit the potential application of non-planar droplet generators (Figure 2.8). Much like traditional microfluidic droplet generators, the novel 3D-printed droplet generator was successfully applied to produce both single (W/O and O/W) and double (W/O/W, O/W/O, and air/O/W) emulsions.

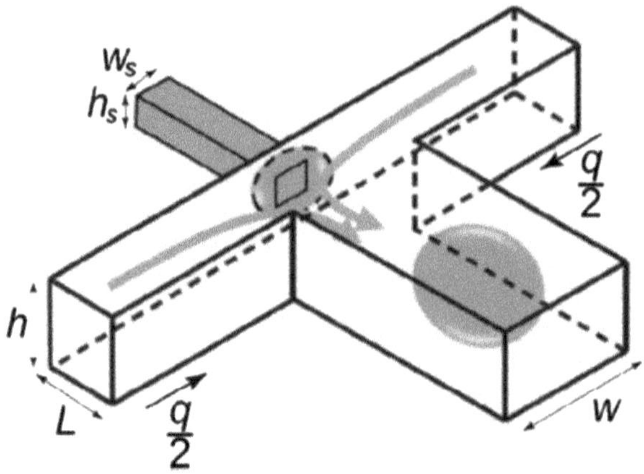


**Figure 2.7** The geometry of the non-planar T-junction based droplet generator. $W_s$ – width of the dispersed phase channel, $h_s$ – height of the dispersed phase channel, $L$ – width of the continuous phase channel, $h$ – height of the continuous phase channel, $W$ – width of the outlet channel, and $q$ – flow rate of the continuous channel. Reproduced from ref. 65 with permission from The Royal Society of Chemistry.

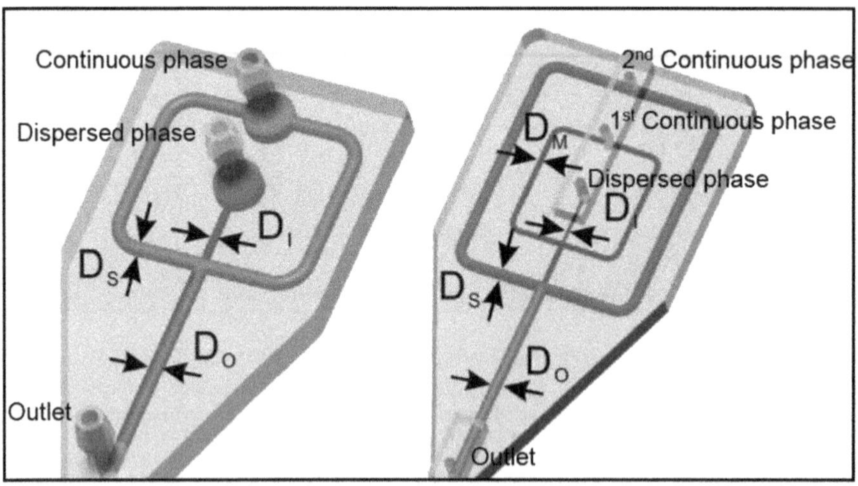


**Figure 2.8** Render of the 3D-printed non-planar droplet generator. Single emulsion generator (left): $D_I$ = 0.6 mm, $D_S$ = 1 mm, and $D_O$ = 1 mm and double emulsion generator (right): $D_I$ = 0.6 mm, $D_M$ = 1 mm, $D_S$ = 2 mm, and $D_O$ = 2 mm. Adapted from ref. 64 with permission from The Royal Society of Chemistry.

In accordance with the channel diameters of the 3D-printed droplet generator shown in Figure 2.8, the original printed device resulted in droplet sizes of *ca.* 1 mm. To generate droplets as small as 50 µm, the terminating channel was replaced with commercial tubing, as the 3D printing of sub

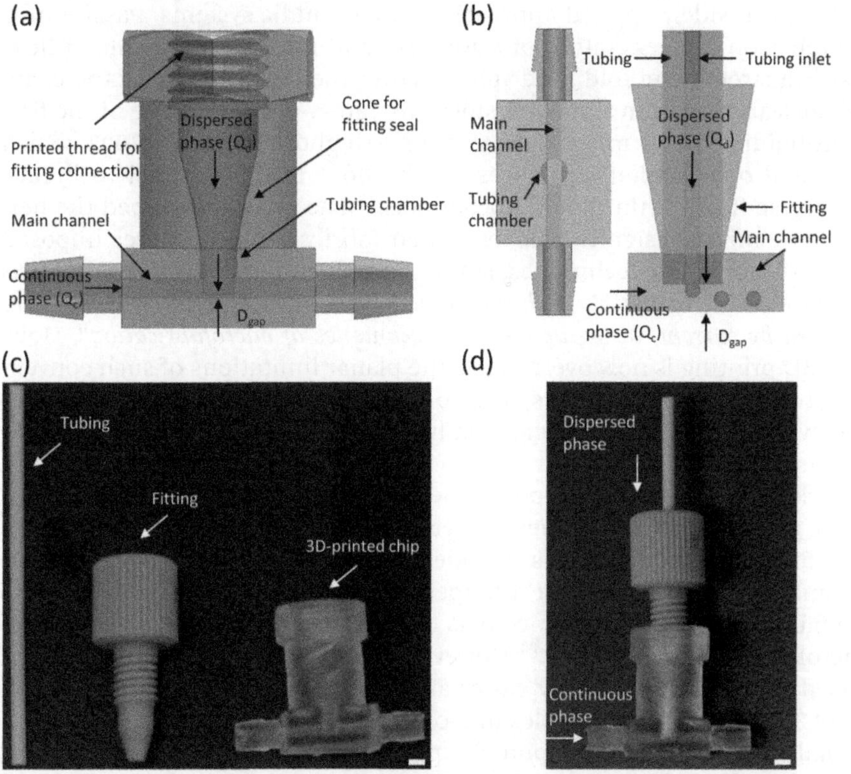


**Figure 2.9** SL-printed polyacrylate based droplet generator with a threaded inlet port: (a) render, (b) bottom view and schematic of droplet generation, (c) discrete parts, and (d) assembled droplet generator. Both scale bars are 1 mm. Reproduced from ref. 66 with permission from The Royal Society of Chemistry.

100 μm closed channels was not possible. The tubing was fed into the droplet generator using standard fittings, *via* a printed and integrated threaded port (Figure 2.9).[66] The ease of design and rapid production of these simple T-junction based droplet generators, from planar to non-planar, with simple integrated connections, highlights the value of 3D printing for device customisation and iteration.

## 2.2.2 Microfluidic Mixers

Microfluidic mixers have been used in various applications, ranging from biochemical assays[67–70] to microscale chemical synthesis.[71–74] However, microfluidic mixing is often challenging due to laminar flow and unfavourable molecular diffusion coefficients. Both active and passive methods have been used to achieve improved mixing in microfluidic devices.[75–77] However, as passive mixing does not require additional external drivers (energy), it is the most widely applied approach in microfluidic systems. Passive mixing relies on the generation of transverse and radial components of flow, that can stretch and fold fluid volumes over the cross-section of the channel. Researchers from the Whitesides laboratory developed one of the first microfluidic passive mixers (staggered herringbone mixer) in 2002, which consisted of bas-relief structures on the floor of the channel to induce transverse flow within the microchannels.[78] The group developed the herringbone mixer design to suit the planar fabrication procedures imposed by soft-lithographic techniques, stating in their study, "*Microfluidic designs should also be compatible with the planar, layer-by-layer geometries that are imposed by current, lithography-based techniques of microfabrication*". However, 3D printing is now overcoming the planar limitations of such conventional fabrication techniques, and so we now begin to see the production and evaluation of three-dimensionally arranged complex passive microfluidic mixers.

Back in 2007, Carrière proposed the structure of a 3D micromixer, which at the time consisted of sequential symmetrical units, based on the Baker's transformation model.[79] This provided mixing by repeatedly splitting and recombining flow streams at a perpendicular angle. As mentioned above, the fabrication of 3D micromixers is challenging when using conventional microfabrication techniques.[80] However, the availability of 3D printers has now delivered a significantly easier and cost-effective means of producing these 3D geometrically complex micromixers.[48]

Shallan *et al.* reported upon the performance of a 3D-printed micromixer, which could be printed in just 50 minutes, with a material cost of only $2, using a low-cost desktop SL printer.[48] Square 500 µm channels were used to form the 3D-printed micromixer (Figure 2.10(a)), and multiple micromixers were used in series to develop a novel five-step gradient generator (*e.g.* for allowing controlled mixing of buffers *etc.*) as shown in Figure 2.10(b).[48]

As discussed in Chapter 1, FDM printing can result in chordal surface defects (Figure 2.11). However, these defects can act as bas-relief structures similar to the herringbone mixers, and so can potentially be used constructively to provide passive mixing within microchannels. With FDM printers, the orientation of these structures with respect to flow, and hence their performance as micromixers, can be controlled by changing filament extrusion angles. Accordingly, the effect of different filament extrusion angles has recently been studied on fluid flow in FDM-printed microfluidic devices.[81] Li *et al.* reported that an extrusion angle of 60° resulted in the highest mixing efficiency in a 3D-printed Y-shaped mixer, because of the resultant rotational motion of the fluid. These FDM-printed Y-mixers were then successfully applied within a platform to perform colourimetric analysis of iron in river water.[81] Extrusion angles of 0° and 90° resulted in laminar flow with minimum fluid mixing. Surprisingly, rough FDM-printed channels with 0° and 90° extrusion angles resulted in a higher degree of laminar flow compared to their relatively smooth PolyJet-printed counterparts.[81]

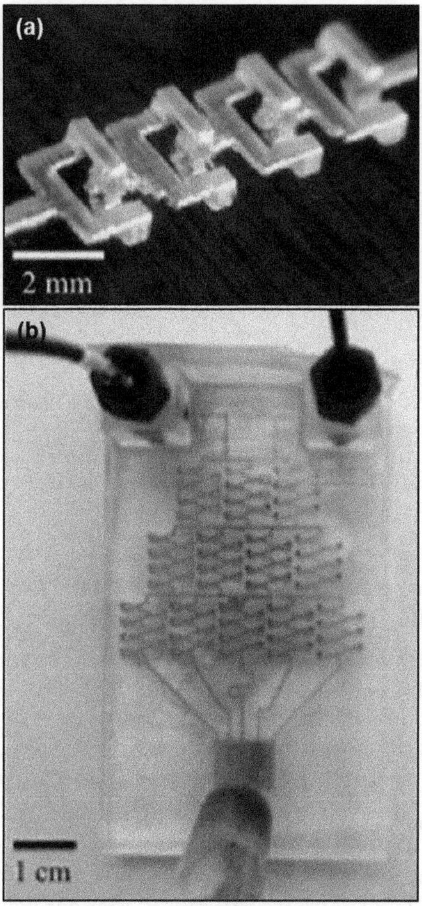


**Figure 2.10** SL-printed polyacrylate composite: (a) micromixer and (b) five-step gradient generator. Reproduced from ref. 48 with permission from American Chemical Society, Copyright 2014.

### 2.2.3 Miscellaneous Microfluidic Elements

In addition to the droplet-generating microfluidic devices and microfluidic mixers, other microfluidics elements have also been 3D-printed that further demonstrate the utility of 3D printing in the field of microfluidics.

For example, Sochol *et al.*[82] have 3D-printed a range of complex microfluidic 'operators', analogous to electrical operators, such as fluidic capacitors, diodes, and transistors, as shown in Figure 2.12. The fabrication of these microfluidic elements would have been practically impossible using conventional microfabrication techniques because of their complex architecture. However, PolyJet printing was used to facilitate their production as robust and single-body designs. Within this work, fluidic capacitors were developed based upon parallel printed diaphragms as shown in Figure

**Figure 2.11** Chordal surface defects on a 0.5 mm × 1 mm section of the FDM-printed channel walls. Reproduced from ref. 47 with permission from Springer Nature, Copyright 2011.

2.12(a). The diaphragms were then modified with a central circular aperture to perform as fluidic diodes, as shown in Figure 2.12(b). Fluidic transistors were fabricated by integrating a piston into the diaphragm centre as shown in Figure 2.12(c) and (d). The fluidic transistors were analogous to the p-channel MOSFET (metal-oxide-semiconductor field-effect transistor) with gate regulation of source to drain fluid flow. Three different diaphragm cross-sections were studied, namely damped sinusoidal, undamped sinusoidal, and planar. The sinusoidal diaphragms resulted in better capacitor performance as compared to the planar diaphragms. The 3D-printed fluidic diodes and transistors were reported to provide a microfluidic diodicity of $80.6 \pm 1.8$ and a pressure gain of $3.01 \pm 0.78$, respectively.[82]

In an equally challenging application, Chen *et al.*[83] developed a 3D-printed electro-millifluidic particle generator with nested channels to produce multi-compartment particles (Figure 2.13(a)).[83] Different particle compartments were filled with model ingredients such as fruit juices, drugs, and magnetic nanoparticles. The 3D-printed particle generator was used to produce different shapes of calcium alginate particles such as concentric core–shells, slice-by-slice multiple lamellae, and petals surrounding the particle centres (Figure 2.13(b)–(d)). These particles can be used for various purposes such as encapsulation, release, and spatially dependent actuation.[83]

## 2.3 Microfluidic Valves

Unger *et al.*[84] introduced the fabrication of microfluidic valves with soft lithographic techniques in early 2000. Since then, microfluidic valves have been extensively used to enable programmability and automation within a wide range of microfluidic devices.[85] However, once again, as above,

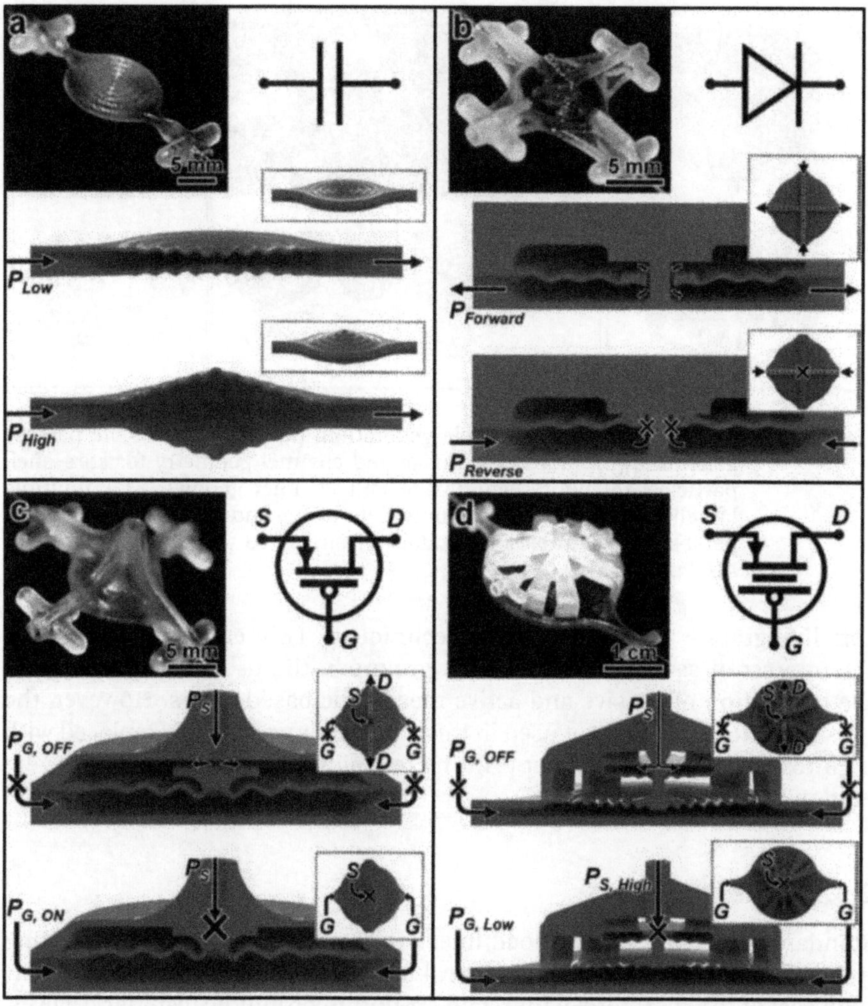


**Figure 2.12** 3D-printed polyacrylate composite microfluidic operators: (a) fluidic capacitors, (b) fluidic diodes, (c) fluidic transistors, and (d) enhanced-gain fluidic transistors. $P$ – pressure inputs, $G$ – gate, $S$ – source, $D$ – drain, and $Q_{SD}$ – fluid flow. Reproduced from ref. 82 with permission from The Royal Society of Chemistry.

soft lithographic fabrication of these complex valves requires multilayer lithographic procedures, with results that are often far from robust and reproducible.

In 2011, Moore *et al.*[47] demonstrated the first successful use of 3D printing to develop functional microfluidic valves. The group used an FDM printer to print passive capillary valves for use within centrifugal microfluidic devices, demonstrating the potential of 3D printing to replace conventional

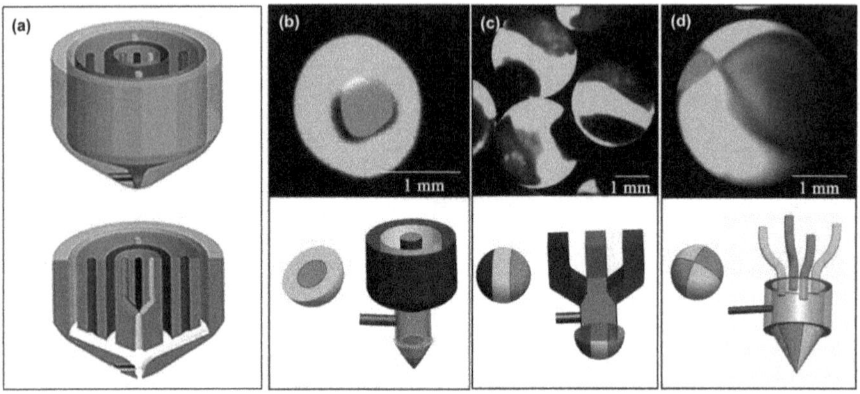


**Figure 2.13** Multi-compartment particle generation: (a) schematic of the particle generator, (b) fluorescent image and channel geometry for core–shell particles, (c) fluorescent image and channel geometry for multiple lamellae particles, and (d) fluorescent image and channel geometry for petal-shaped particles. Reproduced from ref. 83 with the permission of AIP Publishing.

soft lithography and CNC milling techniques. This early report sparked the interest of several other research groups in the use of 3D printing for the fabrication of passive and active membrane-based valves. However, the low-resolution FDM printer used in this early study was quickly replaced with high-resolution SL and PolyJet printing techniques to develop the next generation of valves.

### 2.3.1 Passive Valves

Standard four- and six-port load/inject valves are some of the most commonly used passive valves in analytical laboratories. However, the relatively high cost of the commercial four- and six-port microfluidic (capillary) valves, and their relatively large size, restrict their use in low-cost microfluidic systems. To overcome these limitations, Su *et al.*[86] 3D-printed a functionally similar load/inject valve using an SL printer, and quoted the cost of printing such a device to be less than a dollar. It was fabricated in two parts, namely a base and a rotor. The rotor consisted of two channels that perfectly matched the holes on the base, specifically a straight bypass channel, and a reverse U-shaped sample loop (Figure 2.14). The base and rotor were treated with dry Teflon to provide a watertight seal, which was found to be leak-proof up to 400 kPa. Although the pressure tolerance of the 3D-printed valve was much less than the commercial valves, it was successfully applied within an analytical set-up connected to ICP-MS for the direct and continuous determination of trace elements in microdialysate samples. The simple yet functional 3D-printed valve was trivial to fabricate, could be readily customised, miniaturised, and could be easily integrated into a range of microfluidic devices.

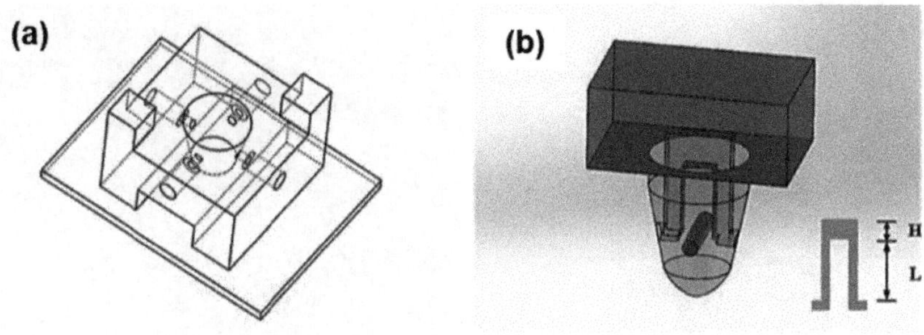


**Figure 2.14** Render of the SL-printed polyacrylate composite load/inject valve: (a) base and (b) rotor. Reproduced from ref. 86 with permission from Elsevier, Copyright 2014.

In a similar work, Chan *et al.*[87] 3D-printed their version of a rotary valve, as shown in Figure 2.15(a) and (b).[87] It was also printed with an SL printer in two parts: a base and a rotor. As with standard rotary valves, after the sample was loaded into the loading channels, the rotor was rotated to make contact between the loading channels and the receiving channels (Figure 2.15(c) and (d)). Once the loading and receiving channels were in contact, the sample passed from the loading channels to the receiving channels (Figure 2.15(e) and (f)). The channel diameters were 500 µm, and the authors reported the use of a thin layer of Vaseline to provide a watertight seal between the base and the rotor.

The same group also developed a simple 3D-printed torque actuated valve to control the opening and closing of a channel.[87] It consisted of a screw with a cylindrical piston attached to its lower surface, which was loosened or tightened to open or close the channel, respectively (Figure 2.16).[87] A thin layer of Vaseline was again applied to the piston surface to provide a watertight seal while closing the channel. In the closed configuration, the valve resisted the passage of fluorescein across the channel for at least 80 minutes.

So-called 'SlipChip' microfluidic devices have been developed to perform multiplexed microfluidic reactions without the use of pumps or valves.[88] However, applying the same operational principle, Chan *et al.*[87] used SL printing to produce a complex multiple pushing valve microfluidic assembly. It consisted of multiple pushing valves and a receiver chip as shown in Figure 2.17. Each pushing valve consisted of sample loading channels and solvent inlets. The receiver chip consisted of pushing valve slots and reaction chambers. During the loading step, the loading channels were filled with a sample solution, and the pushing valves were inserted into the receiver chip (Figure 2.17). To initiate a reaction, the pushing valves were further pushed down to bring the loading channels into contact with the reaction chambers. Once the loading channels and the reaction chambers were in contact, the sample could diffuse from the loading channel into the reaction chamber to start the reaction.

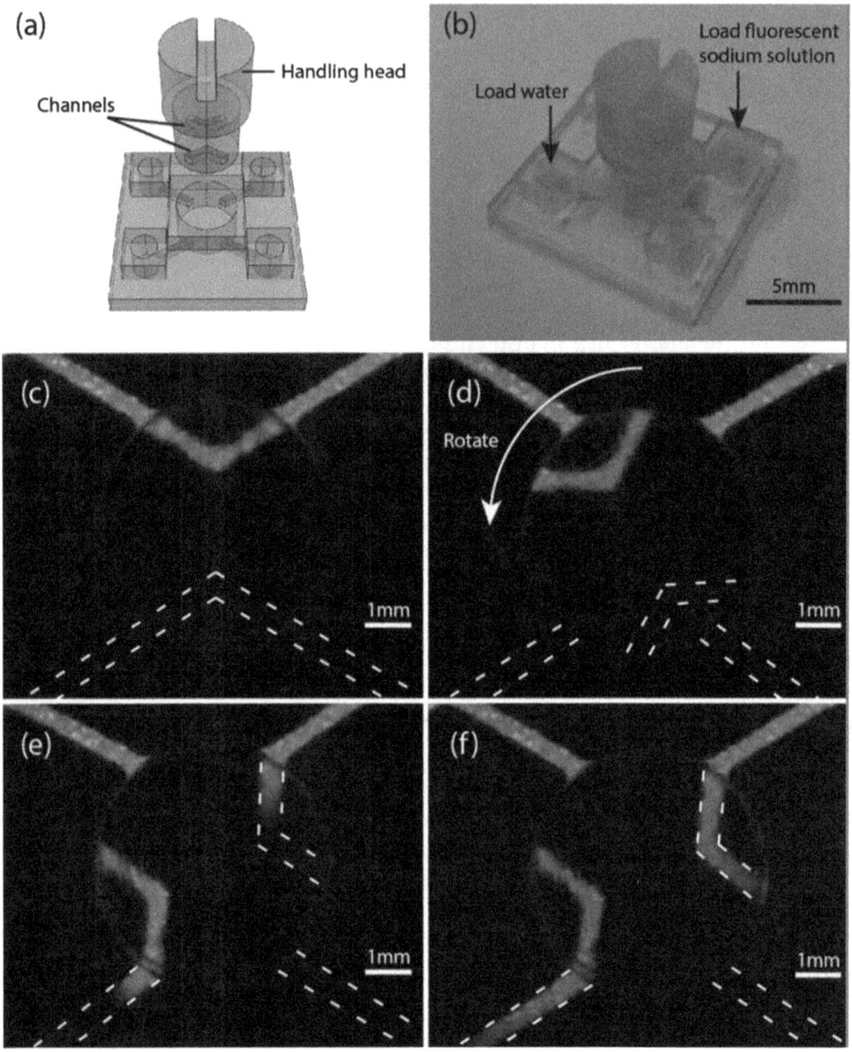


**Figure 2.15** SL-printed polyacrylate composite rotary valve: (a) render and (b) photograph. Fluorescent images of (c) the initial state, (d) 30° rotation, (e) 115° rotation (start of diffusion), (f) 115° rotation (after 20 minutes of diffusion). The dashed lines indicate water filled channels. Reproduced from ref. 87 with permission from American Chemical Society, Copyright 2016.

## 2.3.2 Active Membrane-based Valves

As the name suggests, active membrane-based valves are based on the movement of a membrane to open or close a channel. Typically, these valves consist of a thin membrane between two channels, namely a control channel and a flow channel. The control channel is used to exert pressure on the

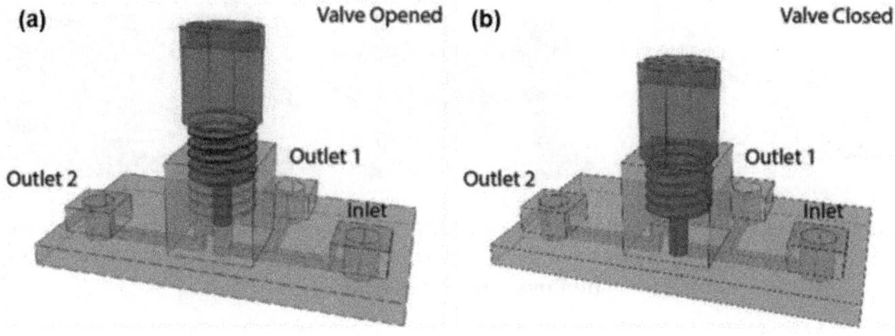


**Figure 2.16** Schematic of the SL-printed polyacrylate composite torque actuated valve in (a) open and (b) closed states. Reproduced from ref. 87 with permission from American Chemical Society, Copyright 2016.

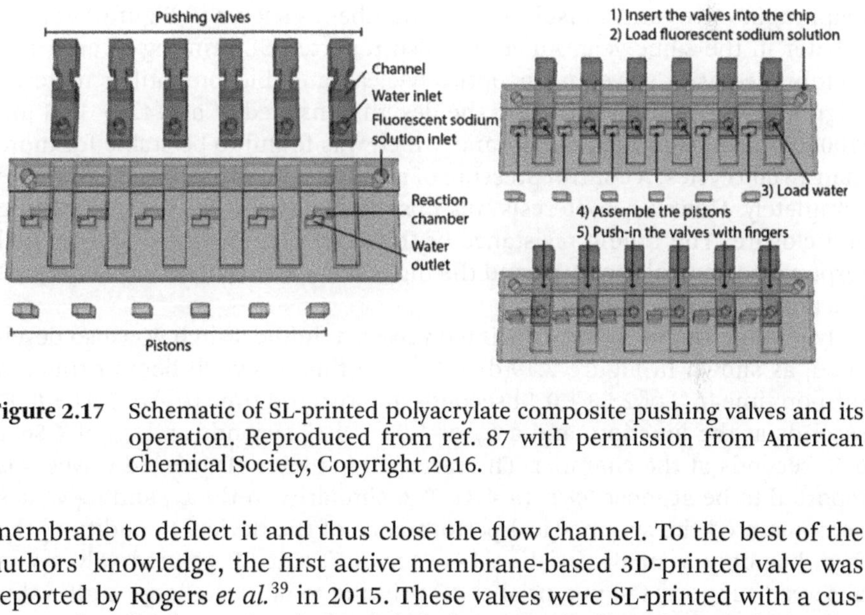


**Figure 2.17** Schematic of SL-printed polyacrylate composite pushing valves and its operation. Reproduced from ref. 87 with permission from American Chemical Society, Copyright 2016.

membrane to deflect it and thus close the flow channel. To the best of the authors' knowledge, the first active membrane-based 3D-printed valve was reported by Rogers *et al.*[39] in 2015. These valves were SL-printed with a custom resin, which consisted of poly(ethylene glycol) diacrylate as the monomer, Irgacure 819 as the photoinitiator, and Sudan I as an absorber dye. The group reported a 100% success rate for printing horizontally arranged rectangular channels (~ 350 μm wide and 250 μm high) and vertically arranged cylindrical channels (~210 μm in diameter) using this resin. The membrane consisted of a single build layer with an exposure time of 5.0 to 5.5 seconds, and was reported to be stable for up to 800 actuations. The group went on later to further reduce the void volume of these valves to 10% of the original volume and increased their stability up to an impressive 1 million actuations.[89] The void volume of the valves was reduced by optimising the resin formulation for high-resolution SL printing,[90] and the mechanical

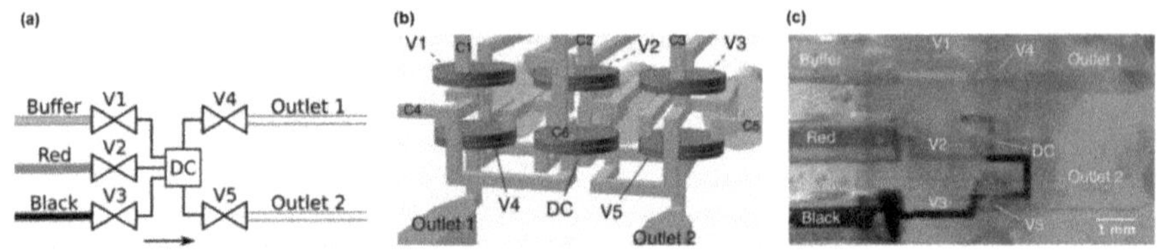


**Figure 2.18** SL-printed poly(ethylene glycol) diacrylate multiplexer: (a) circuit diagram, (b) render, and (c) photograph. Reproduced from ref. 89 with permission from The Royal Society of Chemistry.

robustness of the membrane was improved by incorporating a thermal initiator for post-print baking. The optimised SL resin formulation consisted of 1% (w/w) Irgacure 819, 0.4% (w/w) Sudan I, and 0.01% (w/w) 2,2′-azobisisobutyronitrile in poly(ethylene glycol) diacrylate. These valves were used to generate a complex 3D multiplexer for fluid mixing. It consisted of five diaphragm valves and a displacement chamber as shown in Figure 2.18.

Later in the same year, Au *et al.*[37] also reported SL printing of an active membrane-based valve in an optically-clear and biocompatible material (Figure 2.19(a)–(c)). In this case, the design consisted of a 114.6 ± 15.3 μm (100 μm designed) thick membrane, which was found to be stable for more than 15 000 cycles. A control pressure of *ca.* 41 kPa was able to close the valve completely. However, hysteresis was observed between the valve's opening and closure. The fluidic resistance of the valve demonstrated a sigmoidal response to control pressure, and the highest slope was observed between 3 kPa and 4.5 kPa.

By joining four of these 3D-printed valves, a fluidic switch was also developed, as shown in Figure 2.19(d)–(g).[37] The fluidic switch demonstrated a turn-on time ($t_{on}$) of 2.53 ± 0.14 seconds and turn-off time ($t_{off}$) of 2.41 ± 0.05 seconds at the junction, and a $t_{on}$ of 4.93 ± 0.15 seconds and $t_{off}$ of 4.86 ± 0.33 seconds at the chamber. The variation between 3D-printed valves was reported to be at most 5.8% ($n$ = 3). The similarity in the $t_{on}$ and $t_{off}$ values demonstrated the symmetrical performance of the device. In addition, the fact that only a small variation between different 3D-printed valves was observed supports the potential of 3D printing to fabricate large multi-valve arrays.

The above-mentioned active membrane-based valves were 3D-printed with a single material, and the membranes were only differentiated from the valve body based on their thickness. Hence, actuation of these membranes with fluidic pressure usually results in the deformation of the entire valve structure.[91] With the availability of multi-material PolyJet printers, there now exists the potential to 3D-print such valves with an elastic material for the membrane and a rigid valve body. A wide variety of polyacrylate composite based elastic and rigid materials are now commercially available for PolyJet printing, which can enable this approach. Demonstrating this capability, Keating *et al.*[91] 3D-printed similar multi-material proportional valves

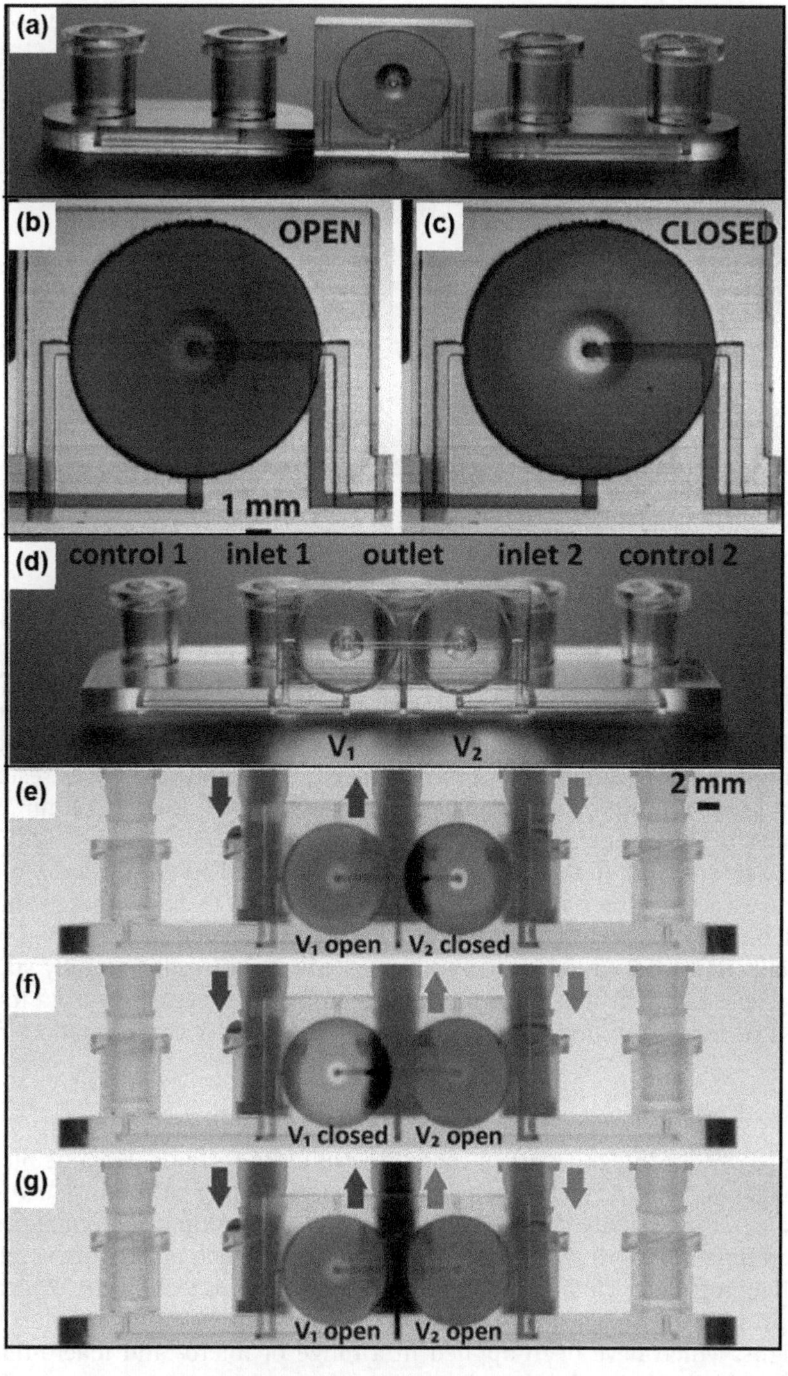


**Figure 2.19** SL-printed active membrane-based valve and fluidic switch: (a) photograph of a single active membrane-based valve, (b) micrograph of the active membrane-based valve in an open state, (c) micrograph of the active membrane-based valve in a closed state, (d) photograph of a two-valve switch, and (e–g) photograph of the switch in three different actuation states as marked. Reproduced from ref. 37 with permission from The Royal Society of Chemistry.

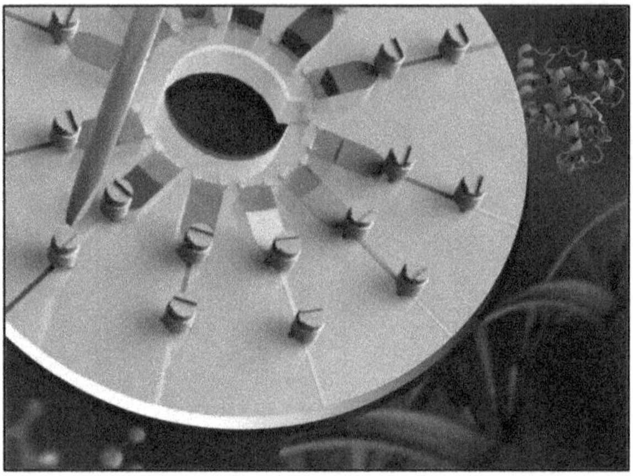

**Figure 2.20** Render of the 3D-printed diaphragm valves. Reproduced from ref. 92 with permission from The Royal Society of Chemistry.

with various elastic membranes of different thicknesses (200 μm to 500 μm) and stiffness (Shore 26–63 Scale A). As expected, thicker and stiffer membranes required higher control pressures to close the valve. However, these multi-material valves did not show any sign of global deformation, whereas similar single material 3D-printed valves clearly showed global deformation during actuation. A repeated proportional control with these valves resulted in an error of only ±4.4 μL $s^{-1}$. The print time for the valve was 25 min, and the cost of the print material was reported as only $1.85.

In a recent work in this area, Kim *et al.*[92] have used an elastic epoxy membrane with 3D-printed adapters to develop push-and-twist diaphragm valves for lab-on-a-disc devices (Figure 2.20). These valves could be addressed individually and were found to be reversible, leak-proof, and thermally stable. They resisted pressures of up to 420 kPa and were successfully used for an enzyme-linked immunosorbent assay and a polymerase chain reaction.[92]

## 2.4 Pumps

Pumps are an integral part of many microfluidic systems, responsible for transporting and delivering small and precise volumes of fluids.[93,94] Microfluidic pumps usually require precise assembly of multiple components and development of both rigid and elastic structures, which once again is rather challenging using more conventional microfabrication techniques. Recently, 3D printing has been used to develop various new pump assemblies and designs, which have been applied in a range of micro- and macro-fluidic applications. Obviously, given the potential complexity of functional microfluidic pump, multi-material PolyJet printing holds significant potential here for producing integrated pump assemblies, allowing the simultaneous printing of the necessary rigid and elastic materials. Rigid materials can be used

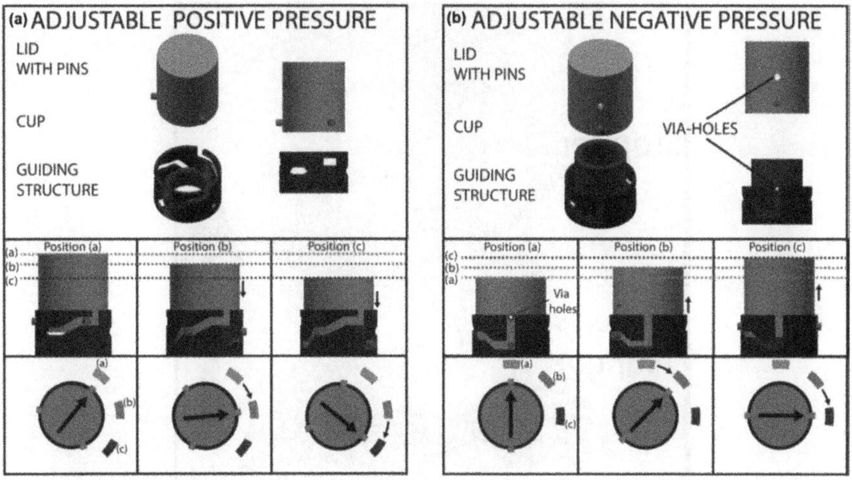


**Figure 2.21** 3D-printed rigid and elastic polyacrylate composite pump based on a pumping lid and cup mechanism: (a) positive pressure control and (b) negative pressure control. Reproduced from ref. 95 with permission from The Royal Society of Chemistry.

to print the pump body parts, such as the barrel and piston for a syringe type pump, with more elastic materials used to print components such as plunger tips, ensuring a hermetic seal between the moving rigid components. As demonstrated below, rigid and elastic multi-material 3D printing has recently enabled the fabrication of integrated, tunable, and positive/negative pressure-generating pumps, with interchangeable and disposable 3D-printed parts.

An excellent example of this potential is the work reported by Begalo *et al.*, in which multi-material 3D-printed pumps were developed, based both on pumping lid and cup assemblies[95] and conventional syringe assemblies.[96] Pumping lids can be moved down and up to generate positive and negative pressures, respectively, as shown in Figure 2.21. The desired pressure can be pre-programmed by adjusting the height of the pumping lid, and it can be further manipulated during experiments. In the 3D-printed pumping lid and cup based pumps, the pressure was generated by either compressing a known gas or by vaporising a volatile solvent.[95] The pressure generated by the pump was found to be solely dependent on the lid and cup dimensions, and was independent of the pumping solution. A hermetic seal was obtained between the pumping lid and cup by using a 1 mm to 1.5 mm thick elastic layer between them. The lid and cups were designed with a 100 μm to 200 μm overlap to ensure an airtight yet easy assembly. In this work, the pump was successfully used for various microfluidic operations, including within droplet generation systems, the control of laminar flow profiles, and the loading of SlipChip devices.[95]

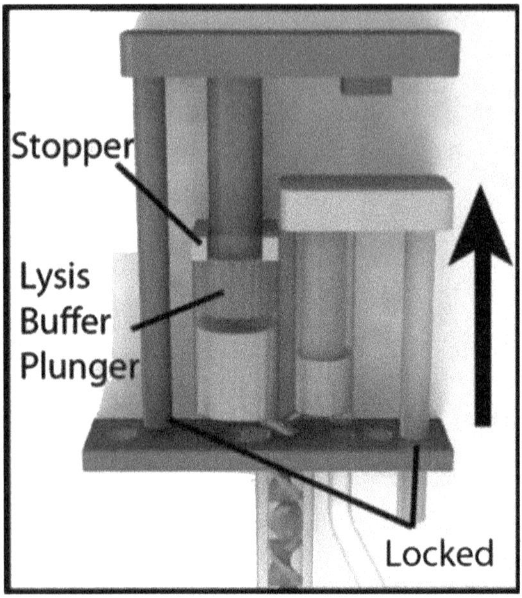


**Figure 2.22** Render of the 3D-printed interlock meter-mix device. Reproduced from ref. 96 with permission from The Royal Society of Chemistry.

The pumping lid and cup based pumps are limited to pneumatic pumping, which often requires re-calibration of the generated pressure with any small change in the local environment. Accordingly, a syringe assembly was developed with multiple 3D-printed rigid and elastic components (Figure 2.22).[96] A hermetic seal was again obtained by using a thin layer of elastic material between the barrel and the plunger. The assembly was used to develop an interlock 'meter-mix' device.[96] It resulted in the accurate aspiration of a predetermined volume with a coefficient of variation of only 0.8%. The performance of the 3D-printed meter-mix device matched the performance of the standard laboratory protocols for urine analysis. Moreover, it significantly reduced the overall analysis time.

Nikola Tesla introduced a new pump assembly in the early 1900s with multiple parallel discs on a rotating shaft.[97,98] Over the years this pump assembly has been successfully used to provide a pulseless flow for various applications, including, for example, artificial hearts.[99,100] Much more recently, Habhab *et al.*[101] have used SL printing to develop a miniaturised version of the Tesla turbine fluidic pump. The pump was 3D-printed in three easily assembled parts, namely the cap holder, rotor, and nozzled housing (Figure 2.23). The 3D-printed Tesla pump achieved a pressure of up to 253 Pa and a flow rate of 12.6 mL min$^{-1}$ at 1200 rpm. In the report, the pump was also successfully integrated with a microfluidic gradient generator.[101]

Low pressure peristaltic pumps are a common sight within many chemistry laboratories. Xiang *et al.*[102] have developed a version of a peristaltic

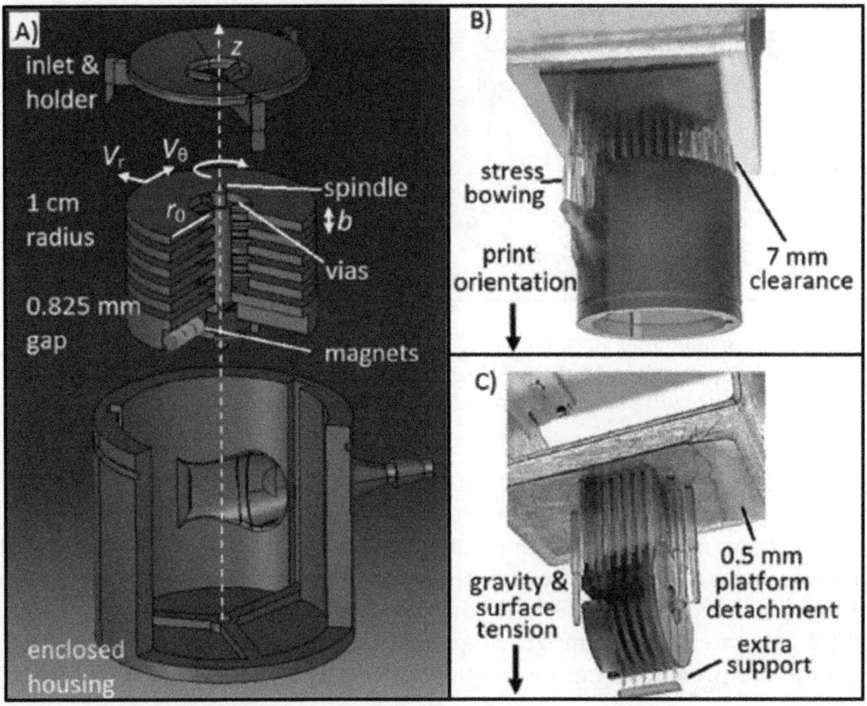


**Figure 2.23** 3D-printed polyacrylate composite micro Tesla pump: (a) render, (b) SL printing of the cup, and (c) SL printing of the rotor. Reproduced from ref. 101 [https://doi.org/10.3390/s16111970] under the terms of a CC BY 4.0 license [https://creativecommons.org/licenses/by/4.0/].

pump using FDM-printed poly(lactic acid) micro-cams. In this case, the pump consisted of a flexible polydimethylsiloxane microfluidic channel and three 3D-printed micro-cams with different phase angles, as shown in Figure 2.24. The pump was shown to be capable of producing a flow rate of 2.74 mL $min^{-1}$ at a driving speed of 170 rpm and resisted backpressure of more than 36 kPa.

As with the previously discussed push-and-twist diaphragm valves, cast membranes can be integrated with bespoke 3D-printed components to develop unique diaphragm pumps. For example, Thomas *et al.*[103] have used a composite polydimethylsiloxane and iron membrane in conjunction with FDM-printed poly(lactic acid) components to develop an electromagnetically actuated closed diffuser nozzle pump. The use of the narrow 3D-printed diffuser elements within this pump generated flow rates between 2.2 to 2.4 μL $min^{-1}$.[103]

Similarly, passive and active membrane-based valves can also be used to develop miniaturised 3D-printed pumps. Chan *et al.*[87] have used a passive torque-actuated valve to develop a torque-actuated pump. The valve screw

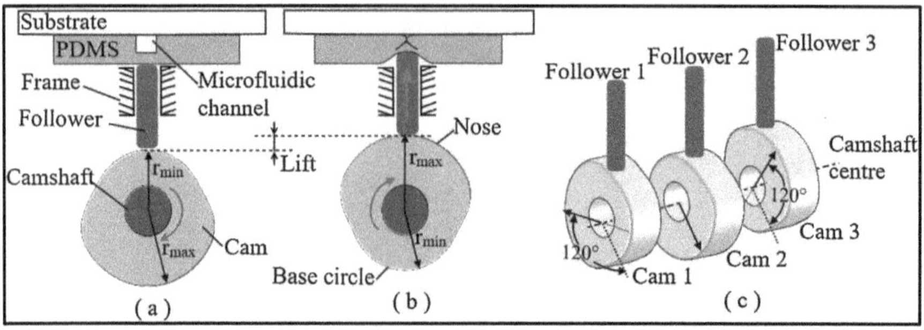


**Figure 2.24** Schematics of a 3D-printed micro-cam actuated peristaltic pump: (a) open state, (b) closed state, and (c) micro-cams. Reproduced from ref. 102 with permission from Elsevier, Copyright 2016.

was tightened or loosened to pump or withdraw the fluid, respectively. The screw can be operated by hand or with a motor, depending on whether a simple operation or an accurate flow rate is required. Au *et al.*[37] have also used active membrane-based valves, in this case to develop a peristaltic pump. The pump consisted of three serially connected active membrane-based valves, which generated a maximum average pumping rate of *ca.* 0.68 mL $min^{-1}$ at $T$ = 75 ms and provided a cut-off frequency of 6.7 Hz.

## 2.5 Plug-and-Play Modules

The production of multifunctional and integrated microfluidic platforms typically requires the arrangement of multiple microfluidic components in different orders and orientations. This obviously requires complex fabrication processes, especially if lithographic and milling techniques are used. Moreover, although built for purpose, such integrated platforms do not allow device flexibility and customisability for additional applications due to the fixed arrangement of the microfluidic functions. To address this inflexibility, the development of so-called microfluidic 'Plug-and-Play' modules has received considerable attention over recent years, as an approach which offers customisability, recyclability, cost-effectiveness, and the opportunity to assemble multi-purpose microfluidic platforms.[104,105]

In 2014, Lee *et al.*[36] published the first report on 3D-printed modular microfluidic components. These modular components were integrated by a combination of O-rings, metal pins, and concave and convex shaped features at the module interfaces (Figure 2.25). A range of standard microfluidic elements were fabricated and assembled in different three-dimensional orientations. The connection of the PolyJet-printed polyacrylate composite modules through O-rings was found to be leak-proof up to 200 kPa. The connection was made leak-proof up to *ca.* 800 kPa by using silicon grease around the O-rings. Paydar *et al.*[106] have also reported the use of 3D-printed O-rings that

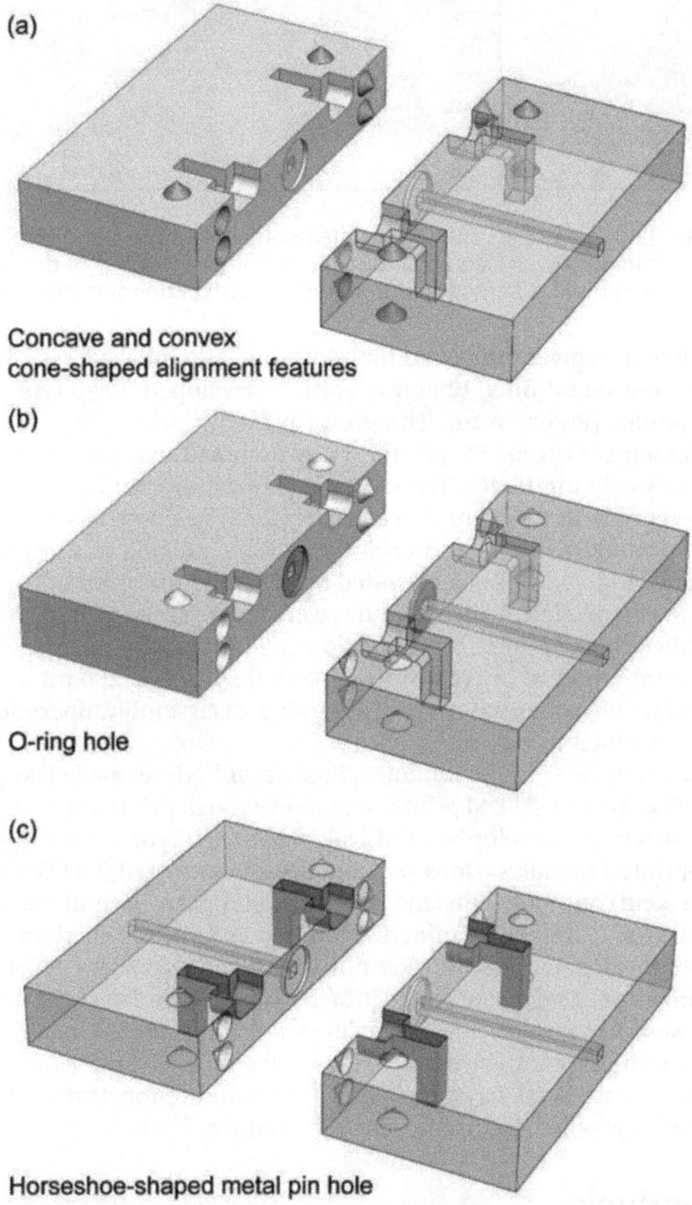


**Figure 2.25** Render of the 3D-printed modules with highlighted (a) conical, (b) O-ring, and (c) metal pin connectors for the assembly. Reproduced from ref. 36 with permission from The Royal Society of Chemistry.

can be simultaneously printed with the modules. These 3D-printed O-rings had an inner diameter of 4 mm, an outer diameter of 7 mm, a measured radial width of 1.5 mm, and a thickness of 2.5 mm, and they provided leak-proof connections up to 416 kPa.

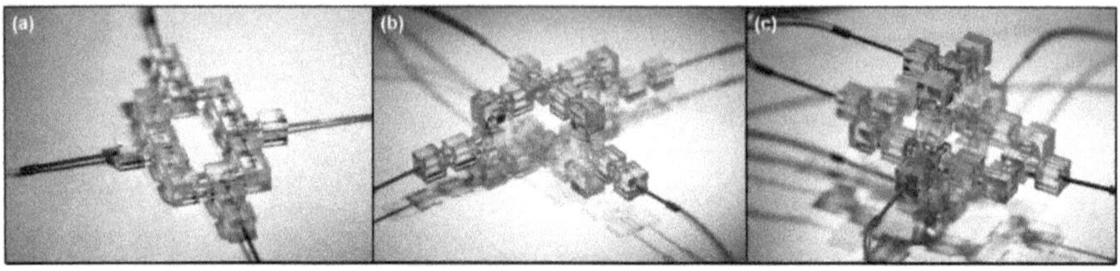


**Figure 2.26** Lego assembly of the 3D-printed polyacrylate composite microfluidic module to develop a microfluidic mixer with (a) two, (b) three, and (c) four outlets. Reproduced from ref. 107 with permission from PNAS.

To further circumvent the need for additional assembly accessories such as O-rings and metal pins, Bhargava *et al.*[107] developed 'Lego-type' microfluidic modules (Figure 2.26). The group printed a wide range of passive fluidic elements with an SL printer, such as junctions, mixers, splitters, and chip-to-world interfaces. These elements were assembled together easily and reversibly in different three-dimensional orientations using interfacing pins and spacers. The interfacing spacers assisted in assembly and disassembly, and additionally provided optical magnification of the fluidic interface for inspection. These modules were successfully used to assemble microfluidic droplet generators, emulsion generators, and mixers. The connections were found to be leak-proof up to a flow rate of 200 mL $h^{-1}$. They resisted up to 28 sequential trials of assembly, disassembly, operation, and cleaning, without any leakage.

Similar to these SL-printed modules, Tsuda *et al.*[108] developed FDM-printed Plug-and-Play modules. FDM printers result in a rough printed surface, which usually restricts the development of leak-proof Lego-type connections with the FDM-printed modules. However, here the group treated the FDM-printed poly(lactic acid) and poly(ethylene terephthalate) parts with dichloromethane vapours for 15 min and 1 min, respectively. The dichloromethane vapour treatment provided a smooth surface finish of the printed parts, which consequently provided a leak-proof connection between them. These FDM-printed modules were used to develop a droplet generator, a sensor-integrated cell growth chamber, and a valve-based flow selector. The FDM-printed parts were further treated with oxygen plasma and (3-aminopropyl)triethoxysilane to bond with the polydimethylsiloxane membranes.[108]

## 2.6 Outlook

Clearly, 3D printing techniques have the potential to largely replace conventional microfabrication techniques in the field of microfluidics going forward. These techniques have already been widely used for rapid and easy fabrication of complex fluid control devices using a variety of materials. To summarise these developments and applications, Table 2.2 lists the different techniques, printers, and print materials that have been used for printing the various micro- and macro-fluidic devices discussed in this chapter.

**Table 2.2** The 3D printing techniques, printers, and print materials that are used for fabricating different micro- and macro-fluidic devices.

| 3DP technique | Printer: Manufacturer | Printer: Model | Print material used | 3D-printed components[a] |
|---|---|---|---|---|
| FDM | Stratasys | Dimensions SST 768 | Acrylonitrile butadiene styrene | Centrifugal valves[47] |
| | ROVA | 3D | Acrylonitrile butadiene styrene | Microfluidic mixer[81] |
| | — | — | Poly(lactic acid) | Plug-and-Play millifluidic devices[108] |
| | — | — | Poly(ethylene terephthalate) | Plug-and-Play millifluidic devices[108] |
| SL | Formlabs | Form 1+ | — | Droplet generator[64] |
| | | | | Plug-and-Play droplet generator[66] |
| | Asiga | PicoPlus 27 | Poly(ethylene glycol) diacrylate monomer based resin | Active membrane-based valves[89] |
| | B9 Creator | V1.1 | Poly(ethylene glycol) diacrylate monomer based resin | Active membrane-based valves[39] |
| | 3D Systems | Viper | Somos® WaterShed XC 11122 | Active membrane-based valves and pumps[37] |
| | — | — | Somos® WaterShed XC 11122 | Discrete microfluidic elements[107] |
| | Miicraft | — | Polyacrylate composite | Microfluidic mixer[48] |
| | | | BV-003 | Pushing valves[87] |
| | | | | Torque actuated valves and pumps[87] |
| | | | | Rotary valves[87] |
| | | | — | Load/inject valve[86] |
| PolyJet printing | Stratasys | Objet Eden 250 | Fullcure® 720 | Homogeniser[46] |
| | | Objet260 Connex | — | Pumping lid and cup based pump[95] |
| | | | VeroClear™ and TangoPlus™ | Interlock meter-mix device[95] |
| | | Objet500 Connex | VeroWhitePlus™ and TangoPlus™ | Multi-material proportional valves[91] |
| | 3D Systems | ProJet 1200 | — | Microfluidic Tesla pump[101] |
| | | ProJet 3000 HD | VisiJet® M3 Crystal | Microfluidic operators[82] |
| | | ProJet 3500 HD | Acrylonitrile butadiene styrene | Droplet generator[63] |
| | | ProJet 3500 HD Plus | VisiJet® M3 Crystal | Modules for integrated microfluidics[36] |
| | | ProJet 3510 HD | — | Push-and-twist diaphragm valves[92] |
| | — | — | TangoBlack™ | O-rings[106] |

[a]The devices have been named in this table as per the names proposed in their respective reports. In the case of the absence of any clearly proposed name, generic names based on their applications have been used.

Future developments in the technology of 3D printers,[109,110] photopolymers,[111] and multi-material SL printing,[109,112] specifically, ever greater print resolution, will certainly rapidly advance the use of 3D printing techniques in the fabrication of more complex (smaller) microfluidic total analysis systems. The SL technique is currently the most commonly used 3D printing technique for fabricating micro- and macro-fluidic devices, because of its higher print resolution compared to the other 3D printing techniques, and the typically lower cost of SL printers compared to PolyJet, SLS, and SLM printers. A further increase in the 3D printing resolution would facilitate the production of smaller channels, better connections between different Plug-and-Play modules, and the reduction of the 'back side'[113] effect during SL printing. The resolution of the SL printers can be further improved by using high definition projectors, reducing the focus depth of the exposure light, and optimising the exposure time, polymerisation wavelength, and printing power.[63] New and non-proprietary print resins should allow better solvent compatibility and desired surface and bulk modification of the printed parts. Development of new dissolvable and liquid support materials[114] for PolyJet printing will provide for the fabrication of smaller and more complex microfluidic channels with ease. Multi-material 3D printing capabilities will continue to develop at a pace and will facilitate the fabrication of Plug-and-Play components that can be hermetically sealed. 3D-printed multi-material active membrane-based valves that do not deform globally can be used to construct elaborate programmable devices by arranging multiple valves in close proximity.[91]

3D printing may hold the solution to the much-awaited commercial and widespread dissemination of customised microfluidic devices. Wider access to low-cost standardised microfluidic devices and capability should allow greater uptake of microfluidic technology through different sections of research, development, and education. The use of various standardised interconnected 3D-printed components will, hopefully, in the words of Chan *et al.*[87] *"facilitate the development of the "lab-on-a-chip" (but not "chip-in-a-lab") devices."*

## References

1. C. D. Chin, T. Laksanasopin, Y. K. Cheung, D. Steinmiller, V. Linder, H. Parsa, J. Wang, H. Moore, R. Rouse and G. Umviligihozo, *Nat. Med.*, 2011, **17**, 1015–1019.
2. A. M. Foudeh, T. F. Didar, T. Veres and M. Tabrizian, *Lab Chip*, 2012, **12**, 3249–3266.
3. S. Mulvaney, K. Myers, P. Sheehan and L. Whitman, *Biosens. Bioelectron.*, 2009, **24**, 1109–1115.
4. M. L. Sin, K. E. Mach, P. K. Wong and J. C. Liao, *Expert Rev. Mol. Diagn.*, 2014, **14**, 225–244.
5. L.-M. Fu, J.-C. Leong, C.-F. Lin, C.-H. Tai and C.-H. Tsai, *Biomed. Microdevices*, 2007, **9**, 405–412.

6. M. T. Koesdjojo, Y. H. Tennico and V. T. Remcho, *Anal. Chem.*, 2008, **80**, 2311–2318.
7. R. A. Maurya, P. H. Hoang and D.-P. Kim, *Lab Chip*, 2012, **12**, 65–68.
8. Z. Meng, X. Zhang and J. Qin, *Nanoscale*, 2013, **5**, 4687–4690.
9. W. Zhan, G. H. Seong and R. M. Crooks, *Anal. Chem.*, 2002, **74**, 4647–4652.
10. L. Gervais, N. De Rooij and E. Delamarche, *Adv. Mater.*, 2011, **23**, H208.
11. A. Hatch, A. E. Kamholz, K. R. Hawkins, M. S. Munson, E. A. Schilling, B. H. Weigl and P. Yager, *Nat. Biotechnol.*, 2001, **19**, 461–465.
12. A. E. Herr, A. V. Hatch, D. J. Throckmorton, H. M. Tran, J. S. Brennan, W. V. Giannobile and A. K. Singh, *PNAS*, 2007, **104**, 5268–5273.
13. S. H. Kim, J. W. Shim and S. M. Yang, *Angew. Chem., Int. Ed.*, 2011, **50**, 1171–1174.
14. W. Wang, M.-J. Zhang and L.-Y. Chu, *Acc. Chem. Res.*, 2013, **47**, 373–384.
15. M. H. Lee, K. C. Hribar, T. Brugarolas, N. P. Kamat, J. A. Burdick and D. Lee, *Adv. Funct. Mater.*, 2012, **22**, 131–138.
16. L. Liu, W. Wang, X.-J. Ju, R. Xie and L.-Y. Chu, *Soft Matter*, 2010, **6**, 3759–3763.
17. Z. Nie, W. Li, M. Seo, S. Xu and E. Kumacheva, *J. Am. Chem. Soc.*, 2006, **128**, 9408–9412.
18. A. Khademhosseini, R. Langer, J. Borenstein and J. P. Vacanti, *PNAS*, 2006, **103**, 2480–2487.
19. C. Xia and N. X. Fang, *Biomed. Microdevices*, 2009, **11**, 1309–1315.
20. P. Neuži, S. Giselbrecht, K. Länge, T. J. Huang and A. Manz, *Nat. Rev. Drug Discovery*, 2012, **11**, 620–632.
21. E. Brouzes, M. Medkova, N. Savenelli, D. Marran, M. Twardowski, J. B. Hutchison, J. M. Rothberg, D. R. Link, N. Perrimon and M. L. Samuels, *PNAS*, 2009, **106**, 14195–14200.
22. M. Chabert and J.-L. Viovy, *PNAS*, 2008, **105**, 3191–3196.
23. S. Nagrath, L. V. Sequist, S. Maheswaran, D. W. Bell, D. Irimia, L. Ulkus, M. R. Smith, E. L. Kwak, S. Digumarthy and A. Muzikansky, *Nature*, 2007, **450**, 1235–1239.
24. L.-Y. Hung, Y.-H. Chuang, H.-T. Kuo, C.-H. Wang, K.-F. Hsu, C.-Y. Chou and G.-B. Lee, *Biomed. Microdevices*, 2013, **15**, 339–352.
25. J. Khandurina, T. E. McKnight, S. C. Jacobson, L. C. Waters, R. S. Foote and J. M. Ramsey, *Anal. Chem.*, 2000, 72, 2995–3000.
26. A. R. Abate, T. Hung, R. A. Sperling, P. Mary, A. Rotem, J. J. Agresti, M. A. Weiner and D. A. Weitz, *Lab Chip*, 2013, **13**, 4864–4869.
27. X. Pan, S. Zeng, Q. Zhang, B. Lin and J. Qin, *Electrophoresis*, 2011, **32**, 3399–3405.
28. S.-m. Park, J.-Y. Ahn, M. Jo, D.-k. Lee, J. T. Lis, H. G. Craighead and S. Kim, *Lab Chip*, 2009, **9**, 1206–1212.
29. F. Moltzahn, A. B. Olshen, L. Baehner, A. Peek, L. Fong, H. Stöppler, J. Simko, J. F. Hilton, P. Carroll and R. Blelloch, *Cancer Res.*, 2011, **71**, 550–560.
30. S. Vorwerk, K. Ganter, Y. Cheng, J. Hoheisel, P. F. Stähler and M. Beier, *New Biotechnol.*, 2008, **25**, 142–149.

31. B. V. Chikkaveeraiah, V. Mani, V. Patel, J. S. Gutkind and J. F. Rusling, *Biosens. Bioelectron.*, 2011, **26**, 4477–4483.
32. M. Hu, J. Yan, Y. He, H. Lu, L. Weng, S. Song, C. Fan and L. Wang, *ACS Nano*, 2009, **4**, 488–494.
33. R. Malhotra, V. Patel, B. V. Chikkaveeraiah, B. S. Munge, S. C. Cheong, R. B. Zain, M. T. Abraham, D. K. Dey, J. S. Gutkind and J. F. Rusling, *Anal. Chem.*, 2012, **84**, 6249–6255.
34. B. L. Ziober, M. G. Mauk, E. M. Falls, Z. Chen, A. F. Ziober and H. H. Bau, *Head Neck*, 2008, **30**, 111–121.
35. P. N. Nge, C. I. Rogers and A. T. Woolley, *Chem. Rev.*, 2013, **113**, 2550–2583.
36. K. G. Lee, K. J. Park, S. Seok, S. Shin, J. Y. Park, Y. S. Heo, S. J. Lee and T. J. Lee, *RSC Adv.*, 2014, **4**, 32876–32880.
37. A. K. Au, N. Bhattacharjee, L. F. Horowitz, T. C. Chang and A. Folch, *Lab Chip*, 2015, **15**, 1934–1941.
38. J. C. McDonald and G. M. Whitesides, *Acc. Chem. Res.*, 2002, **35**, 491–499.
39. C. I. Rogers, K. Qaderi, A. T. Woolley and G. P. Nordin, *Biomicrofluidics*, 2015, **9**, 016501.
40. S. Waheed, J. M. Cabot, N. P. Macdonald, T. Lewis, R. M. Guijt, B. Paull and M. C. Breadmore, *Lab Chip*, 2016, **16**, 1993–2013.
41. A. K. Au, W. Huynh, L. F. Horowitz and A. Folch, *Angew. Chem., Int. Ed.*, 2016, **55**, 3862–3881.
42. C. M. B. Ho, S. H. Ng, K. H. H. Li and Y.-J. Yoon, *Lab Chip*, 2015, **15**, 3627–3637.
43. J. C. McDonald, M. L. Chabinyc, S. J. Metallo, J. R. Anderson, A. D. Stroock and G. M. Whitesides, *Anal. Chem.*, 2002, **74**, 1537–1545.
44. S. Waheed, J. M. Cabot, N. P. Macdonald, U. Kalsoom, S. Farajikhah, P. C. Innis, P. N. Nesterenko, T. W. Lewis, M. C. Breadmore and B. Paull, *Sci. Rep.*, 2017, **7**, 15109.
45. B. C. Gross, J. L. Erkal, S. Y. Lockwood, C. Chen and D. M. Spence, *Anal. Chem.*, 2014, **86**, 3240–3253.
46. A. Bonyár, H. Sántha, B. Ring, M. Varga, J. G. Kovács and G. Harsányi, *Procedia Eng.*, 2010, **5**, 291–294.
47. J. L. Moore, A. McCuiston, I. Mittendorf, R. Ottway and R. D. Johnson, *Microfluid. Nanofluid.*, 2011, **10**, 877–888.
48. A. I. Shallan, P. Smejkal, M. Corban, R. M. Guijt and M. C. Breadmore, *Anal. Chem.*, 2014, **86**, 3124–3130.
49. W. G. Patrick, A. A. Nielsen, S. J. Keating, T. J. Levy, C.-W. Wang, J. J. Rivera, O. Mondragón-Palomino, P. A. Carr, C. A. Voigt and N. Oxman, *PLoS One*, 2015, **10**, e0143636.
50. J. M. Lee, M. Zhang and W. Y. Yeong, *Microfluid. Nanofluid.*, 2016, **20**, 1–15.
51. N. P. Macdonald, J. M. Cabot, P. Smejkal, R. M. Guijt, B. Paull and M. C. Breadmore, *Anal. Chem.*, 2017, **89**, 3858–3866.
52. G. Muschiolik, *Curr. Opin. Colloid Interface Sci.*, 2007, **12**, 213–220.

53. D. Vasiljevic, J. Parojcic, M. Primorac and G. Vuleta, *Int. J. Pharm.*, 2006, **309**, 171–177.
54. I. Tuncer Degim and N. Çelebi, *Curr. Pharm. Des.*, 2007, **13**, 99–117.
55. Q. Xu, M. Hashimoto, T. T. Dang, T. Hoare, D. S. Kohane, G. M. Whitesides, R. Langer and D. G. Anderson, *Small*, 2009, **5**, 1575–1581.
56. H. Becker and U. Heim, *Sens. Actuators, A*, 2000, **83**, 130–135.
57. Y. Xia and G. M. Whitesides, *Annu. Rev. Mater. Sci.*, 1998, **28**, 153–184.
58. A. Utada, E. Lorenceau, D. Link, P. Kaplan, H. Stone and D. Weitz, *Science*, 2005, **308**, 537–541.
59. R. Dangla, F. Gallaire and C. N. Baroud, *Lab Chip*, 2010, **10**, 2972–2978.
60. E. Sollier, C. Murray, P. Maoddi and D. Di Carlo, *Lab Chip*, 2011, **11**, 3752–3765.
61. J. N. Lee, C. Park and G. M. Whitesides, *Anal. Chem.*, 2003, **75**, 6544–6554.
62. K. J. Regehr, M. Domenech, J. T. Koepsel, K. C. Carver, S. J. Ellison-Zelski, W. L. Murphy, L. A. Schuler, E. T. Alarid and D. J. Beebe, *Lab Chip*, 2009, **9**, 2132–2139.
63. L. Donvito, L. Galluccio, A. Lombardo, G. Morabito, A. Nicolosi and M. Reno, *J. Micromech. Microeng.*, 2015, **25**, 035013.
64. J. M. Zhang, E. Q. Li, A. A. Aguirre-Pablo and S. T. Thoroddsen, *RSC Adv.*, 2016, **6**, 2793–2799.
65. A. Rotem, A. R. Abate, A. S. Utada, V. Van Steijn and D. A. Weitz, *Lab Chip*, 2012, **12**, 4263–4268.
66. J. M. Zhang, A. A. Aguirre-Pablo, E. Q. Li, U. Buttner and S. T. Thoroddsen, *RSC Adv.*, 2016, **6**, 81120–81129.
67. G. H. Seong and R. M. Crooks, *J. Am. Chem. Soc.*, 2002, **124**, 13360–13361.
68. H. Roder and M. R. Shastry, *Curr. Opin. Struct. Biol.*, 1999, **9**, 620–626.
69. J. Pihl, J. Sinclair, E. Sahlin, M. Karlsson, F. Petterson, J. Olofsson and O. Orwar, *Anal. Chem.*, 2005, **77**, 3897–3903.
70. S. Sugiura, K. Hattori and T. Kanamori, *Anal. Chem.*, 2010, **82**, 8278–8282.
71. P. Hinsmann, J. Frank, P. Svasek, M. Harasek and B. Lendl, *Lab Chip*, 2001, **1**, 16–21.
72. P. Guichardon and L. Falk, *Chem. Eng. Sci.*, 2000, **55**, 4233–4243.
73. D. Yamamoto, T. Maki, S. Watanabe, H. Tanaka, M. T. Miyahara and K. Mae, *Chem. Eng. J.*, 2013, **227**, 145–150.
74. Y. Song, J. Hormes and C. S. Kumar, *Small*, 2008, **4**, 698–711.
75. N. Nguyen and Z. Wu, *J. Micromech. Microeng.*, 2005, **15**, R1.
76. L. Capretto, W. Cheng, M. Hill and X. Zhang, *Microfluidics*, Springer, 2011, pp. 27–68.
77. C.-Y. Lee, C.-L. Chang, Y.-N. Wang and L.-M. Fu, *Int. J. Mol. Sci.*, 2011, **12**, 3263–3287.
78. A. D. Stroock, S. K. Dertinger, A. Ajdari, I. Mezić, H. A. Stone and G. M. Whitesides, *Science*, 2002, **295**, 647–651.
79. P. Carrière, *Phys. Fluids*, 2007, **19**, 118110.
80. Y. Liao, J. Song, E. Li, Y. Luo, Y. Shen, D. Chen, Y. Cheng, Z. Xu, K. Sugioka and K. Midorikawa, *Lab Chip*, 2012, **12**, 746–749.

81. F. Li, N. P. Macdonald, R. M. Guijt and M. C. Breadmore, *Anal. Chem.*, 2017, **89**, 12805–12811.
82. R. Sochol, E. Sweet, C. Glick, S. Venkatesh, A. Avetisyan, K. Ekman, A. Raulinaitis, A. Tsai, A. Wienkers and K. Korner, *Lab Chip*, 2016, **16**, 668–678.
83. Q. L. Chen, Z. Liu and H. C. Shum, *Biomicrofluidics*, 2014, **8**, 064112.
84. M. A. Unger, H.-P. Chou, T. Thorsen, A. Scherer and S. R. Quake, *Science*, 2000, **288**, 113–116.
85. T. Thorsen, S. J. Maerkl and S. R. Quake, *Science*, 2002, **298**, 580–584.
86. C.-K. Su, S.-C. Hsia and Y.-C. Sun, *Anal. Chim. Acta*, 2014, **838**, 58–63.
87. H. N. Chan, Y. Shu, B. Xiong, Y. Chen, Y. Chen, Q. Tian, S. A. Michael, B. Shen and H. Wu, *ACS Sens.*, 2015, **1**, 227–234.
88. W. Du, L. Li, K. P. Nichols and R. F. Ismagilov, *Lab Chip*, 2009, **9**, 2286–2292.
89. H. Gong, A. T. Woolley and G. P. Nordin, *Lab Chip*, 2016, **16**, 2450–2458.
90. H. Gong, M. Beauchamp, S. Perry, A. T. Woolley and G. P. Nordin, *RSC Adv.*, 2015, **5**, 106621–106632.
91. S. J. Keating, M. I. Gariboldi, W. G. Patrick, S. Sharma, D. S. Kong and N. Oxman, *PLoS One*, 2016, **11**, e0160624.
92. T.-H. Kim, V. Sunkara, J. Park, C.-J. Kim, H.-K. Woo and Y.-K. Cho, *Lab Chip*, 2016, **16**, 3741–3749.
93. M. Pumera, S. Sanchez, I. Ichinose and J. Tang, *Sens. Actuators, B*, 2007, **123**, 1195–1205.
94. C.-H. Wang and G.-B. Lee, *Biosens. Bioelectron.*, 2005, **21**, 419–425.
95. S. Begolo, D. V. Zhukov, D. A. Selck, L. Li and R. F. Ismagilov, *Lab Chip*, 2014, **14**, 4616–4628.
96. E. Jue, N. G. Schoepp, D. Witters and R. F. Ismagilov, *Lab Chip*, 2016, **16**, 1852–1860.
97. N. Tesla, *US Pat.*, 1061142, 1913.
98. N. Tesla, *US Pat.*, 1061206, 1913.
99. G. Miller, A. Sidhu, R. Fink and B. Etter, *Artif. Organs*, 1993, **17**, 590–592.
100. G. Miller, M. Madigan and R. Fink, *Artif. Organs*, 1995, **19**, 680–684.
101. M.-B. Habhab, T. Ismail and J. F. Lo, *Sensors*, 2016, **16**, 1970.
102. J. Xiang, Z. Cai, Y. Zhang and W. Wang, *Sens. Actuators, A*, 2016, **251**, 20–25.
103. D. Thomas, Z. Tehrani and B. Redfearn, *Addit. Manuf.*, 2016, **9**, 30–38.
104. P. K. Yuen, *Lab Chip*, 2008, **8**, 1374–1378.
105. P. K. Yuen, J. T. Bliss, C. C. Thompson and R. C. Peterson, *Lab Chip*, 2009, **9**, 3303–3305.
106. O. Paydar, C. Paredes, Y. Hwang, J. Paz, N. Shah and R. Candler, *Sens. Actuators, A*, 2014, **205**, 199–203.
107. K. C. Bhargava, B. Thompson and N. Malmstadt, *PNAS*, 2014, **111**, 15013–15018.
108. S. Tsuda, H. Jaffery, D. Doran, M. Hezwani, P. J. Robbins, M. Yoshida and L. Cronin, *PLoS One*, 2015, **10**, e0141640.

109. ILIOS RAY, http://www.ilios3d.com/en/shop/shop-3d-printing/ilios-ray/ilios-ray-printer-detail, (accessed 1st April 2016), 2016.
110. OLD WORLD LABS, http://www.oldworldlabs.com/, (accessed 1st April 2016), 2016.
111. X. Wang, X. Cai, Q. Guo, T. Zhang, B. Kobe and J. Yang, *Chem. Commun.*, 2013, **49**, 10064–10066.
112. J.-W. Choi, E. MacDonald and R. Wicker, *Int. J. Adv. Des. Manuf. Technol.*, 2010, **49**, 543–551.
113. A. Waldbaur, H. Rapp, K. Länge and B. E. Rapp, *Anal. Methods*, 2011, **3**, 2681–2716.
114. W. G. Patrick, Masters dissertation, Massachusetts Institute of Technology, 2015.

CHAPTER 3

# 3D-printed Analytical Detectors

## 3.1 Introduction

Point-of-care diagnostic devices and portable miniaturised analysers are becoming increasingly available, arising from application needs across many sectors, including personal healthcare monitoring, biological and chemical hazard checks, online analysis, and environmental monitoring, to name just a few. Miniaturised analytical detection is the essential component of most functional point-of-care devices and portable analytical platforms. The most commonly used analytical detection modes within such devices are optical and electrochemical based, both of which are composed of multiple discrete components within a detector assembly. Optical detectors usually consist of a light source, an optical filter or diffraction module, a detector cell, and a photodetector, whilst electrochemical detectors contain a detector cell, electrode system(s), and the associated electronics/meter. The precise 3D arrangement of components within a detector cell is of course vital to acquiring optimal performance and sensitivity. 3D printing allows these arrangements to be optimised *via* rapid physical iterations with low productions costs.

Over the past few years, 3D printing techniques have been widely applied to the production of various components for both optical and electrochemical detectors, and for their integration into detector assemblies with other off-the-shelf components. These developments, and the performance of 3D-printed detector cells, interfaces, holders, optics, and electrodes within different optical and electrochemical detectors, are described herein.

3D Printing in Chemical Sciences: Applications Across Chemistry
By Vipul Gupta, Pavel Nesterenko and Brett Paull

Published by the Royal Society of Chemistry, www.rsc.org

## 3.2 Optical Detectors

Optical detection is one of the most commonly applied detector techniques across many fields, including general analytical/physical sciences, medicinal science, the life sciences in general, food technology, and the pharmaceutical sciences.[1] Over the past two decades or more, there has been a growing interest in the miniaturisation of optical detectors and the development of optical sensors, for applications in real-time environmental monitoring, forensic investigations, and personal and medical diagnostics. Some of the key requirements of miniaturised optical detectors include portability, low production costs, durability and reliability, and ease of maintenance and fabrication. This is where the advantages of 3D printing techniques, namely ease of use, low-cost manufacturing, rapid prototyping, on-site production capabilities, and fabrication of conventionally impractical designs and architectures, can be of particular utility.

### 3.2.1 Smartphone Optical-sensing Platforms

The current theme of the U.S. Food and Drug Administration: "the right patient with the right drug at the right dose at the right time"[2] has increased the focus on research into some of the key requirements of point-of-care technology. These include those mentioned above, plus reliability, appropriate selectivity, sensitivity and accuracy, and of course delivery of data to individual users and/or their doctors. The smartphone revolution shows no signs of slowing. The computing capabilities of smartphones increase annually, as does the number of applications and functional systems. The use of adapted or adaptive smartphones with, or as, point-of-care devices presents enormous potential. Many reports have already been published on the use of smartphones for colourimetric,[3–5] fluorescence,[6] and label-free[7] determination of various target analytes, based upon the now ubiquitous smartphone cameras. These include biomarkers in body fluids,[4,8–10] cholesterol levels in blood,[10] infectious pathogens in environmental and clinical samples,[6,11–14] viruses,[15] antigens,[16] and vitamins.[8] It was predicted in 2010 that this decade "will probably be known as the era of digital medical devices".[17] Almost a decade later, we can say that indeed there have been significant developments in this area, and that 3D fabrication methods are helping to support this movement.

The development of most smartphone-based optical-sensing platforms requires the development of interfaces and/or sample holders, which can integrate and align the sample cell, the light source, and the other required components with a smartphone camera.[18] The availability of 3D printing to enable the rapid production of the necessary complex and bespoke architectures has led to a plethora of impressive prototype platforms being reported. Moreover, these interfaces can be quickly modified to suit different smartphone models and be customised to enable the integration of different components.

### 3.2.1.1 External Light Sources

Light-emitting diodes (LEDs) and mini-lasers are commonly used miniaturised light sources for optical detection techniques that require a light source to illuminate the sample. 3D-printed interfaces have been used to interface these light sources with a sample cell and a smartphone camera. Such modified smartphones are able to perform qualitative, and in some cases quantitative, measurements by recording the light emitted from, or transmitted through, the sample cell. The captured images are either processed on the phone or transferred wirelessly to a computer for post-processing. Several 3D-printed interfaces and attachments have been developed to isolate assemblies for luminescence-based assays from external and background light sources, and to integrate and align the required optical and analytical components.

Researchers from the Ozcan laboratory have been pioneers in the field of smartphone-based detection, and they have recently reported 3D printing of a low-cost and modified version of their own traditionally fabricated 'optomechanical cell phone attachment'.[19–22] The attachment was 3D-printed with a low-cost and easy to use FDM printer, and its cost was estimated at only a few dollars to produce. It consisted of a 3D-printed cell phone holder, sample tray, and mechanical lid, and it successfully aligned various components such as 14 LEDs, an optical filter, and an aspherical lens with the smartphone camera (Figure 3.1).[23] The 3D-printed attachment provided a portable, lightweight, and low-cost set-up, which was successfully used to perform both brightfield and darkfield microsphere fluorescence immunoassays of anti-recombinant bovine somatotropin (rbST) antibodies in milk extracts.[23] The device when applied was reported to deliver 80% true positive and 95% true negative outcomes. The set-up was further used for the development of a smartphone-based protein microarray fluorescence

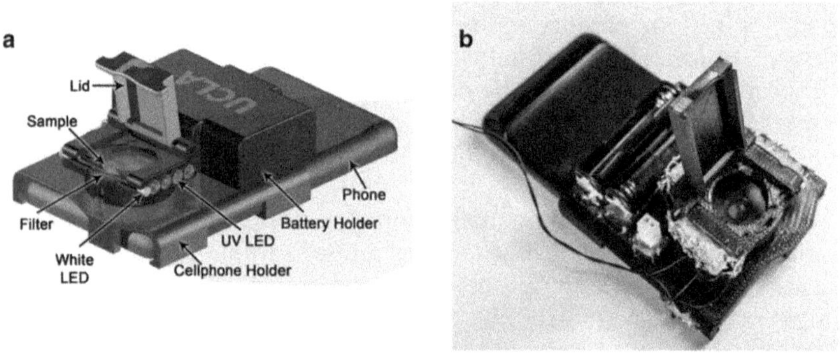


**Figure 3.1** 3D-printed smartphone-based platform for fluorescence immunoassay: (a) render and (b) photograph. Reproduced from ref. 23 with permission from Springer Nature, Copyright 2014.

immunoassay.[24] The immunoassay was utilised for the simultaneous determination of multiple analytes in different complex matrices such as milk, blood, saliva, and urine.

In addition to the smartphone-based fluorescence immunoassays, smartphone-based fluorescence microscopy has also been enabled using 3D-printed attachments.[25] Knowlton *et al.*[25] developed a 3D-printed attachment for a smartphone, which was designed to integrate the various optical components required to perform fluorescence microscopy with magnetic focusing. The printed attachment consisted of a smartphone holder, a two-piece optical component, emission filter holders, and a shade, all of which can be seen within Figure 3.2. All 3D-printed components were designed with tracks to allow easy assembly and disassembly, and cost-effective modification. The cost of materials for 3D printing of the attachment with an SL printer was quoted, rather precisely, as $18.23. The smartphone-based platform successfully magnified microscopic objects and provided images similar to those obtained with conventional fluorescence microscopes. It performed darkfield imaging comparable to the quality of brightfield imaging that had been performed in previously developed magnetic focusing devices.[20,26]

3D-printed interfaces have also been developed to enable smartphone-based measurements in combination with paper-based microfluidic analytical devices, or so-called μ-PADs. Diduk *et al.* 3D-printed a dark box to assist with the readout of a paper-based device that was embedded with fluorescent graphene quantum dots (Figure 3.3).[27] The platform was applied to the determination of phenolic compounds in food and environmental samples, based upon the quenching effect of these phenolic compounds on the fluorescence of quantum dots. The 3D-printed dark box successfully isolated

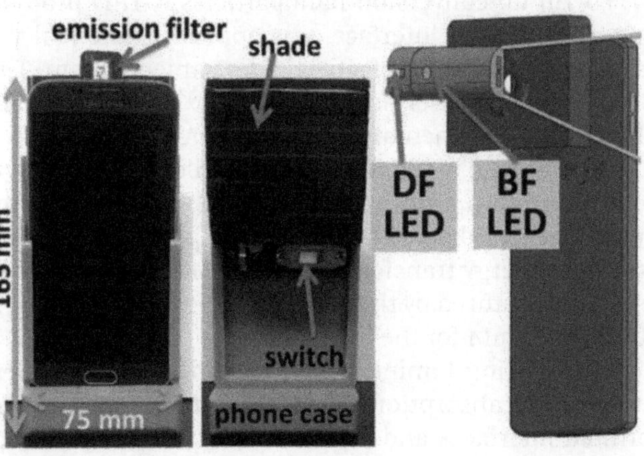


**Figure 3.2** 3D-printed smartphone-based platform for fluorescence microscopy with magnetic focusing. Reproduced from ref. 25 with permission from The Royal Society of Chemistry.

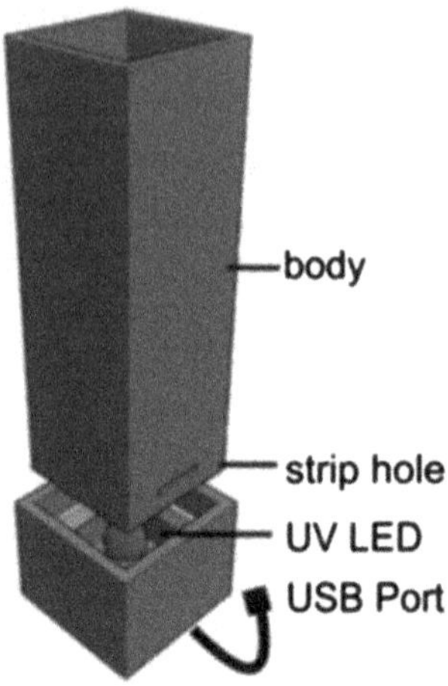


**Figure 3.3** Render of a 3D-printed smartphone-based device for μ-PADs. Reproduced from ref. 27 [https://doi.org/10.1038/s41598-017-01134-3] under the terms of a CC BY 4.0 license [https://creativecommons.org/licenses/by/4.0/].

the paper device from the external light sources and isolated the sample radiation to that produced from a UV ($\lambda$ = 365 nM) LED. The platform was used for screening the flavonoid content in wine samples, and the 4-nitrophenol and the drug paraoxon in seawater samples. The results obtained with the smartphone-based platform correlated well with the standard lab-based fluorimetry results.

A similar 3D-printed interface has also been produced to assist with the smartphone-based measurement from a μ-PAD assay, in this case based upon $NaYF_4$:Yb/Tm up-conversion nanoparticles.[28] This prototype system, incorporating a 3D-printed interface, was applied to the real-time and off-site quantitative determination of a pesticide, namely thiram. The complete set-up consisted of a mini-laser holder, a dark box with an optical filter and a test paper chamber, and the actual phone holder. All of which were 3D-printed using an FDM printer and made from acrylonitrile butadiene styrene. In the demonstrated application, thiram was detected by quenching the blue luminescence of the laser-irradiated $NaYF_4$:Yb/Tm up-conversion nanoparticles, through resonance energy transfer. Measurement of the blue channel intensities of the images captured by the smartphone camera resulted in a limit of detection (LOD) of 0.1 μM for the target analyte.

In addition to imaging luminescence, modified smartphones have also been used to perform absorption-based photometric measurements. Once again, 3D-printed interfaces and attachments have been developed to integrate various optical and analytical components with the smartphone to provide this capability.

Monitoring of the food supply chain is becoming increasingly important, as exemplified by multiple worldwide food adulteration incidents in recent years.[29,30] In the area of food safety, 3D-printed components have once again been designed and developed to enable smartphones to provide a portable detection option for food-safety related markers, such as aflatoxin B1[31] and streptomycin.[32] Li *et al.* FDM-printed a smartphone adapter to develop a so-called 'Smartphone-App-Chip System' for colourimetric detection of aflatoxins.[31] The 3D-printed adapter was designed to align and uniformly illuminate an assay chip for imaging by a smartphone camera as shown in Figure 3.4. The images were directly processed on the phone using a custom-developed Android application. The platform was used to study spiked and mould infected corn samples, resulting in an LOD of 3 ± 1 ppb and a dynamic range upto 250 ppb. The results were consistent with the results obtained by conventional enzyme-linked immunosorbent assay (ELISA) and high performance liquid chromatography (HPLC) methods.

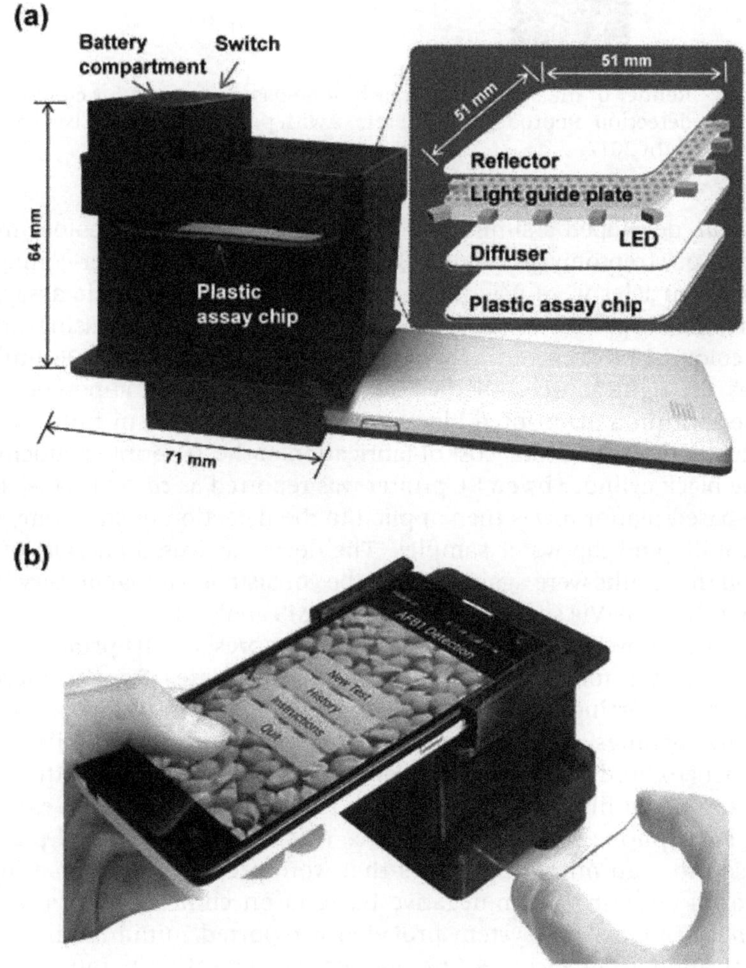


**Figure 3.4** 3D-printed smartphone-based platform for imaging chip-based immunoassay: (a) schematic, and (b) platform in operation. Reproduced from ref. 31 with permission from American Chemical Society, Copyright 2017.

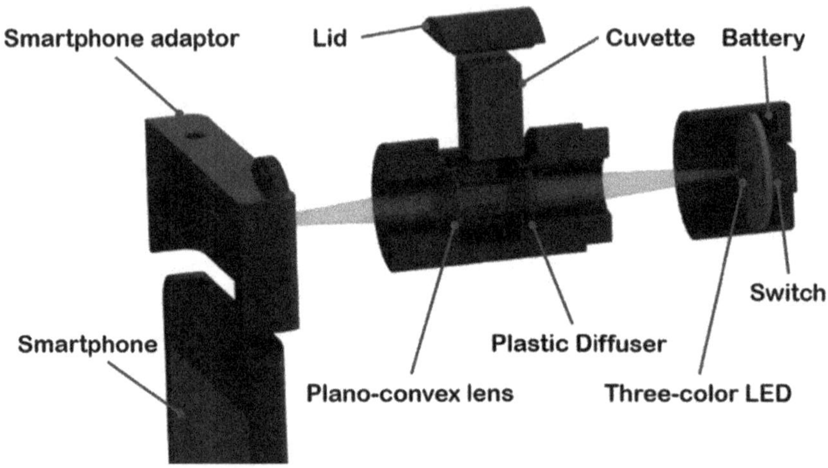


**Figure 3.5** Render of the 3D-printed smartphone-based platform for colourimetric detection. Reproduced from ref. 32 with permission from Elsevier, Copyright 2017.

Liu *et al.* developed a similar 3D-printed accessory for the colourimetric detection of streptomycin (Figure 3.5).[32] In this case, aptamer-conjugated gold nanoparticles were used as the basis of the colourimetric assay, and a smartphone applied as the light capturing and data processing unit. A 'three-colour' LED (centre wavelengths 473 nm, 520 nm, and 625 nm) was used as the light source. All the electrical and optical components were enclosed within a 3D-printed black cylinder to isolate them from external light fluctuations. The total cost of fabrication of the 3D-printed attachment and the black cylinder by an SL printer was reported as *ca.* $13. This smartphone-based platform was then applied to the detection of streptomycin in honey, milk, and tap water samples. The device delivered an LOD of 12.3 nM, and the results were again found to be consistent with laboratory-based methods, here UV-Vis spectroscopic and LC-MS analysis.

Feng *et al.* have developed a particularly impressive 3D-printed smartphone interface for reading 96-well microtiter plates for antimicrobial susceptibility testing.[33] The assembled device is shown in Figure 3.6. The assembly includes a 3D-printed attachment (no printer details given), which acts to hold the microtiter plates and illuminate them with an LED array. An optical fibre array was used to transmit the light from each well to the smartphone camera for capture. This smartphone-based platform was used to read microtiter plates that were prepared with 17 different antibiotics targeting Gram-negative bacteria on clinical isolates of *Klebsiella pneumoniae*. The system provided a reported minimum-inhibitory concentration accuracy of 95.12% and a drug susceptibility interpretation accuracy of 99.23%.

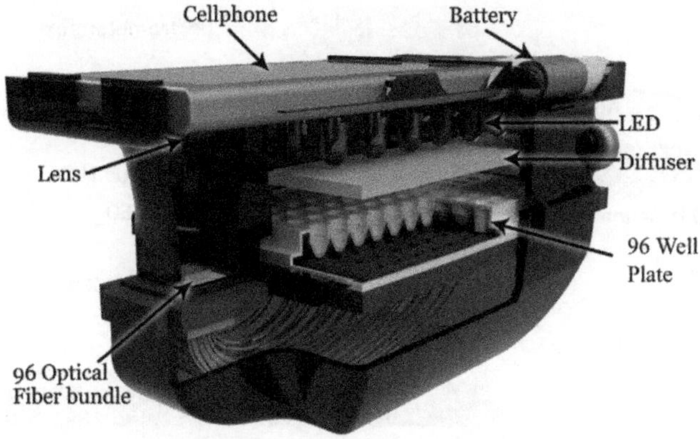


**Figure 3.6** Render of the smartphone-based platform for reading 96-well microtiter plates. Reproduced from ref. 33 [https://doi.org/10.1038/srep39203] under the terms of a CC BY 4.0 license [https://creativecommons.org/licenses/by/4.0/].

Finally, Wang *et al.* reported upon the use of a digital versatile disc diffraction grating which was incorporated into a 3D-printed smartphone interface to perform visible spectroscopy.[34] The diffraction grating was used to split the collimated light from a white light LED into wavelengths ranging from 400 nm to 700 nm. An FDM-printed sample holder was used for the sample illumination. This smartphone-based platform was used to assay paraoxons using two cholinesterase enzymes – acetylcholinesterase and butyrylcholinesterase, with an LOD of 2.9 nM and a dynamic range of 5 nM to 25 μM.

### 3.2.1.2 *Smartphone Flashlight as a Light Source*

To circumvent the need for an external light source in various modes of optical detection, novel 3D-printed interfaces have been developed to exploit the smartphone flashlight as a light source. For most smart devices, the flashlight can be controlled independently of the camera using either the inbuilt, or a third-party software. Where this is possible, the use of the smartphone flashlight to replace the external light source can both lower the cost and complexity of the final device. Again, the bespoke nature of 3D printing allows the development of holders and interfaces that can effectively integrate smartphone flashlights with other optical components, such as filters, diffraction grating, *etc.*

Hossain *et al.* have recently produced a smartphone-based spectrometer for both absorption and fluorescence based detection by developing a novel 3D-printed interface.[35] The 3D-printed attachment was made from acrylonitrile butadiene styrene, and was printed using an FDM printer. It was used to integrate the white flashlight of the smartphone, orthogonally positioned

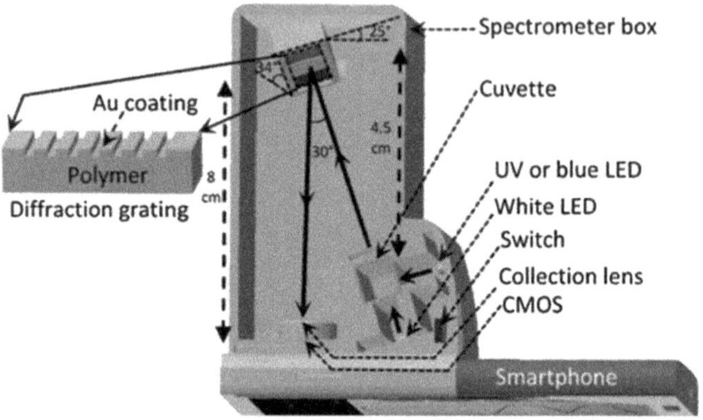


**Figure 3.7** Render of the 3D-printed smartphone-based spectrometer for absorption and fluorescence detection. Reproduced from ref. 35 with permission from The Optical Society.

UV ($\lambda_{ex}$ = 370 nm) and blue ($\lambda_{ex}$ = 450 nm) LEDs, and a gold-coated nano-imprinted diffraction grating (Figure 3.7). This integrated platform was powered by the smartphone battery. The smartphone-based device was used to obtain absorbance and fluorescence spectra of molecular probes for both pH and $Zn^{2+}$ concentrations in water samples. This simple yet effective platform was able to perform detection over a 300 nm span with a 0.42 nm per pixel resolution.

Similarly, Petryayeva *et al.* recently reported a 3D-printed interface to perform all-in-one smartphone-based excitation and imaging of photoluminescence.[36] The relatively simple set-up is shown in Figure 3.8. The 3D-printed interface was again printed in acrylonitrile butadiene styrene with an FDM printer, and it consisted of a blackout box, a box lid, and two plastic reflectors. It was used for a protein binding assay and a Förster resonance energy transfer assay of proteolytic activity in serum. This platform provided the proof-of-concept for a simple, miniaturised, lightweight, and low-cost system for all-in-one sample excitation and photoluminescence measurements.

### 3.2.1.3 *Luminescence-based Assays*

Modified smartphones with 3D-printed attachments have been applied in luminescence assays, such as bioluminescence and chemiluminescence, which circumvent the need for either an internal or external light source. Obviously, elimination of any light source from the smartphone-based platform considerably simplifies its design, production, and use. The following are examples of where simple 3D-printed housings and attachments have been developed to deliver smartphone-based bioluminescence and chemiluminescence assays.

Roda *et al.* have utilised 3D-printed cartridges and smartphone housings to perform bioluminescence assays of total bile acid,[9] a chemiluminescence

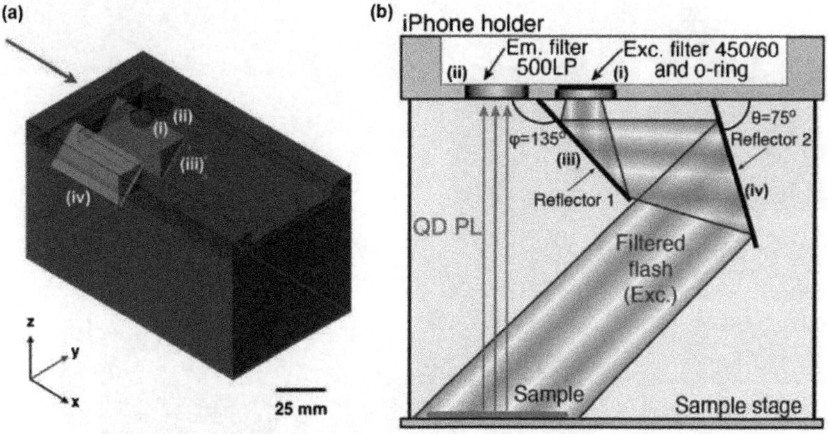


**Figure 3.8** 3D-printed smartphone accessory for all-in-one excitation and imaging of photoluminescence. (a) Render with filter positions marked as (i) and (ii), and reflector supports marked as (iii) and (iv). (b) Schematic of the path of excitation light. Reproduced from ref. 36 with permission from Springer Nature, Copyright 2016.

assay of total cholesterol in serum,[9] chemiluminescence assays of total lactate in sweat and oral fluids,[37] and whole cell toxicity studies.[38] Initially, in 2014, the group reported the FDM printing of an acrylonitrile butadiene styrene mini-cartridge and smartphone accessory to quantify target analytes in biological fluids (Figure 3.9). The quantification was based on biochemiluminescence-coupled biospecific enzymatic reactions. They performed two assays as a proof-of-concept, namely a bioluminescence assay for total bile acid and a chemiluminescence assay for total cholesterol. The total bile acid was measured using 3α-hydroxyl steroid dehydrogenase co-immobilised with a bacterial luciferase system. The total cholesterol was measured using cholesterol esterase/cholesterol oxidase coupled with luminol-$H_2O_2$-horseradish peroxidase. The smartphone-printed housing was designed to serve two functions. Firstly, to provide a dark space by eliminating the ambient light, and secondly, to host the mini-cartridge. The smartphone-based platform delivered an analysis time of just 3 minutes and an LOD of 0.5 μM and 20 mg $dL^{-1}$ for total bile acid and total cholesterol, respectively.

In the same year, the same group also reported the 3D printing of a disposable mini-cartridge for a chemiluminescence assay of lactate in sweat and oral fluids.[37] The disposable cartridges were again printed in acrylonitrile butadiene styrene using an FDM printer. The lactate was measured by coupling lactate oxidase with horseradish peroxidase. The smartphone-based platform consisted of a similar disposable analytical mini-cartridge, mini-dark box, and smartphone adapter. The smartphone adapter also consisted of a plano-convex plastic lens to focus the reaction chamber image of the cartridge onto the smartphone camera. The platform delivered an LOD of

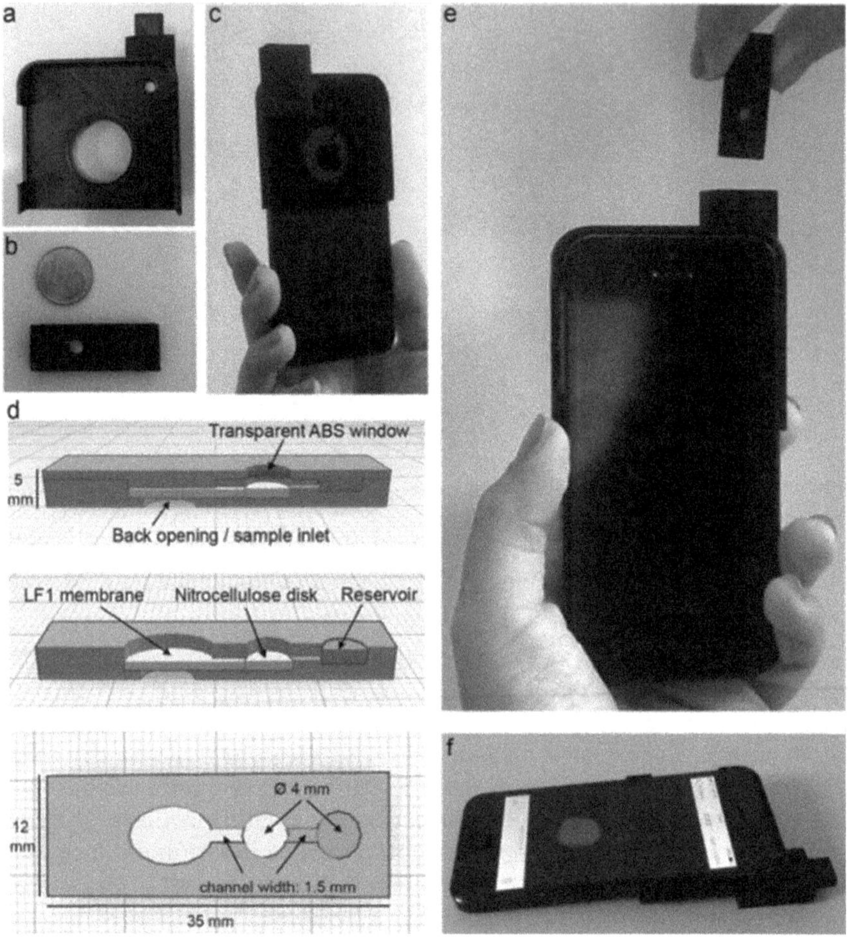


**Figure 3.9** 3D-printed smartphone based biochemiluminescence platform: (a) photograph of the 3D-printed accessory, (b) photograph of the 3D-printed mini-cartridge, (c) photograph of the assembled platform, (d) schematic of the mini-cartridge, (e) assembly of the mini-cartridge, and (f) photograph of the smartphone-based detection. Reproduced from ref. 9 with permission from American Chemical Society, Copyright 2014.

0.5 mM and 0.1 mM for lactate in oral fluids and sweat, respectively. It was successfully used to monitor the endurance of athletes.

Later in 2016, the group further explored applications of their set-up, and reported upon the FDM printing of mini-cartridges and a smartphone holder to integrate an array of bioluminescent cells.[38] The FDM-printed holder was designed for holding a pre-loaded cell mini-cartridge, providing a dark box, integrating a reagent reservoir and disposable droppers, and assisting light acquisition from the transparent windows of a sliding lid, as shown in Figure 3.10. This complex platform was used to perform smartphone-based

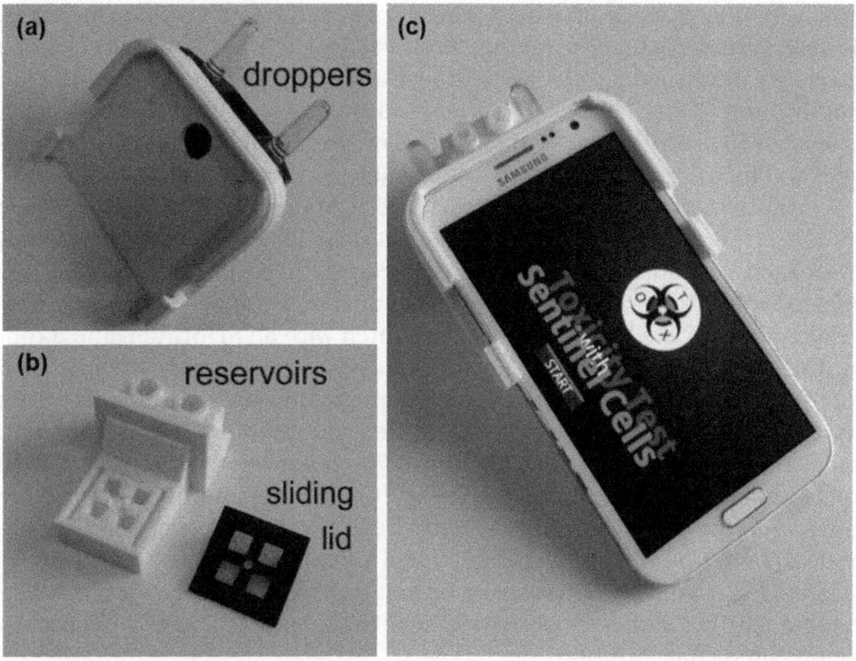


**Figure 3.10** 3D-printed smartphone-based platform for bioluminescence detection: (a) 3D-printed adapter, (b) 3D-printed cartridge, and (c) assembled opto-sensing platform. Reproduced from ref. 38 with permission from Elsevier, Copyright 2016.

bioluminescence whole-cell toxicity studies. The test was based on the difference in the light emitted from the bacteria, with and without toxicant exposure. Genetically engineered human embryonic kidney cells that expressed a dominant green-emitting luciferase mutant were used as 'sentinel cells'. Ubiquitous products such as tap water, moisturiser, stain removers, mineral and synthetic oils, bio-degreaser, floor and toilet cleaners, and an alcohol-based aftershave lotion were used as samples. The analysis time was less than 30 minutes, and the results were found to correspond well with results obtained from laboratory-scale instruments and with commercially available toxicity assays.

These studies successfully demonstrate the application of low-cost FDM-printed smartphone interfaces to deliver sensitive luminescence assays that require complete isolation from the ambient light, and perform quantitative assays using only a low-sensitivity smartphone camera.

### 3.2.1.4 Universal Smartphone Interfaces

The smartphone interfaces mentioned above were each specifically fabricated for one particular smartphone model, due to their closed and complex architecture. One would require access to the design files to modify the dimensions in order to use these interfaces with other smartphone models. However, recently, 3D printing has been used to develop open smartphone interfaces that can be used with almost any smartphone, without further modification. These simple interfaces, and their use within standard smartphone-based platforms can result in nearly anyone developing their own portable and functional smartphone-based analyser.

To the best of the authors' knowledge, Comina *et al.*[18] developed the first 3D-printed universal smartphone interface for optical detection. They 3D-printed a simple focusing lens that can be used universally with any smartphone's front camera to provide optical detection for various microfluidic systems. The lens itself was 3D-printed with an SL printer, and a thin layer of uncured resin left on the surface of the polymerised lens was then post-print cured by UV exposure to provide a smooth surface finish. The polymerisation of the excess resin provided the desired magnification and transparency to the lens. This 3D-printed lens was used with a 3D-printed single-bodied lab-on-a-chip device to facilitate the readout by a smartphone camera (Figure 3.11). The lab-on-a-chip device consisted of calibration references, finger pumps, and unidirectional valves for glucose monitoring as described in Chapter 4. This combined smartphone-based platform resulted in the quantification of glucose with an error of only 1.22% with respect to the prepared concentration.

Wang *et al.*[39] developed a similarly simple single-strip microtiter plate reader that could also be interfaced with any smartphone's front camera, as opposed to the above-mentioned more complex 96-well plate reader. The

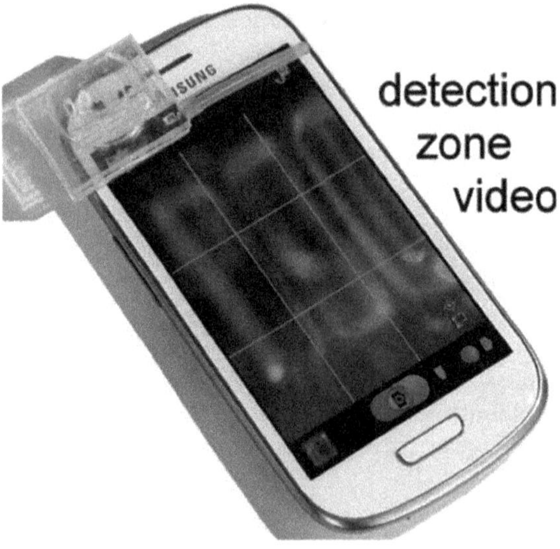


**Figure 3.11** 3D-printed universal smartphone interface for optical detection. Reproduced from ref. 18 with permission from John Wiley and Sons, Copyright © 2015 WILEY-VCH Verlag GmbH & Co. KGaA, Weinheim.

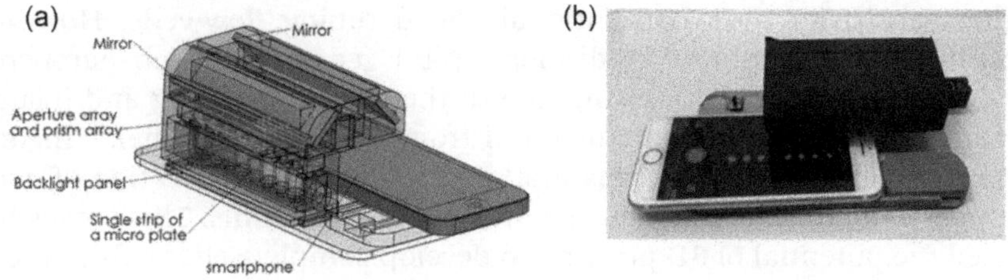


**Figure 3.12** 3D-printed smartphone-based platform for reading single-stripe microlitre plates: (a) render and (b) photograph. Reproduced from ref. 39 with permission from American Chemical Society, Copyright 2017.

smartphone-based single-stripe reader used FDM-printed acrylonitrile butadiene styrene accessories and single-strip plates for in-field colourimetric detection of 2,4-dichlorophenoxyacetic acid. It was able to screen eight samples simultaneously as shown in Figure 3.12. The device was used to analyse spiked environmental (water) and biological samples (rat and human plasma/serum) resulting in an LOD for this herbicide of less than 1 ppb and a linear range from 1 ppb to 80 ppb.

### 3.2.2 Flow-cells for Mixing, Reaction, and Detection

Miniaturised high sensitivity and/or high sample capacity/volume analytical flow-cells require the production of precise channels of set dimensions and designs, specific to the particular application. For example, in the case of detection cells for luminescence measurements, this could be long and narrow channels that are compactly wrapped or arranged within small footprints. Conventional fabrication techniques struggle with the production of such arrangements, especially when working within the limitations of two-dimensions. Fabrication of three-dimensionally aligned channels by conventional means involves labour-intensive alignment of multiple layered components and sealing of open channels. With these complications in mind, it's not surprising that 3D printing has been recently applied to deliver complex arrangements of complete microchannels within high performance miniaturised optical flow-cells.

High-sensitivity chemiluminescence flow-cells require sufficiently long and precisely arranged microchannels to deliver the required hold-up time and volume to allow maximum emission to be captured by the attached detector. With regard to these requirements, in 2014, Spilstead *et al.*[40] fabricated the first 3D-printed chemiluminescence detection flow-cell with features that were not attainable using conventional tube coiling methods. The flow-cells were printed with a PolyJet printer in a polyacrylate composite material, and used within a flow injection analysis (FIA) manifold, and with HPLC. The developed methods were applied to the rapid chemiluminescent-based detection of morphine and other adrenergic phenolic amines. The use of 3D-printed flow-cells with a reflective housing gave higher sensitivity than conventional coiled tubing flow-cells. However, these flow-cells employed traditional spiral geometry, which hampered the PolyJet post-print processing due to the time-consuming and insufficient removal of the support material from the spiral channel.[40] Incomplete removal of the support material necessitated the fabrication of open channels, which were later sealed with epoxy-acetate films. This somewhat missed the potential of 3D printing to develop complete channels that can be directly used without any post sealing.

3D printing of the first closed channel chemiluminescence flow-cell was reported later by Gupta *et al.*[41] Within this work, a new radial flow-cell geometry was developed by diverging multiple linear channels from a common central port, as shown in Figure 3.13. The complex cell could be successfully printed with both PolyJet and FDM printers in polyacrylate composite and acrylonitrile butadiene styrene materials, respectively. PolyJet-printed radial flow-cells required only 10 hours of post-print processing time as compared to 360 hours of processing time required for their 3D-printed spiral counterparts. Moreover, an FDM printer that had failed to fabricate the spiral flow-cell resulted in the successful fabrication of the counterpart radial flow-cell. The 3D-printed radial chemiluminescence flow-cell was compared with the 3D-printed and coiled-tubing spiral flow-cells for chemiluminescent-based detection of hydrogen peroxide. The radial flow-cell resulted in *ca.* 60% increase in the peak height, *ca.* 90% increase in the peak area, and *ca.* 40% increase in the signal duration as compared to its spiral flow-cell equivalent.

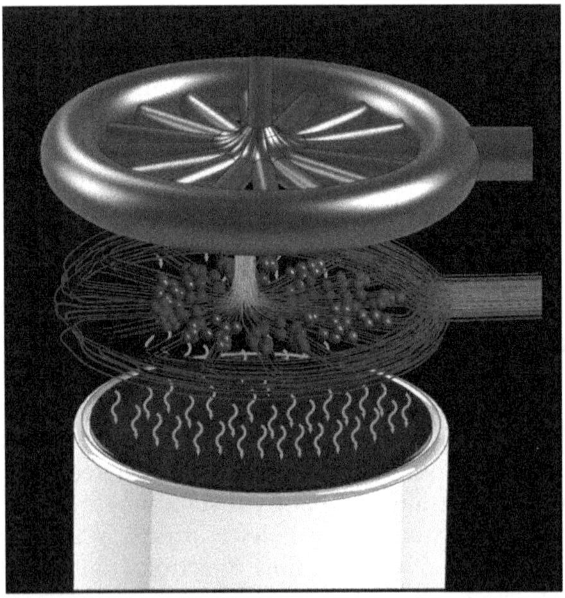

**Figure 3.13** 3D-printed radial flow-cell for chemiluminescence detection. Reproduced from ref. 41 with permission from Elsevier, Copyright 2018.

The 3D-printed radial flow-cell was applied within an ion chromatography method, using chemiluminescense detection of hydrogen peroxide in urine and coffee extract samples.

3D-printed flow-cells have also been applied in immunomagnetic assays involving functionalised magnetic nanoparticles, which have been utilised for the rapid detection of bacteria.[42] Conventional immunomagnetic assays suffer from various limitations, including the capacity to handle only a small sample volume, inefficient binding between the antibody-functionalised magnetic nanoparticles and the pathogenic bacteria, and the low concentrations of both magnetic nanoparticles and pathogenic bacteria in the sample solutions.[42] These limitations increase the analysis time and reduce the sensitivity of conventional immunomagnetic assays.

To address these difficulties, Lee *et al.* developed a novel 3D immunomagnetic flow-based assay, employing a 3D-printed microfluidic flow-cell and a luminescent ATP detection kit.[42] The microfluidic flow-cell was printed with an SL printer in a polyacrylate composite material, and was used for the rapid and sensitive detection of salmonella bacteria in relatively large volume samples (foodstuffs and extracts). Salmonella bacteria were extracted from the lettuce wash by binding with antibody-functionalised magnetic nanoparticle clusters. The nanoparticle clusters were magnetically immobilised on the surface of 3D-printed cylindrical microchannels as shown in Figure 3.14. Stacked permanent magnets with spacers were used inside the cylinder to generate high magnetic fields. Magnetic concentration within the narrow microchannels provided the capture of a high percentage of the nanoparticle–bacteria complexes. High magnetic fields allowed a post-rinsing step at high flow rates, which reduced the analysis time for large volume samples. The immunomagnetic flow assay resulted in an LOD of less than 10 colony-forming units per mL.

Thin layer spectroelectrochemistry is based upon the investigation of electrochemically generated species through absorption and fluorescence spectroscopy.[43–45] Spectroelectrochemistry measurements require the presence of a thin layer of sample between optically transparent electrodes to study

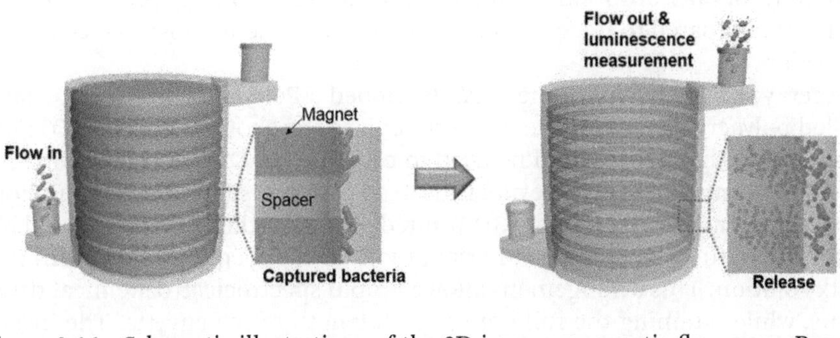


**Figure 3.14** Schematic illustrations of the 3D immunomagnetic flow assay. Reproduced from ref. 42 with permission American Chemical Society, Copyright 2014.

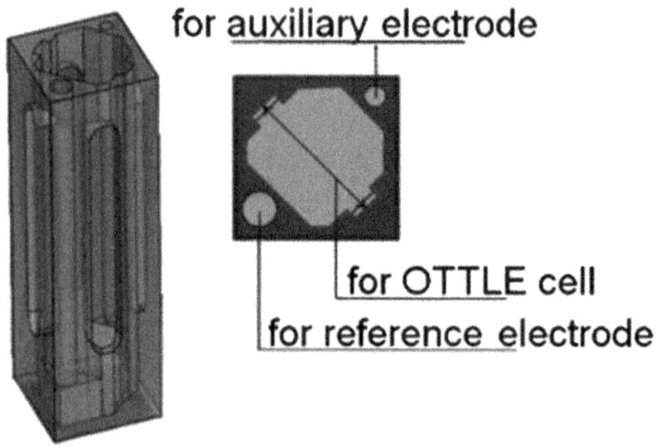


**Figure 3.15** Render of the 3D-printed OTTLE cell. Reproduced from ref. 47 with permission from John Wiley and Sons, Copyright © 2010 WILEY-VCH Verlag GmbH & Co. KGaA, Weinheim.

both electrochemical and optical properties. This set-up requires the construction of a custom-built optically transparent thin layer electrochemical (OTTLE) cell (Dutton cell).[46] In 2010, Wilson *et al.*[47] used 3D printing to construct such OTTLE cells to facilitate spectroelectrochemical measurements in conjunction with standard bench-top absorbance and fluorescence spectrophotometers. The assembly consisted of two parts, namely an OTTLE cell and a cell holder. The cell holder was designed to align the cell with a reference electrode and an auxiliary electrode as shown in Figure 3.15. The material cost of the holder was quoted as being less than $10. The assembly was used to study the spectroelectrochemistry of tris-(2,2′-bipyridine) ruthenium(II) chloride ($Ru\text{-}(bpy)_3Cl_2$) and 1-hydroxypyrene. A good correlation was observed between the results and the known properties of these analytes. The printed assembly could be used with any commercial optical platform that measures 1 cm above the base of the cuvette, such as most standard absorbance and fluorescence spectrophotometers. The ability to use this assembly for both absorbance and fluorescence spectroscopy provided lower LODs as compared to the cells that were only suitable for absorbance-based detection.

A few years later, Brisendine *et al.* developed a PolyJet-printed microchannelled polyacrylate composite insert to replace the above-mentioned OTTLE assembly.[48] This insert could be used to establish an OTTLE cell within any standard 1 cm-square cuvette and thus used with any standard transmission spectrometer. The inserts were 3D-printed such that the electrodes could be placed in optimal positions and orientation for rapid equilibration with the bulk solution. This arrangement allowed rapid spectroelectrochemical titrations, while retaining the full optical path length of the cuvette. The insert was successfully used for the determination of reduction potentials for proteins and small molecules. This 3D-printed insert could be quickly and inexpensively manufactured, and it combined the best features from both the standard Dutton cells[46] and the above OTTLE devices.

The above flow-cells were typically printed with PolyJet and SL printers, because FDM-printed devices are usually opaque, as discussed in Chapter 1. However, Morgan *et al.*[49] have been able to develop transparent FDM-printed microfluidic devices. The devices were printed with poly(lactic acid), transparent poly(ethylene terephathalate), modified acrylonitrile butadiene styrene, and transparent poly(lactic acid). The transparency of the FDM-printed devices was improved by minimising under-extrusion by reducing the layer thickness and the speed of the printing. A reduction in under-extrusion also improved the fidelity and the pressure resistance of the FDM-printed devices by ensuring a better seal between the printed layers. These FDM-printed devices were found to be leak-proof up to a pressure of at least 2 MPa.[49]

### 3.2.3 Miscellaneous Optical Devices

Spectroscopy and microscopy in their various modes are found in almost all analytical laboratories. Outside of the laboratory, as with the plethora of smartphone-based devices, portable spectrometers and microscopes for various applications are now widespread, including hand-held and miniaturised optical devices. With regard to the latter, in recent years 3D printing has been used to produce new components and devices to deliver optical detectors that are low-cost, lightweight, miniaturised, and easy to repair/replace. 3D printing can provide customised spatial arrangements, and the developed 3D printing designs can be shared between research groups allowing easy transfer of knowledge.[50] The use of 3D printing techniques to fabricate various vital components for microscopes, on-capillary detectors, and spectrometers is described below.

#### *3.2.3.1 Microscopes*

Microscopy is a crucial technique for the bioanalytical sciences. However, for many laboratories, especially those in resource-limited countries, the cost and maintenance of commercial microscopes is problematic. In addition, most commercial conventional microscopes and accessories are not easily customised or adapted to specific applications. Recently, however, 3D printing has been applied to provide some novel solutions, and to tailor microscope platforms that are low-cost, lightweight, and customisable for different microscopes.

Time-lapse imaging microscopy is a widely recognised technique for studying live cells and microorganisms. It allows the study of cellular behaviour and dynamics over an extended period. However, the high cost of these imaging systems presents a workflow bottleneck even in well-funded institutions.[51] Recently, in 2016, Vera *et al.* developed a modular and affordable time-lapse

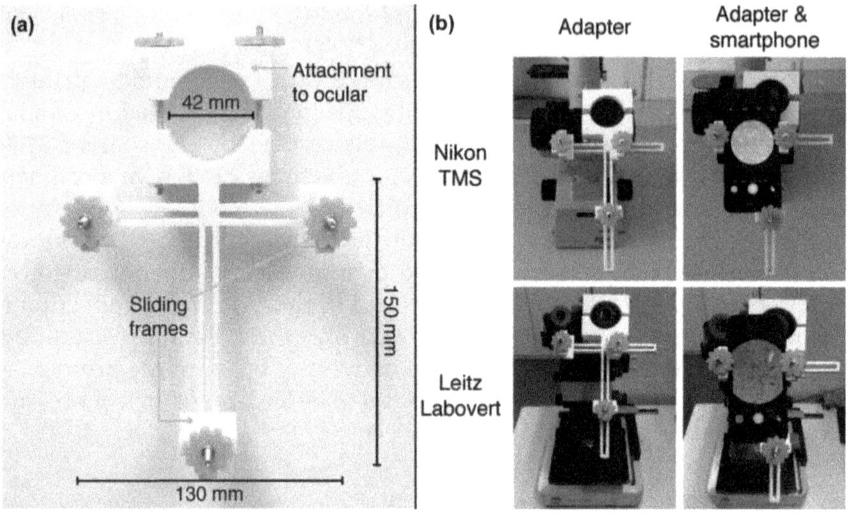


**Figure 3.16** Time-lapse imaging and incubation system (ATLIS): (a) 3D-printed smartphone holder, and (b) ATLIS assembly on two different microscopes as indicated. Reproduced from ref. 51 [https://doi.org/10.1371/journal.pone.0167583] under the terms of a CC BY 4.0 license [https://creativecommons.org/licenses/by/4.0/].

imaging and incubation system (ATLIS) using FDM-printed poly(lactic acid) parts, a smartphone, and off-the-shelf electronic components (Figure 3.16).[51] The so-called ATLIS set-up consists of four modular units that transform a simple inverted microscope into a live cell imaging system, these being an imaging unit, a heating unit, an onstage incubator, and a control unit. The combined system was successfully used to obtain time-lapse images of traditional and microfluidic cell culture systems.

Multi-cellular aggregates are used as *in vitro* tumour models.[52] High throughput screening of drugs on these 3D cell cultures requires multi-position imaging of thick samples in three-dimensions over an extended period. The three-dimensional imaging requires customisation of traditional observation platforms. Accordingly, Alessandri *et al.* developed a multi-use observation chamber for live 3D bio-imaging, which was named UniverSlide.[53] The frame of the UniverSlide was SL-printed in a polyacrylate composite material and sealed with a removable lid. The SL-printed frame was autoclavable and found to be suitable for use in controlled cell culture environments. UniverSlide facilitated the imaging of 3D cultures and organisms irrespective of sample shape and microscope orientation (upright or inverted).

In 2015, Ozbay *et al.*[54] developed a lightweight and compact confocal fluorescence microscope for 3D fluorescence imaging of thick tissues. In order to reduce the weight of the microscope and facilitate its installation onto the head of a freely moving mouse, they used a 3D-printed adapter for integrating various optical components of the microscope.

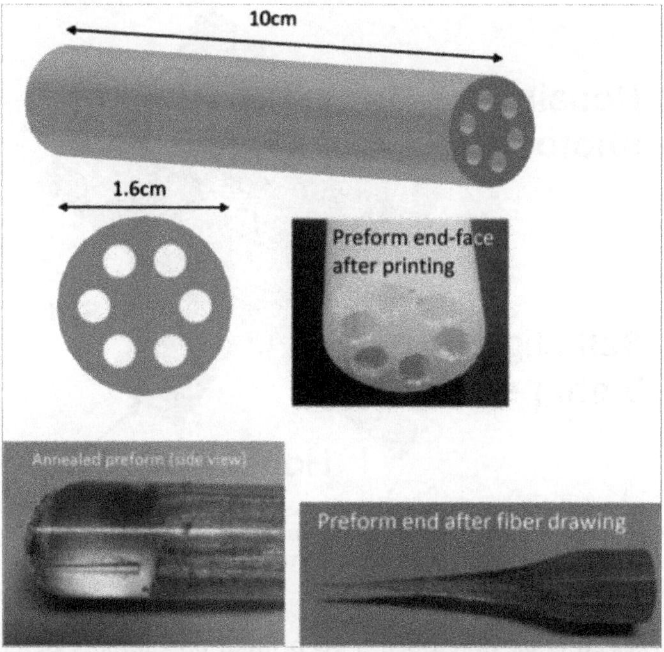


**Figure 3.17** Render and photographs of the 3D-printed styrene–butadiene-copolymer and polystyrene mixture preform as indicated. Reproduced from ref. 55 with permission from The Optical Society.

Optical fibres are being increasingly used in the construction of optical detectors and analytical platforms. They allow easy and efficient coupling between different components of an optical detector, such as a light source, a flow-cell, and a photodetector. In 2015, Cook *et al.*[55] reported the development of the first 3D-printed optical fibres (Figure 3.17). They drew optical fibres from FDM-printed preforms of a propriety polystyrene mixture containing styrene–butadiene-copolymer and polystyrene. This polystyrene mixture offered transparency, a high degree of flexibility, and did not suffer from discolouration under stress. This first report was later followed by multiple papers from the same and other groups on the development of other similar 3D-printed optical fibres.[56–60]

### *3.2.3.2 On-capillary Detectors*

Recently, a number of papers have been presented on the use of 3D printing to produce customised arrangements for on-capillary optical detection. Various 3D-printed devices have been used to integrate and align the capillaries with the essential optical components to provide miniature on-capillary detection for capillary scale flow analysis and analytical separations (*e.g.* capillary electrophoresis or capillary chromatography).

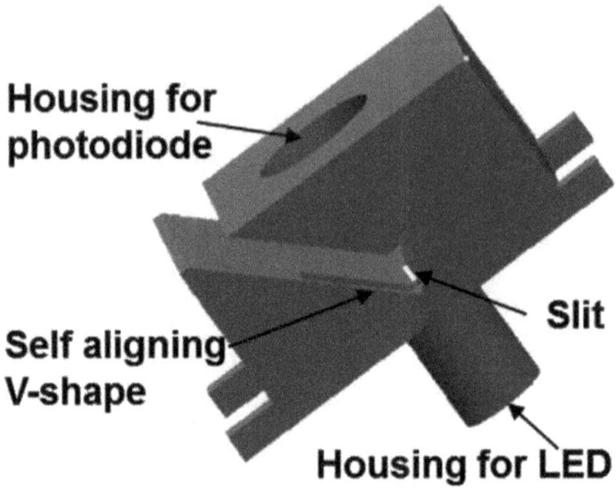


**Figure 3.18** Render of the 3D-printed on-capillary detector platform. Reproduced from ref. 61 with permission from Elsevier, Copyright 2017.

Recently, Cecil *et al.* developed a 3D-printed on-capillary detector platform designed to integrate and align an LED, used as a light source, and a photodiode for detection, across various types of capillaries for on-capillary photometric detection.[61] The interface was printed with an FDM printer using poly(lactic acid). It consisted of a 70 μm slit and a V-shaped opening to precisely align the capillaries (Figure 3.18). The size of the slit was limited by the resolution (50 μm) of the FDM printer. The slit resulted in a stray light estimation of 3.8% and 50% for 500 μm and 50 μm capillaries, respectively. The robustness of the device was tested by repeat repositioning of the capillary, which resulted in an RSD of only 1.9% ($n$ = 10). The on-capillary detector, fitted with a 470 nm LED light source, was applied to capillary electrophoretic separations of coloured metal-chelate complexes, and provided an LOD of 6.8 μM and 4.5 μM for $Zn^{2+}$ and $Cu^{2+}$, respectively. The performance of the 3D-printed on-capillary detector was found to be comparable to the performance of similar commercial conventionally fabricated assemblies.

Along similar lines, Prikryl *et al.*[50] developed a slightly more elaborate FDM-printed poly(lactic acid) on-capillary detection arrangement, again for capillary based separations, as shown in Figure 3.19. It consisted of a 3D-printed detector head and cell for LED-induced fluorescence detection. The detector head was designed for an epifluorescence arrangement. The cost of 3D printing the detector head and cell was quoted as only $5. Optical fibres were used to couple the detector with the source LED and the detection unit. The detector was used with both semiconductor photodiodes and photomultiplier tubes. It was successfully used as an on-capillary detector

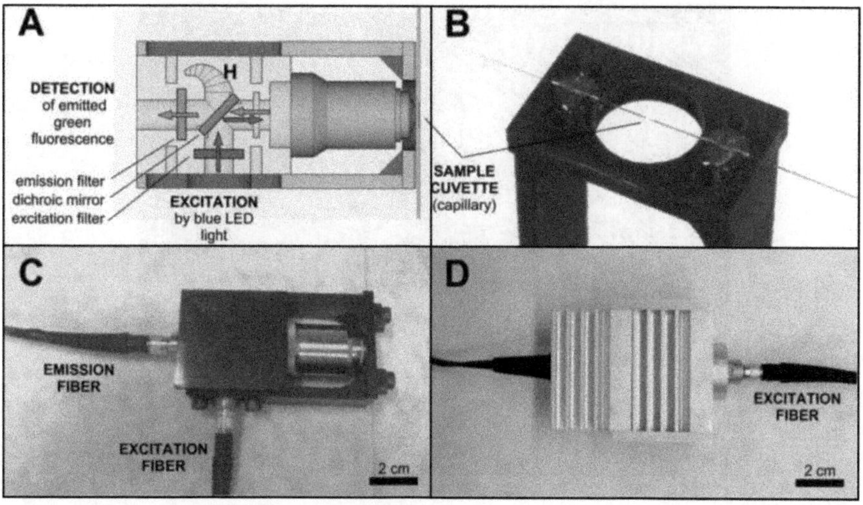


**Figure 3.19** 3D-printed LED-induced fluorescence detector: (A) schematic of the detector head, (B) photograph of the 3D-printed detector cell, (C) assembled detector, and (D) LED housing. Reproduced from ref. 50 with permission from American Chemical Society, Copyright 2014.

for capillary electrophoretic separations of oligosaccharides labelled with 8-aminopyrene-1,2,3,6-trisulfonate.

#### 3.2.3.3 Simple Spectrometers

A key aspect of a basic spectrometer is diffraction of light, which in many instances is achieved using a simple prism. Conventional prisms are typically rather expensive and not readily customised. However, Hinman *et al.* have recently described an SL-printed polymethacrylate prism and its use as a light-guiding sensing platform (Figure 3.20).[62] Several different prism geometries were printed, demonstrating the ease of customising such printed prisms. Simple benchtop polishing procedures were used to polish the prism faces. The smoothing process achieved high transparency across the prism with a surface roughness of only 23 nm. For sensing purposes, the smooth prism surface was further coated with a gold layer, which was modified with 3-mercaptopropanol for application in Kretschmann-configured plasmonic sensing of bacterial toxins. A 50 nm gold film was evaporated on to the prism surface, and it was immersed in a 3-mercaptopropanol solution to form a self-assembled monolayer.

Going further, Gao *et al.* developed an inexpensive optical particle spectrometer with an SL-printed enclosure and optical mounts.[63] Since most of its structural and component parts were 3D-printed, it was named by Gao and coworkers as a 'printed optical particle spectrometer'. As it only weighed

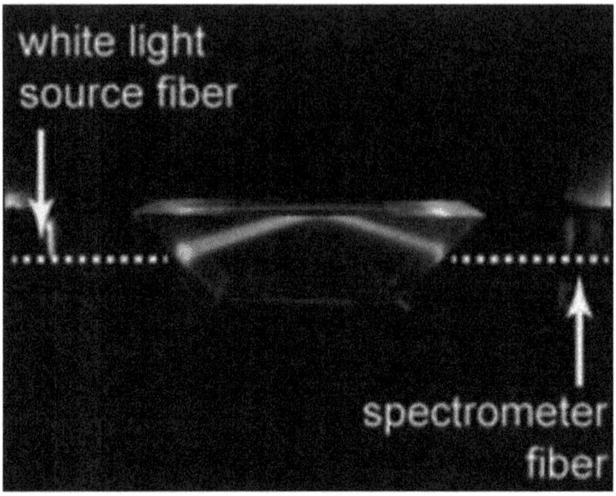


Figure 3.20 A photograph of the 3D-printed methacrylate prism. Reproduced from ref. 62 with permission from American Chemical Society, Copyright 2017.

800 grams, it was easily mounted on unmanned aerial vehicles and balloon sondes. It was successfully used to measure aerosol number, concentration, and size.

## 3.3 Electrochemical Detectors

Electrochemical detectors typically offer both selectivity and sensitivity, and are readily miniaturised without significant loss in either case. These qualities have made electrochemical detectors a popular choice for portable and point-of-care analytical devices. The four most commonly used electrochemical detection techniques are amperometry,[64,65] voltammetry,[66,67] electrochemiluminescence,[68,69] and coulometry.[70] Over the past few years, 3D printing techniques have been used to fabricate low-cost, customised, and modular components for most of these electrochemical detection modes.

### 3.3.1 Electrochemical Detector Cells

To the best of the authors' knowledge, the first 3D-printed integrated electrochemical detector was reported by Erkal *et al.*[71] in 2014. The group 3D-printed flow-cells with a PolyJet printer in a polyacrylate composite material to facilitate both amperometric and chronoamperometric electrochemical detection. The amperometric detector flow-cell consisted of a 0.5 mm high, 0.5 mm wide, and 30 mm long straight channel and threaded female connection ports. The 3D-printed threaded ports allowed easy integration of electrodes through standard fittings, as shown in Figure 3.21. This flow-cell was used for the determination of dopamine and nitric oxide.

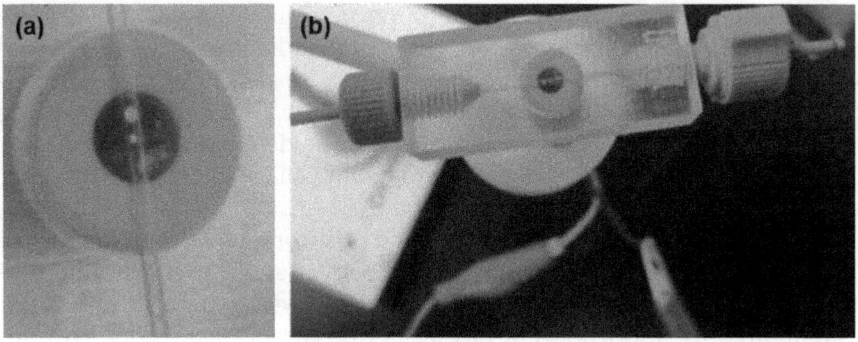

**Figure 3.21** 3D-printed amperometric detector flow-cell with modular electrode connection ports: (a) electrode arrangement, and (b) assembled flow-cell. Reproduced from ref. 71 with permission from The Royal Society of Chemistry.

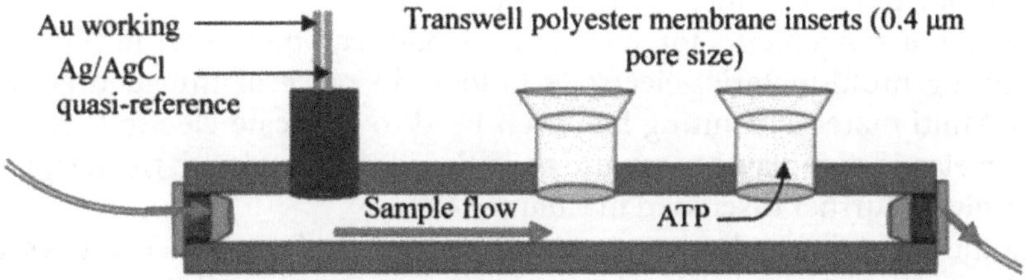


**Figure 3.22** Schematic of the 3D-printed chronoamperometric detector flow-cell with Transwell® membrane inserts. Reproduced from ref. 71 with permission from The Royal Society of Chemistry.

The chronoamperometric detector flow-cell consisted of a 0.5 mm high, 3 mm wide, and 7 mm long straight channel with the same connection ports as mentioned above. It also consisted of additional holders for Transwell® membrane inserts as shown in Figure 3.22. The detector was used for the determination of oxygen concentration and its effect on the release of adenosine triphosphate (ATP). The low oxygen permeability of the polyacrylate 3D-printed parts minimised the background noise during oxygen measurements. Both flow-cells were designed with modular components for easy assembly and disassembly. These simple 3D-printed flow-cells were robust, could be used multiple times, and could be readily customised with regard to connections and electrodes.

The above flow-cells were produced using advanced PolyJet printers. Cheaper and simpler FDM printers have also been used to fabricate electrochemical flow-cells with similar threaded ports for easy integration of electrodes. Bishop *et al.*[72] FDM-printed threaded ports with acrylonitrile butadiene styrene material, which successfully allowed the integration of electrodes with standard fittings. The electrochemical flow-cell consisted of a linear 800 μm square conduit channel. It was FDM-printed with clear poly(ethylene terephathalate), and were found to be semi-transparent. The flow-cells were printed in less than 50 minutes, and they resulted in a material cost of only \$0.30 to \$0.60. These FDM-printed flow-cells were said to be leak-proof up to a flow rate of 2 mL min$^{-1}$. They were used for the electrochemical sensing of hydrogen peroxide, based upon amperometric detection with a Prussian blue nanoparticles-modified gold electrode. An LOD of 100 nM and a linear response from 100 nM to 20 μM was obtained.

The same research group later continued this work, using an SL printer to produce similar electrode connections as part of an electrochemiluminescence detector.[73] The use of an SL printer to print with a methacrylate-based resin allowed the fabrication of flow-cells with 90% transparency (610 nm wavelength) as opposed to the above-mentioned semi-transparent flow-cells fabricated by an FDM printer. The SL-printed flow-cells required *ca.* 3.5 hours of print time at 50 μm print resolution and resulted in a material cost of *ca.* \$2.50 each. The higher material transparency allowed electrochemiluminescence detection of tris(2,2′-bipyridyl)dichlorruthenium(II) hexahydrate with an LOD of 7 μM and a linear range of 9 to 900 μM.

Modular arrangements for smaller electrodes can be further improved by fabricating multi-material electrode holders in place of threaded connections. Multi-material printing has been used to fabricate electrode holders with an elastic outer layer to ensure their liquid-tight seal with the rest of the assembly, as further discussed in Chapter 4.[74]

Anodic stripping voltammetry (ASV) is a technique which allows the rapid, sensitive, and selective detection of trace metals.[75–77] Thin-layer cells for ASV offer various advantages such as low sample volume, high sensitivity, and improved reproducibility.[78,79] With these advantages in mind, Sun *et al.*[80] developed a 3D-printed thin-layer flow-cell for continuous flow accumulation square-wave ASV. The flow-cell was printed with a PolyJet printer in a polyacrylate composite material and allowed for the electrochemical determination of lead ions with an LOD of 0.2 ppb and a linear range of 0.5 to 100 ppb, and a 95.5 to 106.5% recovery. The ASV results using the printed cell were found to be in-line with those from atomic fluorescence spectroscopy.

Wall-jet based electrochemical flow cells typically provide higher sensitivity for electrochemical detection as compared to the thin layer electrode assemblies, as they can deliver improved analyte mass transport between the carrier solution and the electrode.[81] In the past, the two-dimensional nature of conventional production methods has limited the design of wall-jet electrode assemblies.[82] However, 3D printing can be used to easily produce the desired three-dimensional complex geometries, to maximise the performance of the detector and further explore structure-response relationships.

Munshi *et al.*[82] developed a simple wall-jet electrode assembly-based electrochemical flow-cell, as shown in Figure 3.23. The flow-cell was printed with a PolyJet printer in a polyacrylate composite material. It consisted of 375 μm square cross-section channels and the above-mentioned threaded

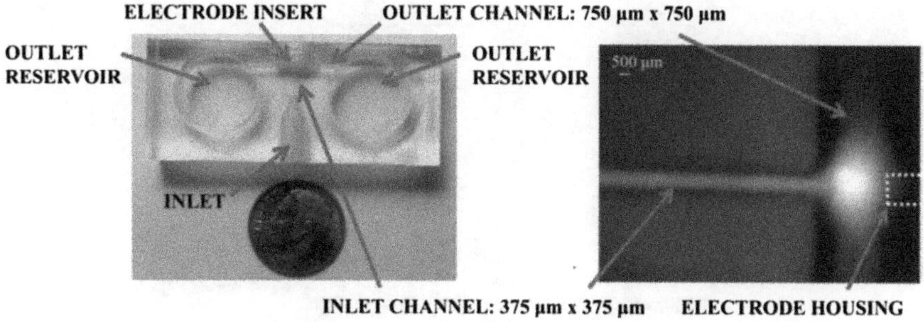


**Figure 3.23** 3D-printed wall-jet electrode assembly based electrochemical flow-cell. Reproduced from ref. 82 with permission from The Royal Society of Chemistry.

connections for electrodes. The performance of the 3D-printed wall-jet electrode and thin layer electrode assembly were compared for the determination of a model compound, namely catechol.[82] The wall-jet electrode design provided 16-times increased sensitivity as compared to the thin layer type electrochemical cell. The printed wall-jet flow-cell provided an LOD of 500 nM, whereas the thin layer electrode counterpart provided an LOD of only 6 µM. Regarding application, the 3D-printed wall-jet electrochemical cell was later successfully used in conjunction with HPLC.[82]

Dias *et al.*[83] took the idea of the 3D-printed wall jet electrode assembly further to develop an electrochemical batch injection analysis cell. Batch injection analysis cells allow direct introduction (with a pipette) of a sample onto electrodes. The direct sample introduction in the batch injection analysis system circumvents the need for pumps and valves that are required for flow injection analysis (FIA) based systems. The batch injection analysis cell was 3D-printed in less than 4 hours with an FDM printer, and the cost of printing in acrylonitrile butadiene styrene material was quoted as just $5. The 3D-printed batch injection analysis cell was used for the amperometric determination of glucose. It resulted in an LOD of 0.11 mM, a linear response from 1 mM to 10 mM, and an RSD ($n$ = 6) of 1.5 to 2.8%. The results were found to be in agreement with the reference methods.

In 2015, Schimo *et al.*[84] produced an SL-printed flow-cell for *in situ* electrochemical hydrogen loading and cyclic voltammetry. It was used for scanning Kelvin probe measurements of the hydrogen diffusion coefficient in palladium. Production using 3D printing allowed easy fabrication of non-planar shaped electrolyte channels, such as rectangular, V-shaped, and U-shaped. Moreover, each channel had a so-called 'breaker' (as shown in Figure 3.24) within it to force the solution to pass close to the surface. The complex arrangement of the cell, enabled by 3D printing, allowed different electrochemical measurements without removing the sample from the cell. Undisturbed sample and cell position ensured uniform measurement conditions throughout the experiment.

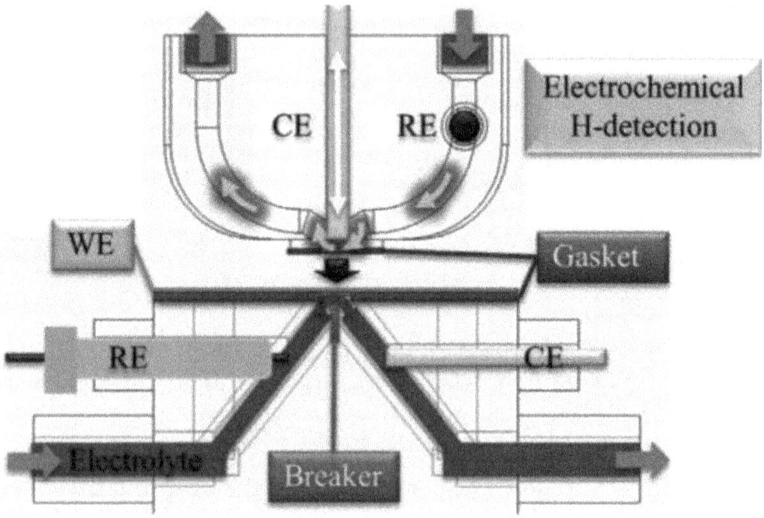


**Figure 3.24** Schematic of the 3D-printed flow-cell for *in situ* electrochemical hydrogen loading and cyclic voltammetry. CE: counter electrode, RE: micro-reference electrode, WE: working electrode. Reproduced from ref. 84 with permission from Elsevier, Copyright 2015.

### 3.3.2 Printed Electrodes

3D printing can easily be applied to the production of bespoke and complex three-dimensional electrodes and electrode arrays, for use within electrochemical detectors and indeed a multitude of other electrochemical devices. In recent years, both polymer and metal 3D printing techniques have been used to fabricate such electrodes, which have been demonstrated for a variety of applications.

To the best of the authors' knowledge, Salvo *et al.*[85] reported the first 3D-printed electrodes in 2011, which were produced by the metallisation of PolyJet-printed polyacrylate composite structures. These electrodes consisted of 180 conical needles (3 mm high, 600 μm diameter base, and 100 μm diameter tip) on a truncated conical support (0.5 mm high, 15 mm diameter base, and 13 mm diameter top). The polyacrylate composite structures were metallised by a two-step process that involved titanium sputter coating and gold evaporative coating. The electrodes were used for ECG/EEG recording and did not require any skin preparation or electrolytic gel. The 3D-printed dry electrodes performed similarly to the planar wet Ag/AgCl electrodes traditionally used for recording ECG/EEG.

Later in 2014, Zhao *et al.*[86] reported the direct 3D metal printing of electrodes to circumvent the need for post-printing metallisation as mentioned above. Titanium alloy (Ti-6Al-4V) interdigitated electrodes were developed using the SLM technique. The electrodes were electrodeposited with polypyrrole for their use as capacitive electrodes. They were used with a poly(vinyl alcohol)-$H_3PO_4$ polymer electrolyte to form an interdigitated

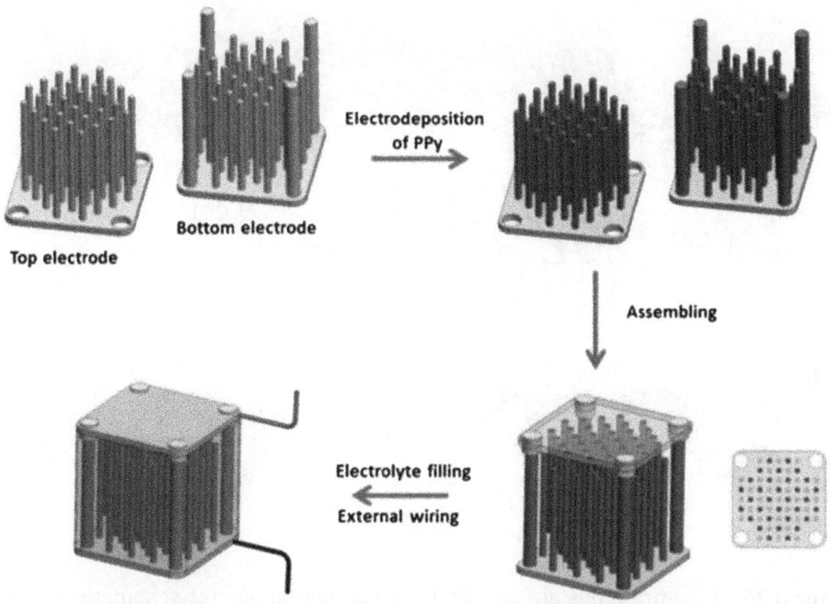


**Figure 3.25** The steps needed to assemble the supercapacitor made from 3D-printed titanium alloy interdigitated electrodes. Reproduced from ref. 86 with permission from Elsevier, Copyright 2014.

supercapacitor, which consisted of top and bottom sections as shown in Figure 3.25. The top section was designed with 32 posts, and the bottom section with 33 posts. The bottom section also consisted of four alignment posts (10 mm high and 1 mm diameter), to correspond with four holes in the top section (*ca.* 1.5 mm diameter). Each post was 8 mm in height and 500 μm in diameter. The 3D-printed electrode-based supercapacitor provided a volumetric capacitance of 2.4 F $cm^{-3}$ and 0.98 F $cm^{-3}$ at current densities of 3.74 mA $cm^{-3}$ and 37.4 mA $cm^{-3}$, respectively, resulting in an energy density of 213.5 Wh $m^{-1}$ and a power density of 15 kW $m^{-3}$. The 3D-printed electrodes gave a higher areal energy density ($2.98 \times 10^{-6}$ Wh $cm^{-2}$) as compared to their planar counterparts. These electrodes also provided higher cycling stability as compared to the previous three-dimensional carbon electrode supercapacitors.

In addition to titanium alloy, SLM technique has also been used to 3D-print electrodes in other metals, such as stainless steel.[87] Schematic designs for SLM-printed stainless steel helical electrodes are shown in Figure 3.26. The stainless steel electrodes were electrodeposited with iridium(IV) oxide. These 3D-printed electrodes gave a pseudo-capacitance of 14 mF $cm^{-2}$ at a current of 0.1 mA and resulted in the Nernstian behaviour for electrochemical pH sensing. They also offered good catalytic activity for an oxygen evolution reaction that enabled the oxidation of water at a much lower potential as compared to the glassy carbon electrodes.

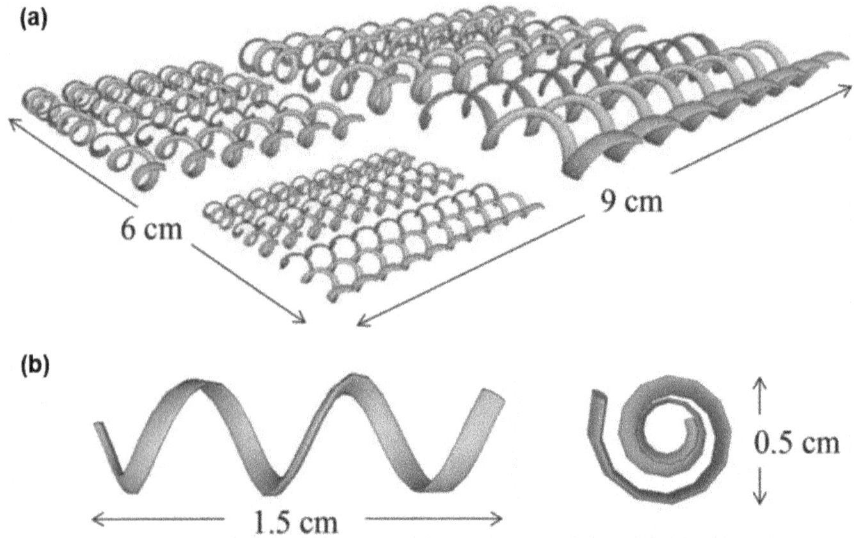


**Figure 3.26** SLM-printed stainless steel helical electrodes: (a) schematic and (b) dimensions. Reproduced from ref. 87 with permission from John Wiley and Sons, Copyright © 2015 WILEY-VCH Verlag GmbH & Co. KGaA, Weinheim.

### 3.3.3 Miscellaneous Electrochemical Devices

The development of any in-house built or prototype analytical platform typically requires numerous bespoke components. Machining such parts or customising off-the-shelf components has been in the past both costly and time-consuming. As seen throughout this chapter, the advent of 3D printing has obviously drastically changed this process. Some final examples of such low-cost and bespoke electrochemical arrangements are presented below.

Meloni *et al.* have 3D-printed linear stages for electrochemical-scanning probe microscopes.[88] Individual stages were 3D-printed with an FDM printer in poly(lactic acid) for each axis of movement. The use of 3D-printed components along with microcontroller boards and off-the-shelf parts allowed the construction of a low-cost and flexible electrochemical-scanning probe microscope (Figure 3.27). This simple set-up could be used for teaching exercises and as a basis to build more elaborate systems.[88]

Cabot *et al.*[89] developed a 3D-printed platform for thread-based electrophoresis. The platform consisted of modular bespoke components, including stage and buffer reservoirs. The buffer reservoirs consisted of a hoop to tie the thread, a cylinder to hold the electrodes, horizontal rollers to guide the thread, and a basin to hold the buffer as shown in Figure 3.28. All components were 3D-printed with a PolyJet printer in a polyacrylate composite material. 3D-printed components enabled rapid assembly and disassembly, and the platform offered great potential for multiplexed analysis.

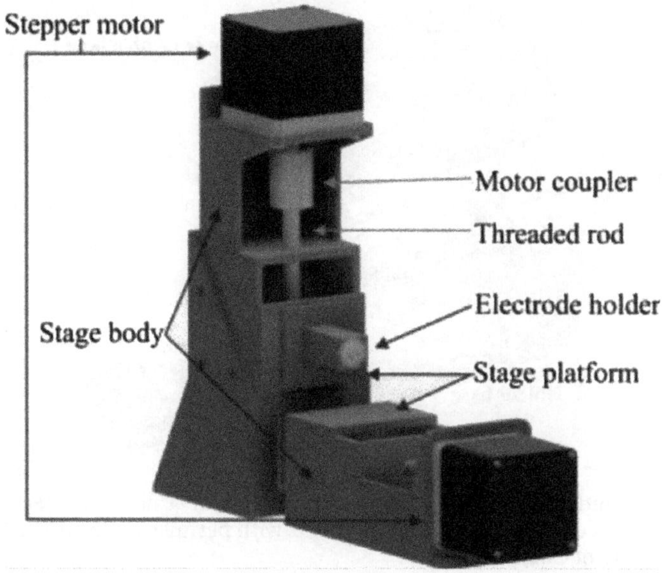


**Figure 3.27** Render of the electrochemical-scanning probe microscope with 3D-printed components. Reproduced from ref. 88 with permission from American Chemical Society, Copyright 2017.

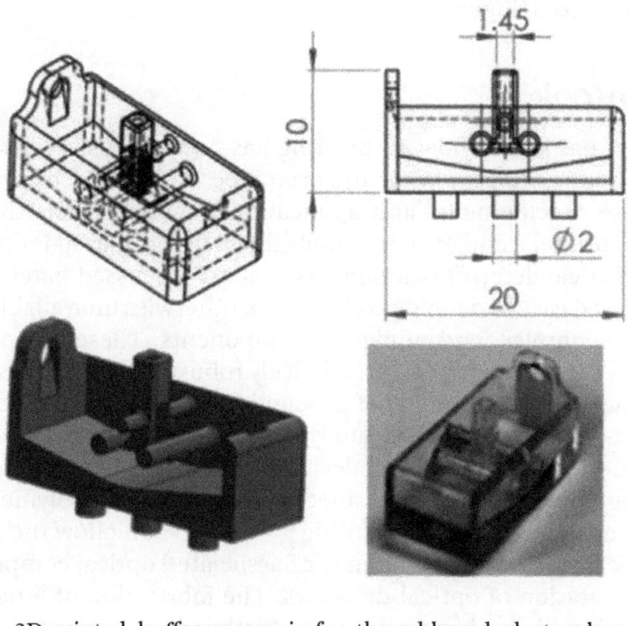


**Figure 3.28** 3D-printed buffer reservoir for thread-based electrophoresis. Reproduced from ref. 89 with permission from Elsevier, Copyright 2018.

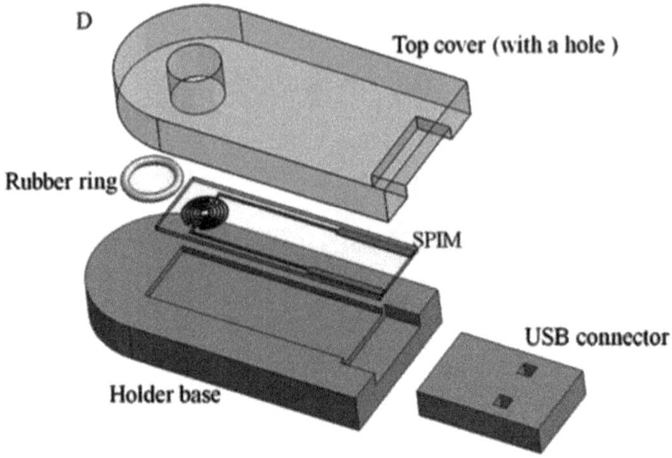


**Figure 3.29** Render of the 3D-printed USB holder for a portable electrochemical detector. Reproduced from ref. 92 with permission from The Royal Society of Chemistry.

3D-printed holders and enclosures have also been developed to integrate different components of electrochemical detectors. A modular holder has been 3D-printed to integrate electrodes, electrical connectors, and a switchable magnet.[90] The assembled electrochemical detector was used for amperometric detection of myeloperoxidase in human plasma. Similar enclosures have also been developed for electrochemical cells[91] and portable electrochemical detectors (Figure 3.29).[92]

## 3.4 Outlook

This chapter has shown how 3D printing has been extensively used to fabricate various components of new and prototype analytical detectors. To summarise these developments and applications, Table 3.1 lists the different techniques, printers, and print materials that have been used for printing the various analytical detector components that are discussed herein. 3D printing has allowed easy and rapid production of otherwise unavailable, three-dimensionally complex, and miniature components. These components are typically low-cost and lightweight, relatively robust and easy to iterate or customise, which is why this mode of production is so attractive when developing the next-generation portable analytical detectors.

However, current 3D-printed devices usually suffer from low optical transparency and relatively low mechanical strength (polymer devices). The recent advent of glass 3D printing[93,94] could soon allow the printing of robust optical flow-cells, and a range of associated optical components, for the next generation of optical detectors. The fabrication of long and complex flow-cells will also benefit from the further development of more readily soluble and liquid PolyJet printing support materials.[95] The ability to use

**Table 3.1** The 3D printing techniques, printers, and print materials that are used for fabricating different analytical detector components.

| 3DP technique | Printer: Manufacturer | Printer: Model | Print material used | 3D-printed components[a] |
|---|---|---|---|---|
| FDM | 3D Systems | CubePro | Acrylonitrile butadiene styrene | Smartphone adapter for photoluminescence[36] |
| | 3Dfactories | Easy3DMaker | Poly(lactic acid) | On-capillary fluorescence detector[50] |
| | Deltasine | Redback-2 | Styrene–butadiene-copolymer and polystyrene mixture | Optical fibre[55] |
| | Cliever | CL1 Black | Poly(lactic acid) | Scanning electrochemical microscope[88] |
| | Felix | 3 | Acrylonitrile butadiene styrene | Radial chemiluminescence flow-cell[41] |
| | | | Poly(lactic acid) | On-capillary photometric detector[61] |
| | | | | Microdroplet generator[98] |
| | MakerBot | Replicator 2X | Acrylonitrile butadiene styrene | Minicartridge for smartphone-based biochemiluminescence detection[9] |
| | | | | Device for smartphone-based chemiluminescence[37] |
| | | | | Smartphone adapter for bioluminescence detection[38] |
| | | | | Smartphone spectrometer[35] |
| | | | | Smartphone adapter for luminescence detection[28] |
| | | | Poly(ethylene teraphathalate) and acrylonitrile butadiene styrene | Electrochemical detector flow-cell[72] |
| | Makerfarm | Prusa | Poly(lactic acid) | Components for an affordable time-lapse imaging and incubation system[51] |
| | | Prusa i3 | Acrylonitrile butadiene styrene | Batch injection analysis cell[83] |
| | Stratasys | uPrint SE | — | Smartphone-based single-stripe microplate reader[39] |
| | Zortax | M200 | Acrylonitrile butadiene styrene | Smartphone optosensing platform[34] |
| Polyjet printing | 3D Systems | ProJet 3000 HD | VisiJet® EX200 | Chemiluminescence flow-cell[40] |
| | Stratasys | Objet Eden 333 | Fullcure® 720 | Spectroelectrochemical cell insert[48] |
| | | Objet Eden 260VS | VeroClear™ | Radial chemiluminescence flow-cell[41] |
| | | | | Platform for thread-based microfluidics[89] |
| | | Objet Eden 350V | Fullcure® 720 | Electrodes[85] |
| | | Objet350 Connex | VeroClear™ | Electrochemical detector flow-cell[71] |
| | | Objet Eden 260V | Fullcure® 720 | Wall-jet electrode detector[82] |
| | — | — | Polyacrylate composite resin | Thin layer flow-cell for anodic stripping voltammetry[80] |

*(continued)*

**Table 3.1** (*continued*)

| | Printer | | | |
|---|---|---|---|---|
| 3DP technique | Manufacturer | Model | Print material used | 3D-printed components[a] |
| SL | 3D Systems | Viper | Somos® WaterShed XC 11122 | High-capacity efficient magnetic O-shaped separator[42] |
| | | Projet 1500 | — | Components for a spectrometer[63] |
| | envisionTEC | Micro Plus Hi-Res | HTM140 | UniverSlide[53] |
| | Formlabs | Form 1+ | Polyacrylate composite resin | Smartphone adapter for fluorescence microscopy[25] |
| | | | — | Smartphone-based platform for colourimetric detection[32] |
| | | | Methacrylate-based resin | Electrochemical detector flow-cell[73] |
| | — | — | Methacrylate-based resin | Prism for surface plasmon resonance[62] |
| SLM | ConceptLaser GmbH | Concept Laser | Stainless steel | Electrodes[87] |
| | RealizerGmbH | SLM50 | Ti-6Al-4V | Electrodes[86] |

[a]The devices have been named in this table as per the names proposed in their respective reports. In the case of the absence of any clearly proposed name, generic names based on their applications have been used.

3D printing preforms for producing optical fibres would further reduce the cost of developing elaborate optical platforms, such as the one shown in Figure 3.6.

Most 3D printing techniques currently lack adequately conductive print materials, which restrict the direct multi-material printing of 3D-printed electrodes within electrochemical detectors. The advent of new conductive 3D-printable materials such as copper-based[96] and carbon-based[97] inks would assist the production of low-cost, high resolution, and customised electrodes. Moreover, the ability to use conductive materials with multi-material printers would allow the fabrication of integrated electrochemical detectors. As with optical detectors, the advent of glass 3D printing would also assist the development of electrochemiluminescence detectors.

## References

1. C. P. Bacon, Y. Mattley and R. DeFrece, *Rev. Sci. Instrum.*, 2004, **75**, 1–16.
2. L. A. Simmons, M. A. Dinan, T. J. Robinson and R. Snyderman, *Pers. Med.*, 2012, **9**, 85–91.
3. L. Shen, J. A. Hagen and I. Papautsky, *Lab Chip*, 2012, **12**, 4240–4243.
4. V. Oncescu, M. Mancuso and D. Erickson, *Lab Chip*, 2014, **14**, 759–763.
5. B.-Y. Chang, *Bull. Korean Chem. Soc.*, 2012, **33**, 549–552.
6. E. Petryayeva and W. R. Algar, *Anal. Chem.*, 2014, **86**, 3195–3202.
7. D. Gallegos, K. D. Long, H. Yu, P. P. Clark, Y. Lin, S. George, P. Nath and B. T. Cunningham, *Lab Chip*, 2013, **13**, 2124–2132.
8. S. Lee, V. Oncescu, M. Mancuso, S. Mehta and D. Erickson, *Lab Chip*, 2014, **14**, 1437–1442.
9. A. Roda, E. Michelini, L. Cevenini, D. Calabria, M. M. Calabretta and P. Simoni, *Anal. Chem.*, 2014, **86**, 7299–7304.
10. V. Oncescu, D. O'Dell and D. Erickson, *Lab Chip*, 2013, **13**, 3232–3238.
11. J. Jiang, X. Wang, R. Chao, Y. Ren, C. Hu, Z. Xu and G. L. Liu, *Sens. Actuators, B*, 2014, **193**, 653–659.
12. T. San Park, W. Li, K. E. McCracken and J.-Y. Yoon, *Lab Chip*, 2013, **13**, 4832–4840.
13. G.-H. Chen, W.-Y. Chen, Y.-C. Yen, C.-W. Wang, H.-T. Chang and C.-F. Chen, *Anal. Chem.*, 2014, **86**, 6843–6849.
14. C. C. Stemple, S. V. Angus, T. San Park and J.-Y. Yoon, *J. Lab. Autom.*, 2014, **19**, 35–41.
15. T. Laksanasopin, T. W. Guo, S. Nayak, A. A. Sridhara, S. Xie, O. O. Olowookere, P. Cadinu, F. Meng, N. H. Chee and J. Kim, *Sci. Transl. Med.*, 2015, 7, 273re271.
16. A. I. Barbosa, P. Gehlot, K. Sidapra, A. D. Edwards and N. M. Reis, *Biosens. Bioelectron.*, 2015, **70**, 5–14.
17. E. J. Topol, *Sci. Transl. Med.*, 2010, **2**, 16cm14.
18. G. Comina, A. Suska and D. Filippini, *Angew. Chem., Int. Ed.*, 2015, **54**, 8708–8712.

19. H. Zhu and A. Ozcan, *J. Visualized Exp.*, 2013, e50451.
20. H. Zhu, S. Mavandadi, A. F. Coskun, O. Yaglidere and A. Ozcan, *Anal. Chem.*, 2011, **83**, 6641–6647.
21. H. Zhu, U. Sikora and A. Ozcan, *Analyst*, 2012, **137**, 2541–2544.
22. H. Zhu, O. Yaglidere, T.-W. Su, D. Tseng and A. Ozcan, *Lab Chip*, 2011, **11**, 315–322.
23. S. K. Ludwig, H. Zhu, S. Phillips, A. Shiledar, S. Feng, D. Tseng, L. A. van Ginkel, M. W. Nielen and A. Ozcan, *Anal. Bioanal. Chem.*, 2014, **406**, 6857–6866.
24. S. K. Ludwig, C. Tokarski, S. N. Lang, L. A. van Ginkel, H. Zhu, A. Ozcan and M. W. Nielen, *PLoS One*, 2015, **10**, e0134360.
25. S. Knowlton, A. Joshi, P. Syrrist, A. F. Coskun and S. Tasoglu, *Lab Chip*, 2017, **17**, 2839–2851.
26. M. M. Gong, B. D. MacDonald, T. V. Nguyen, K. Van Nguyen and D. Sinton, *Lab Chip*, 2014, **14**, 957–963.
27. R. Álvarez-Diduk, J. Orozco and A. Merkoçi, *Sci. Rep.*, 2017, 7, 976.
28. Q. Mei, H. Jing, Y. Li, W. Yisibashaer, J. Chen, B. N. Li and Y. Zhang, *Biosens. Bioelectron.*, 2016, **75**, 427–432.
29. J. Zhang, X. Zhang, L. Dediu and C. Victor, *Food Control*, 2011, **22**, 1126–1135.
30. K. Everstine, J. Spink and S. Kennedy, *J. Food Prot.*, 2013, **76**, 723–735.
31. X. Li, F. Yang, J. X. H. Wong and H.-Z. Yu, *Anal. Chem.*, 2017, **89**, 8908–8916.
32. Z. Liu, Y. Zhang, S. Xu, H. Zhang, Y. Tan, C. Ma, R. Song, L. Jiang and C. Yi, *Anal. Chim. Acta*, 2017, **966**, 81–89.
33. S. Feng, D. Tseng, D. Di Carlo, O. B. Garner and A. Ozcan, *Sci. Rep.*, 2016, **6**, 39203.
34. L.-J. Wang, Y.-C. Chang, X. Ge, A. T. Osmanson, D. Du, Y. Lin and L. Li, *ACS Sens.*, 2016, **1**, 366–373.
35. M. A. Hossain, J. Canning, S. Ast, K. Cook, P. J. Rutledge and A. Jamalipour, *Opt. Lett.*, 2015, **40**, 1737–1740.
36. E. Petryayeva and W. R. Algar, *Anal. Bioanal. Chem.*, 2016, **408**, 2913–2925.
37. A. Roda, M. Guardigli, D. Calabria, M. M. Calabretta, L. Cevenini and E. Michelini, *Analyst*, 2014, **139**, 6494–6501.
38. L. Cevenini, M. M. Calabretta, G. Tarantino, E. Michelini and A. Roda, *Sens. Actuators, B*, 2016, **225**, 249–257.
39. Y. Wang, M. M. Zeinhom, M. Yang, R. Sun, S. Wang, J. N. Smith, C. Timchalk, L. Li, Y. Lin and D. Du, *Anal. Chem.*, 2017, **89**, 9339–9346.
40. K. B. Spilstead, J. J. Learey, E. H. Doeven, G. J. Barbante, S. Mohr, N. W. Barnett, J. M. Terry, R. M. Hall and P. S. Francis, *Talanta*, 2014, **126**, 110–115.
41. V. Gupta, P. Mahbub, P. N. Nesterenko and B. Paull, *Anal. Chim. Acta*, 2018, **1005**, 81–92.
42. W. Lee, D. Kwon, B. Chung, G. Y. Jung, A. Au, A. Folch and S. Jeon, *Anal. Chem.*, 2014, **86**, 6683–6688.
43. W. R. Heineman, *Anal. Chem.*, 1978, **50**, 390A–402A.

44. K. M. Kadish, *Electrochemical and Spectrochemical Studies of Biological Redox Components*, American Chemical Society, Washington DC, 1st edn, 1982.
45. S. Dong, J. Niu and T. M. Cotton, *Methods Enzymol.*, 1995, **246**, 701–732.
46. P. L. Dutton, *Methods Enzymol.*, 1977, **54**, 411–435.
47. R. A. Wilson, T. S. Pinyayev, N. Membreno and W. R. Heineman, *Electroanalysis*, 2010, **22**, 2162–2166.
48. J. M. Brisendine, A. C. Mutter, J. F. Cerda and R. L. Koder, *Anal. Biochem.*, 2013, **439**, 1–3.
49. A. J. Morgan, L. H. San Jose, W. D. Jamieson, J. M. Wymant, B. Song, P. Stephens, D. A. Barrow and O. K. Castell, *PLoS One*, 2016, **11**, e0152023.
50. J. Prikryl and F. Foret, *Anal. Chem.*, 2014, **86**, 11951–11956.
51. R. H. Vera, E. Schwan, N. Fatsis-Kavalopoulos and J. Kreuger, *PLoS One*, 2016, **11**, e0167583.
52. R. M. Sutherland, *Science*, 1988, **240**, 177–184.
53. K. Alessandri, L. Andrique, M. Feyeux, A. Bikfalvi, P. Nassoy and G. Recher, *Sci. Rep.*, 2017, 7, 42378.
54. B. N. Ozbay, J. T. Losacco, R. Cormack, R. Weir, V. M. Bright, J. T. Gopinath, D. Restrepo and E. A. Gibson, *Opt. Lett.*, 2015, **40**, 2553–2556.
55. K. Cook, J. Canning, S. Leon-Saval, Z. Reid, M. A. Hossain, J.-E. Comatti, Y. Luo and G.-D. Peng, *Opt. Lett.*, 2015, **40**, 3966–3969.
56. K. Cook, G. Balle, J. Canning, L. Chartier, T. Athanaze, M. A. Hossain, C. Han, J.-E. Comatti, Y. Luo and G.-D. Peng, *Opt. Lett.*, 2016, **41**, 4554–4557.
57. J. Canning, M. A. Hossain, C. Han, L. Chartier, K. Cook and T. Athanaze, *Opt. Lett.*, 2016, **41**, 5551–5554.
58. Q. Zhao, F. Tian, X. Yang, S. Li, J. Zhang, X. Zhu, J. Yang, Z. Liu, Y. Zhang and T. Yuan, *Optik*, 2017, **133**, 60–65.
59. T. H. Marques, B. M. Lima, J. H. Osório, L. E. da Silva and C. M. Cordeiro, 3D printed microstructured optical fibers, Brazil, 2017.
60. J. Kaufman, C. Bow, F. Tan, A. Cole and A. Abouraddy, 3D Printing Preforms for Fiber Drawing and Structured Functional Particle Production, Australia, 2016.
61. F. Cecil, M. Zhang, R. M. Guijt, A. Henderson, P. N. Nesterenko, B. Paull, M. C. Breadmore and M. Macka, *Anal. Chim. Acta*, 2017, **965**, 131–136.
62. S. S. Hinman, K. S. McKeating and Q. Cheng, *Anal. Chem.*, 2017, **89**, 12626–12630.
63. R. Gao, H. Telg, R. McLaughlin, S. Ciciora, L. Watts, M. Richardson, J. Schwarz, A. Perring, T. Thornberry and A. Rollins, *Aerosol Sci. Technol.*, 2015, **50**, 88–99.
64. W. R. Vandaveer, S. A. Pasas, R. S. Martin and S. M. Lunte, *Electrophoresis*, 2002, **23**, 3667–3677.
65. M. R. Hardy, R. R. Townsend and Y. C. Lee, *Anal. Biochem.*, 1988, **170**, 54–62.
66. A. M. Bond, *Analyst*, 1994, **119**, 1R–21R.

67. O. Niwa, Y. Xu, H. B. Halsall and W. R. Heineman, *Anal. Chem.*, 1993, **65**, 1559–1563.
68. X.-B. Yin and E. Wang, *Anal. Chim. Acta*, 2005, **533**, 113–120.
69. H. Wei and E. Wang, *Luminescence*, 2011, **26**, 77–85.
70. O. Spreux-Varoquaux, G. Bensimon, L. Lacomblez, F. Salachas, P. F. Pradat, N. Le Forestier, A. Marouan, M. Dib and V. Meininger, *J. Neurol. Sci.*, 2002, **193**, 73–78.
71. J. L. Erkal, A. Selimovic, B. C. Gross, S. Y. Lockwood, E. L. Walton, S. McNamara, R. S. Martin and D. M. Spence, *Lab Chip*, 2014, **14**, 2023–2032.
72. G. W. Bishop, J. E. Satterwhite, S. Bhakta, K. Kadimisetty, K. M. Gillette, E. Chen and J. F. Rusling, *Anal. Chem.*, 2015, **87**, 5437–5443.
73. G. W. Bishop, J. Satterwhite-Warden, I. Bist, E. Chen and J. F. Rusling, *ACS Sens.*, 2015, **1**, 197–202.
74. S. A. Gowers, V. F. Curto, C. A. Seneci, C. Wang, S. Anastasova, P. Vadgama, G.-Z. Yang and M. G. Boutelle, *Anal. Chem.*, 2015, **87**, 7763–7770.
75. P. Figura and B. McDuffie, *Anal. Chem.*, 1980, **52**, 1433–1439.
76. J. Wang, J. Lu, S. B. Hocevar, P. A. M. Farias and B. Ogorevc, *Anal. Chem.*, 2000, **72**, 3218–3222.
77. W. Joseph, *Electroanalysis*, 2005, **17**, 1341–1346.
78. M. A. T. Gilmartin and J. P. Hart, *Analyst*, 1994, **119**, 2331–2336.
79. N. Gharib Naseri, S. J. Baldock, A. Economou, N. J. Goddard and P. R. Fielden, *Anal. Bioanal. Chem.*, 2008, **391**, 1283–1292.
80. Q. Sun, J. Wang, M. Tang, L. Huang, Z. Zhang, C. Liu, X. Lu, K. W. Hunter and G. Chen, *Anal. Chem.*, 2017, **89**, 5024–5029.
81. D. T. Chin and C. H. Tsang, *J. Electrochem. Soc.*, 1978, **125**, 1461–1470.
82. A. S. Munshi and R. S. Martin, *Analyst*, 2016, **141**, 862–869.
83. A. A. Dias, T. M. Cardoso, R. M. Cardoso, L. C. Duarte, R. A. Muñoz, E. M. Richter and W. K. Coltro, *Sens. Actuators, B*, 2016, **226**, 196–203.
84. G. Schimo, W. Burgstaller and A. W. Hassel, *Electrochem. Commun.*, 2015, **60**, 208–211.
85. P. Salvo, R. Raedt, E. Carrette, D. Schaubroeck, J. Vanfleteren and L. Cardon, *Sens. Actuators, A*, 2012, **174**, 96–102.
86. C. Zhao, C. Wang, R. Gorkin, S. Beirne, K. Shu and G. G. Wallace, *Electrochem. Commun.*, 2014, **41**, 20–23.
87. A. Ambrosi, J. G. S. Moo and M. Pumera, *Adv. Funct. Mater.*, 2015, **26**, 698–703.
88. G. N. Meloni, *Anal. Chem.*, 2017, **89**, 8643–8649.
89. J. M. Cabot, M. C. Breadmore and B. Paull, *Anal. Chim. Acta*, 2018, **1000**, 283–292.
90. J. Barallat, R. Olivé-Monllau, J. Gonzalo-Ruiz, R. Ramírez-Satorras, F. X. Muñoz-Pascual, A. G. Ortega and E. Baldrich, *Anal. Chem.*, 2013, **85**, 9049–9056.
91. H. V. Nguyen, K. Tmejova, L. Krejcova, D. Hynek, P. Kopel, J. Kynicky, V. Adam and R. Kizek, *Int. J. Electrochem. Sci.*, 2014, **9**, 3364–3373.
92. Z. Li, Z. Ye, Y. Fu, Y. Xiong and Y. Li, *Anal. Methods*, 2016, **8**, 548–553.

93. M. Grant, P. Renuka, S. Duane and G. Mark, *Rapid Prototyping J.*, 2011, **17**, 187–194.
94. F. Kotz, K. Arnold, W. Bauer, D. Schild, N. Keller, K. Sachsenheimer, T. M. Nargang, C. Richter, D. Helmer and B. E. Rapp, *Nature*, 2017, **544**, 337.
95. W. G. Patrick, Masters dissertation, Massachusetts Institute of Technology, 2015.
96. B. K. Park, D. Kim, S. Jeong, J. Moon and J. S. Kim, *Thin Solid Films*, 2007, **515**, 7706–7711.
97. S. J. Leigh, R. J. Bradley, C. P. Purssell, D. R. Billson and D. A. Hutchins, *PLoS One*, 2012, **7**, e49365.
98. P. Oomen, J. Mulder, E. Verpoorte and R. Oleschuk, *Anal. Chim. Acta*, 2017, **988**, 50–57.

CHAPTER 4

# 3D Printing in Analytical Chemistry Methods and Applications

## 4.1 Introduction

The use of 3D printing in the development of micro- and macro-fluidic devices and miniaturised analytical detectors has been presented within Chapters 2 and 3, respectively. This chapter describes the use of 3D printing to develop analytical devices, materials and methods for solid phase extraction and solute isolation, chromatographic separations, mass spectrometry interfaces and detection, and advanced lab-on-a-chip applications.

Much of what is discussed within the following chapter relies upon the material properties of the printed substrates and devices. In those methods and applications where it is necessary for the target solute to physically contact or in some way chemically interact with the printed material, then the nature of that material and its surface are of course of the upmost importance. In the following sections, investigation and various analytical applications of printed materials with unique bulk and surface properties, such as mechanical strength, pressure and temperature stability, porosity, thermal and electrical conductivity, hydrophobicity or surface charge, are discussed.

## 4.2 Extraction Methods

Analysis of complex environmental and biological samples typically requires some form of sample clean-up, matrix elimination, solute extraction and/or pre-concentration. A variety of extraction methods have been developed to

3D Printing in Chemical Sciences: Applications Across Chemistry
By Vipul Gupta, Pavel Nesterenko and Brett Paull

Published by the Royal Society of Chemistry, www.rsc.org

address these requirements, including for example, solid-phase extraction (SPE), solid-phase microextraction (SPME), liquid–liquid extraction (LLE), centrifugation, distillation, *etc.* Over the last few years, 3D printing has begun to see extensive application in this area and has been used to produce a wide variety of microfluidic platforms, reactors, and holders for these and other related extraction methods. The use of modern printing technology enables the prototyping of extraction devices with three-dimensional complexity, integration, and developed surface areas, in materials which can have inherent or functionalisable surface selectivity. A number of small and low-cost 3D-printed devices produced to-date have shown potential as extraction phases in real-world applications, as discussed below.

### 4.2.1 Solid-phase Extraction

Solid-phase extraction (SPE), as the name suggests, utilises a solid sorbent to extract target analytes from liquid or gaseous samples. When compared to liquid–liquid extraction, SPE typically provides lower solvent consumption, can be quicker and to some extent automated. These advantages have seen SPE established as the standard sample preparation and solute pre-concentration procedure in most routine analytical laboratories. Solid-phase extraction can be based upon a wide variety of extraction materials, in various formats. However, commonly used extraction phases are typically particle based or porous media (*e.g.* monolithic materials), and either packed or otherwise immobilised within a cartridge or column. The use of 3D printing to replicate these structures is a challenge, most notably in terms of print resolution and material properties, but significant advances have been made towards this goal.

#### *4.2.1.1 Packed and Supported Sorbents*

As mentioned above, SPE can be carried out based upon packed sorbents in cartridges, columns, capillaries, or other microfluidic channels. SPE carried out in packed micro-columns or microfluidic devices is termed micro-solid phase extraction (μ-SPE).[1] Wang *et al.*[2] have developed a μ-SPE device by packing an SL-printed polyacrylate composite microfluidic device with polyaniline-modified magnetite nanoparticles. The device consisted of a single rectangular channel (0.8 mm × 3.0 mm × 50 mm) with 0.35 mm thick walls (Figure 4.1). It was designed to place two rectangular magnets close to each other to hold the bed of magnetite nanoparticles in place. It was used for the extraction of alkyl benzoate analogues and triclosan from human saliva and urine samples. The 3D-printed extraction device was coupled online to a HPLC system in an online fashion. The device was successfully reused more than 20 times with an RSD of less than 15%. A recovery of 84% to 117% was observed from the spiked saliva and urine samples, and the enhancement factors ranged from 16 to 25.

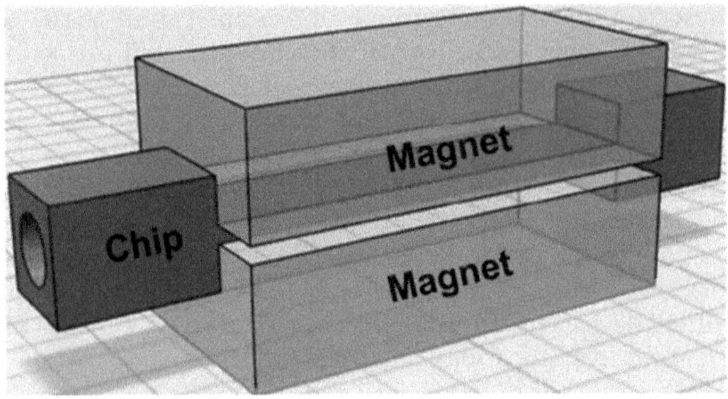


**Figure 4.1** Render of the magnetic μ-SPE chip. Adapted from ref. 2 with permission from American Chemical Society, Copyright 2017.

Kataoka *et al.*[3] produced another 3D-printed μ-SPE device, using an FDM printer to print a microfluidic chip, containing channels that were subsequently packed with a flux-calcined diatomaceous earth material, namely Celite® 545. The packed micro-channel was used for demulsification and deasphalting of petroleum samples for gas chromatography–mass spectrometry and Fourier transform ion cyclotron resonance mass spectrometry, respectively. The 3D-printed microfluidic chip consisted of a cylindrical sample compartment, a rectangular stationary phase container, and two stainless steel connection ports (Figure 4.2). The microfluidic device was printed with poly(lactic acid) material and was reported to be stable while exposed to nonpolar solvents for at least 20 minutes, and was successfully used for fast emulsion breaking and solvent deasphalting of petroleum. According to the authors a 98% recovery was achieved without compromising maltene integrity. It also delivered a ten-fold reduction in the preparation time for petroleum samples, as compared to centrifugation. The cost of fabrication of the 3D-printed microfluidic chip was stated as only *ca.* $0.50, and was thus suitably disposable.

The majority of polymeric 3D-printed devices lack tolerance to high-pressures, specifically for example, those devices containing internal channels for use in flow-based applications. In the latter instance, this restricts the use of densely packed particle beds, or the packing of small particles, which is typically the case in SPE for obtaining efficient extraction. To help circumvent this limitation, SPE discs have been incorporated into 3D-printed microfluidic devices, which do not present significant flow restriction and back-pressure. Calderilla *et al.*[4] developed a 3D-printed modular device to hold SPE discs to perform various processes, including automated analyte oxidation, disc-based extraction, and analyte complexation. The device was composed of two modular components with chelating disks between them. Part 1 consisted of two confluence points and a

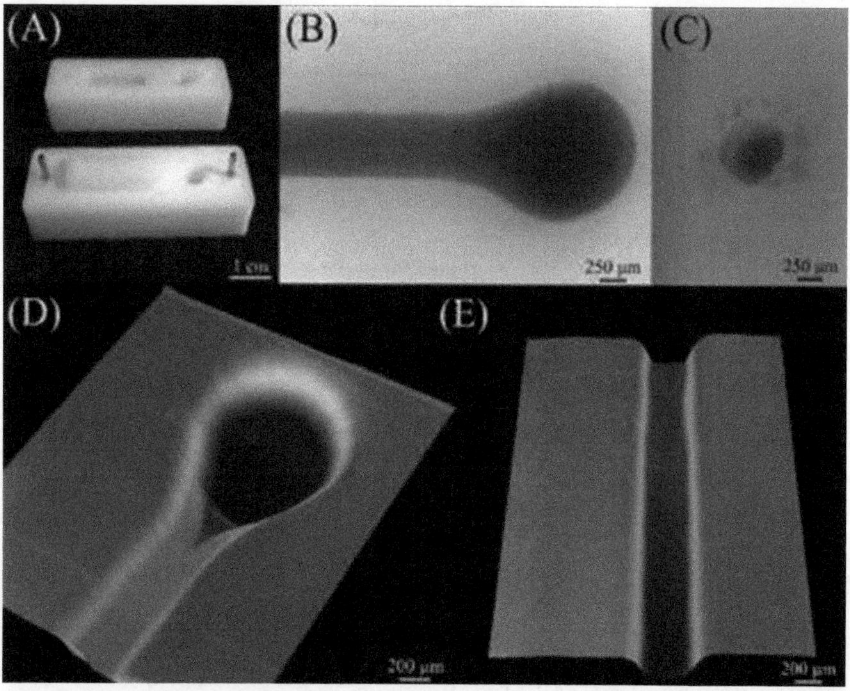


**Figure 4.2** 3D-printed μ-SPE device for sample preparation of petroleum. (A) Photograph of the device, (B) photograph of the half-printed microchannel, (C) cross-section of circular channel, (D) profilometry image of the half-printed base component of the device, and (E) profilometry image of the microchannel. Reproduced from ref. 3 with permission from American Chemical Society, Copyright 2017.

mixer, and Part 2 consisted of a support for the sorbent disk, a confluence point, and a mixer as shown in Figure 4.3(a). The mixer was composed of a matrix of 0.4 mm cubes as shown in Figure 4.3(b) and (c) with a separation distance of 0.3 mm and a superposition of 0.05 mm. The disk support was a truncated cone as shown in Figure 4.3(d), and it consisted of a matrix of 0.9 mm cubes with a separation distance of 0.5 mm and a superposition of 0.2 mm. Part 1 and Part 2 were integrated through threaded connectors. The 3D-printed device was successfully used in a multi-syringe based FIA system for the extraction, speciation, and spectrophotometric detection of iron in water samples. The developed method delivered an LOD of 7 ng and a linear range from 22 ng to 2400 ng for a 3 mL sample. Although the device was printed using an SL printer in a polyacrylate composite, there was no significant adsorption of Fe(III) on the 3D-printed walls. The 3D-printed device was reported to give an 85% greater peak height for the determination of total iron as compared to the conventional tube and connector based manifolds.

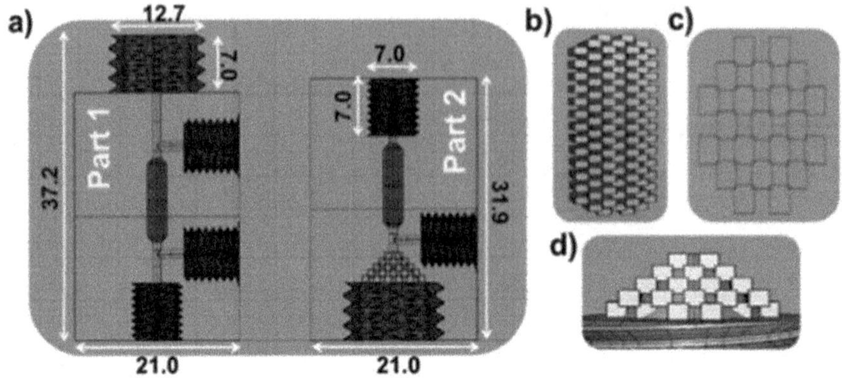


**Figure 4.3** Render of the 3D-printed solid phase extraction device: (a) two modular components (dimensions in mm), (b) mixer, (c) arrangement of cubes in the mixer, and (d) sorbent disk support. Reproduced from ref. 4 with permission from Elsevier, Copyright 2017.

### *4.2.1.2 Surface Modification of Printed Structures*

Structures for use within SPE devices can be printed to provide developed surface areas which can be then modified (surface coated or bonded) with various selective chemistries/biochemistries. This approach has been explored by several groups recently, printing either regularly structured porous beds, or other high (relatively) surface area support structures. However, for most of the 3D-print resins, the exact composition is proprietary, which greatly hinders the process of surface functionalisation. However, some success has been reported in the functionalisation of 3D-printed features with both enzymes and metal–organic frameworks, as discussed below.

In 2016, Su *et al.* reported upon the immobilisation of enzymes, such as glucose oxidase and lactate oxidase, on 3D-printed acrylonitrile butadiene styrene supports to form a simple bioreactor.[5] The printed flow bioreactor consisted of two parallel reaction chambers and two female fittings (as seen in Figure 4.4). Each reaction chamber was composed of 118 ordered cuboids (1 mm × 1 mm × 0.8 mm) with an overlapping square of 0.1 mm × 0.1 mm at each edge. This geometrically 3D complex reactor design was printed using a low-cost FDM printer. The reactor surface was activated by glutaraldehyde polymerisation for the immobilisation of glucose oxidase. Under testing, the immobilised glucose oxidase maintained its activity for up to 42 days at 4 °C, whereas a similarly immobilised lactate oxidase lost 60% of its activity within seven days. The 3D-printed flow bioreactors were reported to deliver recoveries of 105.2% and 94% for glucose and lactate, respectively. An RSD of less than 15% was observed in the analytical performance of four similarly printed bioreactors.

The application and recycling of metal–organic frameworks (MOFs) for the adsorption and removal of organic solutes has typically involved time and labour-intensive manual processes, such as centrifugation and filtration.

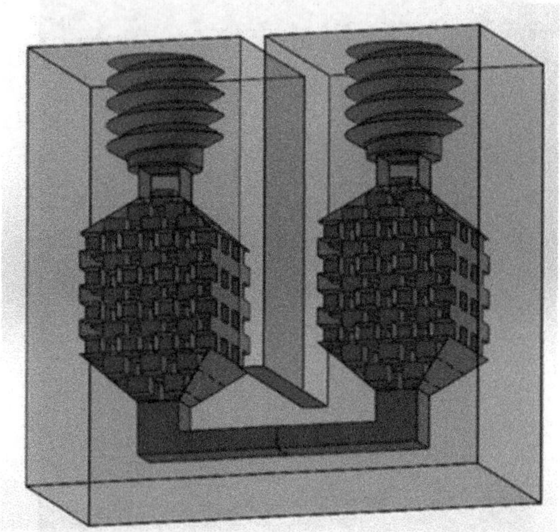

**Figure 4.4** Render of the 3D-printed structure for use as a surface immobilised flow bioreactor based upon 118 ordered cuboids. Reproduced from ref. 5 with permission from American Chemical Society, Copyright 2016.

However, recent work has focused upon the immobilisation of MOFs onto high surface area solid supports to facilitate flow-based extraction. Wang *et al.*[6] have recently immobilised HKUST-1 metal–organic framework material onto the surface of an FDM-printed acrylonitrile butadiene styrene porous support structure (as shown in Figure 4.5) for the extraction of methylene blue. In this work, Cu-BTC particles ranging from 200 nm to 900 nm were successfully deposited onto the 3D-printed support using a step-by-step *in situ* growth process. When applied to extraction, the immobilised MOFs resulted in recoveries of 93.3% and 98.3% for 10 mg $L^{-1}$ and 5 mg $L^{-1}$ methylene blue solutions, respectively. The Cu-BTC particle modified printed support was readily regenerated by soaking in methanol for repeat applications.

In a similar work, Shi *et al.* produced HKUST-1 modified poly(lactic acid)-based sorbents to extract malachite green from wastewater.[7] Similar solid supports were once again FDM-printed, and the support was recovered by soaking the composite in 1 M HCl solution, which removed Cu-MOFs. These two studies suggest that the FDM-printed acrylonitrile butadiene styrene and poly(lactic acid) supporting structures are reasonably stable in solvents such as methanol and 1 M HCl, respectively, allowing their use and regeneration in multiple applications.

### 4.2.1.3 Native (Unmodified) Substrates

3D printing has already been used to deliver porous materials of enhanced surface area, as mentioned above. An ability to 3D-print such substrates with materials that are inherently selective towards target species obviously

**Figure 4.5** 3D-printed Cu-BTC particle modified support. Adapted from ref. 6 with permission from Springer Nature, Copyright 2014.

eliminates the need for additional sorbent packings and/or functionalisation steps in the delivery of simple SPE devices. Although advances in this approach are limited by the availability and variety of functional 3D print materials, some reports have been presented that have used polyacrylate composite resins and similar new porous thermoplastic materials to directly print SPE sorbents/devices.

Su *et al.*[8] have used an SL printer to directly print and apply a polyacrylate-based composite preconcentrator phase for the extraction of trace elements. The work exploits the electron donor groups, such as carbonyl and ester, which are inherently present on the printed polyacrylate surface to attract and retain (complex) metal ions.[9–11] The design of the preconcentrator was very similar to the above-mentioned 3D-printed reactor (Figure 4.4), however, it was printed using a higher resolution SL printer, which allowed smaller and more closely packed cuboids to be printed. The SL-printed preconcentrator consisted of 526 cuboids with dimensions of 0.4 mm × 0.4 mm × 0.2 mm (l × l × h), arranged in 29 layers. This preconcentrator column could be printed in <40 minutes, and the cost of fabrication was said to be only $1.50. In terms of its performance, the printed preconcentrator was applied to the concentration of trace divalent cations such as Mn, Ni, Cu, Cd, Zn, and Pb in solutions of up to 4.5% sodium chloride. Enrichment factors and LODs ranging from 11.2 to 107.9 and 0.3 ng $L^{-1}$ to 18.0 ng $L^{-1}$, respectively, were reported when

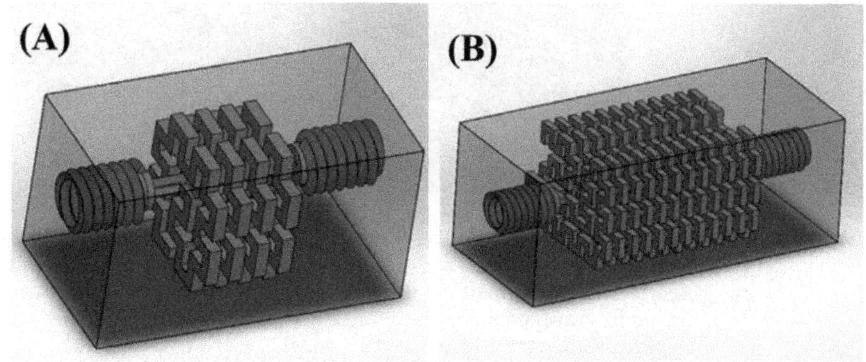


**Figure 4.6** Render of the 3D-printed knotted reactors: (A) 128 turns and (B) 768 turns. Reproduced from ref. 12 with permission from Elsevier, Copyright 2016.

used in conjunction with a quadrupole-based, inductively-coupled plasma mass spectrometer (ICP-MS). The printed phase was highly permeable, could be rapidly regenerated with dilute acids, and the device was monolithic and frit free.

In a related study, the same group also developed SL-printed knotted reactors for the extraction and differentiation of $Ag^+$ ions and Ag nanoparticles in wastewater samples.[12] The polyacrylate composite knotted reactors functioned through the generation of centrifugal forces within the flow stream, which acted to enhance adsorption of neutral metal complexes on the inner walls of the reactors.[13–15] The 3D-printed knotted reactor consisted of angle-defined square flow channels and two 10–32 flat bottom female connectors (Figure 4.6). The fabrication times for 128-turn and 768-turn knotted reactors were 56 minutes and 85 minutes, respectively. The extraction efficiencies of the knotted reactors increased with an increase in the number of turns, although this effect levelled off at 768 turns for 1 mL (5 $\mu g\ L^{-1}$ Ag) samples, providing LODs of 0.86 ng $L^{-1}$ and 0.52 ng $L^{-1}$, and recoveries of 54.5% and 32.3% for $Ag^+$ ions and Ag nanoparticles, respectively, when used in conjunction with an ICP-MS. The 3D-printed knotted reactors resulted in extraction efficiencies of 3.3-fold higher for $Ag^+$ ions, and 4-fold higher for Ag nanoparticles, as compared to PTFE tubing-based counterparts.

Recently, a number of new thermoplastic materials for 3D printing have become available, which following printing, and post-print treatments (washing, soaking, *etc.*), are inherently porous. Examples include 'Lay-Fomm' and 'Lay-Felt' filaments from the Poro-Lay portfolio of print materials, which have been developed for use with FDM printers. These materials consist of two major components; a rubber-elastomeric polymer and a water-soluble poly(vinyl alcohol) component that can be dissolved in water after printing. A number of research groups have been exploring these new materials for the

**Figure 4.7** 3D-printed porous tube-shaped sorbent. (a) Photograph of the sorbent: (I) before washing with water, and (II) after washing with water, (b) scanning electron micrograph of the sorbent before washing, and (c) scanning electron micrograph of the sorbent after washing. Reproduced from ref. 16 with permission from American Chemical Society, Copyright 2017.

direct 3D printing of both solid-phase microextraction sorbents and porous membranes. For example, Belka *et al.*[16] have used Lay-Fomm 60 material to FDM-print a tube-shaped sorbent (as shown in Figure 4.7) for the extraction of small molecules, specifically in the reported application, glimepiride from water samples. Also in the area of separations, although here membrane-based, Li *et al.*[17] have used Lay-Felt material to FDM-print membranes within simple polymer microfluidic chips for the selective passage of nitrate ions from soil samples into a receiving reagent solution for low-cost and portable analysis.[17] Within this work, it was reported that the FDM-printed Lay-Felt membrane had an average pore diameter of 12 nm after washing with water.

Furthermore, Kalsoom *et al.*[18] used the Lay-Felt material to develop a passive sampling device. The device consisted of two interlocking frames with integrated membranes, which could be efficiently sealed together without the need for additional O-rings (Figure 4.8). The frames were 3D-printed with a multi-material FDM printer. Poly(lactic acid) was used to print the body and Lay-Felt was used to print the membrane. In this case, the authors observed an average pore size of 30 nm after washing the membrane with water. The membranes allowed selective permeation of anionic and neutral solutes, preventing any transfer of cations. A 10 mm overlap between the membrane and the holder body was found to be sufficient to provide a strong attachment of the membrane. The device was filled with a microparticulate sorbent to extract atrazine from sea water. It resulted in a depletion value of 87% and a sampling rate of 0.19 litres per day, which was found to be comparable to the depletion value (92%) and sampling rate (0.20 litres per day) observed with the sampling devices fitted with commercial PES membranes.

### 4.2.2 Centrifugation and Distillation Devices

In a very practical and cost-saving application of 3D printing, Desai *et al.*[19] developed customisable analytical ultra-centrifugation centrepieces to replace otherwise expensive commercial counterparts. The commercially available centrepieces were reported to cost more than $1000 each and

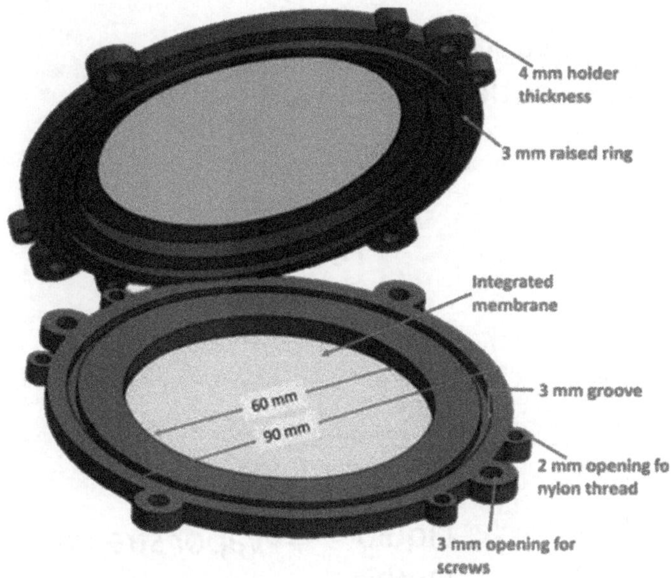


**Figure 4.8** Render of the 3D-printed passive sampler. Reproduced from ref. 18 with permission from American Chemical Society, Copyright 2018.

are considered by the users as consumable items, due to their short lifespan. In addition, these ultra-centrifugation centrepieces are required in various designs and arrangements for different experiments. The group reported on the performance of various centrepieces, 3D-printed using PolyJet and SL printers, and prepared from different materials, including polyacrylate-based, carboxylate-based, polypropylene carbonate, and acrylonitrile butadiene styrene-like resins. All of the centrepieces were found to provide a similar performance. The authors reported that the cost of PolyJet printing these centrepieces in a polyacrylate material was between 1 and 10% of the cost of commercial analytical ultra-centrifuge centrepieces, with the cost even lower for SL printing. The PolyJet-printed centrepieces were used for analytical ultra-centrifugation and were found to withstand gravitational forces of more than 300 000 *g*, provided precise sedimentation equilibrium and velocity experiments with a statistical error of less than 1%.

In the area of distillation, Mardani *et al.* have developed a coiled and modular polyacrylate-based distillation column, produced using SL printing.[20] The column was designed in a coiled shape to provide a sufficiently long column length, and hence a larger number of equilibrium stages, while minimising its space requirements. The column was designed and printed in multiple modular sections, connected through flange connections because of the limited build area of the SL printer used (Figure 4.9).

The group also printed a column packing material, which consisted of cubic shapes with a maximum width of 4 mm. The column was used for

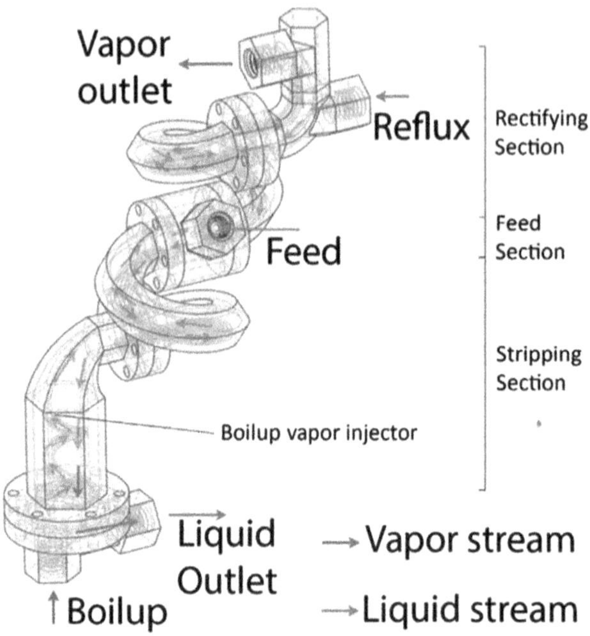


**Figure 4.9** Render of the 3D-printed distillation column. Reproduced from ref. 20 with permission from Elsevier, Copyright 2016.

the distillation of a binary mixture of hexane and cyclohexane. Steady-state conditions were achieved, and distillations continued for up to 9 hours. The column was also successfully used for multiple distillation runs. However, prolonged use of the column resulted in the swelling and breakdown of the printed polyacrylate material due to high temperature and exposure to solvent vapours.

## 4.3 Analytical and Preparative Separations

Chromatography is one of the most widely applied analytical separation techniques, and exists in a number of well-established and advanced modes, for the separation and analysis of both liquid and gaseous samples, originating from a huge variety of sources. Electrophoresis, both capillary and gel-based, is closely related to chromatography, and is also an established separation technique, particularly in the area of bioseparations, and so also considered herein. As so-called 'gold standard' laboratory techniques, the market for both liquid and gas chromatography consumables is extremely large and profitable. This has inevitably seen 3D printing gain a foothold in this field, as users seek to source lower cost and disposable chromatographic devices, including columns and stationary phases, and other related components.

The use of 3D printing to develop fluid handling platforms/devices and analytical detectors, both of which can, and indeed already have, found application within chromatography and electrophoretic separation systems, is already described in Chapters 2 and 3, respectively. The printing of porous and surface modified substrates for use in SPE has also been discussed above. The use of 3D printing in the field of capillary electrophoresis is covered in Section 4.5 (lab-on-a-chip). The following section is therefore limited to the production of columns and stationary phases which have been produced to deliver actual chromatographic separations.

## 4.3.1 Liquid Chromatography

Column-based liquid chromatographic techniques are typically classified according to the dominant retention mechanism or solute interactions taking place, such as reversed-phase, ion-exchange, hydrophilic interaction, and size-exclusion. In the case of liquid chromatographic techniques, they are of course applied extensively to the separation of non-volatile solutes. The most important component within any liquid chromatograph is the separation column. Herein the selectivity towards the solute(s), which provides the means to achieve separations is expressed. The column itself consists of two major elements, the actual column body or housing, and the internal stationary phase. In recent years, there have been several attempts to rethink the nature (and future) of the chromatographic column and the stationary phase within, materialised through the application of 3D printing.[21]

### *4.3.1.1 Chromatographic Columns*

The chromatographic column (housing) has a significant role in the ultimate performance of the separation method, and indeed its physical footprint. Traditional 'long' (25 to 30 cm) tubular liquid chromatographic columns, typical of standard HPLC systems of the 1970s, 1980s, and 1990s, have largely been replaced with shorter 5 to 15 cm columns, housing particle packed stationary phases, which have decreased proportionally over these decades, from 20 to 10 to 5 µm and smaller particles. To deviate from the classic straight tube-based columns, and to produce chromatographic columns with smaller footprints, requires the introduction of column coiling, or 'turns' more typical of what has been achieved within microfluidic platforms in the past. With conventional manufacturing techniques restricted to two dimensions, the production of coiled columns is certainly a challenge, particularly in non-planar formats. However, 3D printing can be used to fabricate high-aspect ratio coiled columns rather easily. Further still, 3D printing offers the 'fail fast and often' approach for rapid optimisation of column design, which allows more detailed exploration of column geometry effects upon both flow and dispersion-based phenomena within.

The most commonly used 3D printers are produced to print a limited range of commercial polymeric materials. However, these commercial materials, and their subsequent printed structures, lack sufficient structural and thermal stability to be used under the high pressures and temperatures associated with modern liquid, and indeed gas, chromatography. Further, for application in most modes of liquid chromatography, solvent compatibility is also critical, and once again most print polymers currently available have very limited resistance to common organic solvents. However, printing in a metal can potentially overcome each of the above limitations. The first 3D-printed metal–liquid chromatographic column was reported by Sandron *et al.*[22] in 2014. The coiled column body was 3D-printed with stainless steel (316L) build material using an SLM metal 3D printer. The printed column was formed within a single metal printed unit, containing a 600 mm long and 0.9 mm I.D. (designed as 1 mm) internal channel, which was wrapped in a double-handed spiral configuration, all within a footprint of just 5 mm × 30 mm × 30 mm, as shown in Figure 4.10. The column channel ends had integrated 3D-printed threaded ports, to accommodate standard HPLC fittings. In this relatively early work, the printed column was slurry packed with a range of octadecyl silica (ODS) particles as the stationary phase and applied to simple separations in reversed-phase HPLC. Most significantly, the 3D-printed stainless steel column was found to be stable up to at least 28 MPa of operating pressure and was fully solvent compatible. However, the print resolution of the SLM printer used resulted in an internal wall roughness of *ca.* 50 μm, which hinders efficient packing of small particles, and hence greatly limits the achievable chromatographic efficiency of the coiled column.

**Figure 4.10** Render of the 3D-printed stainless steel (316L) liquid chromatographic column. Reproduced from ref. 22 with permission from The Royal Society of Chemistry.

The adverse effects of wall roughness in the above-printed columns can be to some extent circumvented by exploiting *in situ* polymerised wall bonded monoliths as the stationary phases. The bonding of such organic monoliths to inorganic surfaces usually requires pre-modification of the walls with organosilanes.[23] However, in the past, the silanisation of stainless steel surfaces has proven to be hydrolytically unstable.[23] Accordingly, the above work was extended by Gupta *et al.*[24] to produce the SLM-printed metal column in a titanium alloy (Ti-6Al-4V). The resultant oxidised titanium surface was relatively simple to silanise and offered greater hydrolytic stability. The final column was also solvent, pressure and temperature stable, and relatively chemically inert.[23,25] The 3D-printed titanium column housed an *in situ* polymerised and wall-bonded poly(butyl methacrylate-co-ethylene glycol dimethacrylate) monolith. The monoliths were stable up to an operating pressure of at least 20 MPa, and the column was successfully applied to the reversed-phase liquid chromatographic separations of various proteins and peptides. Moreover, the thermal stability of the titanium alloy printed column was successfully tested through the application of rapid and bi-directional thermal gradients for liquid chromatographic separations.

Exploiting the unique capabilities of metal 3D printing further, Gupta *et al.* also studied the effect of column 3D geometry and coiling on chromatographic efficiency.[26] Conventional fabrication techniques have previously limited detailed experimental investigations into the effects of complex column geometry in liquid chromatography, and hence these effects have been debated for many decades. However, metal 3D printing provided the opportunity to experimentally study these potentially significant effects for a range of column formats/designs. Three different columns with different aspect ratio turns were designed, and printed, with each containing an *in situ* polymerised wall-bonded poly(butyl methacrylate-co-ethylene glycol dimethacrylate) monolithic stationary phase. These were then characterised to study the effects in reversed-phase HPLC. The columns, in increasing order of aspect ratio turns, were 2D serpentine, 3D spiral, and 3D serpentine in shape, as shown in Figure 4.11. Experimental and computational

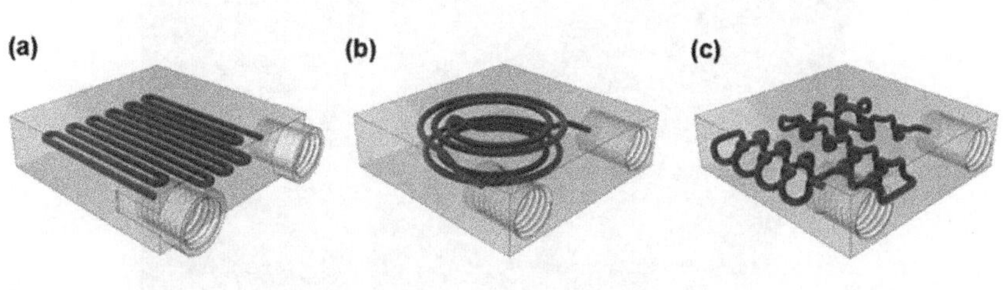


**Figure 4.11** Render of the 3D-printed titanium alloy (Ti-6Al-4V) liquid chromatographic columns: (a) 2D serpentine, (b) 3D spiral, and (c) 3D serpentine. Reproduced from ref. 26 with permission from American Chemical Society, Copyright 2018.

fluid dynamic studies indicated that a potential increase in the chromatographic efficiency could be gained with an increase in the aspect ratio of the turns, especially at higher linear velocities. The high aspect ratio 3D serpentine column resulted in a 58% shorter analysis time for the isocratic separations of small molecules, and 74% higher peak capacity in the gradient separations of proteins, with respect to the low aspect ratio 3D spiral column. This study presented the first experimental evidence for the higher chromatographic efficiency of high aspect ratio coiled liquid chromatographic columns.

The generation of polymer monoliths in columns of I.D. greater than a few hundred μm can be problematic and subject to voids and detachment from the channel walls. In a further example of the use of 3D metal printing in the area, Vonk *et al.* used 3D-printed titanium columns, which have internal 3D 'scaffold' structures to prevent such issues and avoid the 'collapse' of *in situ* polymerised organic monoliths.[25] The internal scaffold consisted of 110 μm × 110 μm pillars, which were regularly spaced at 225 μm from each other, as shown in Figure 4.12. The scaffold was generated within a 50 mm long square channel, 1.3 mm across, and was used to support a poly(styrene-co-divinylbenzene) monolith. The complete monolithic column was reported to be stable up to an operating pressure of at least 80 MPa.

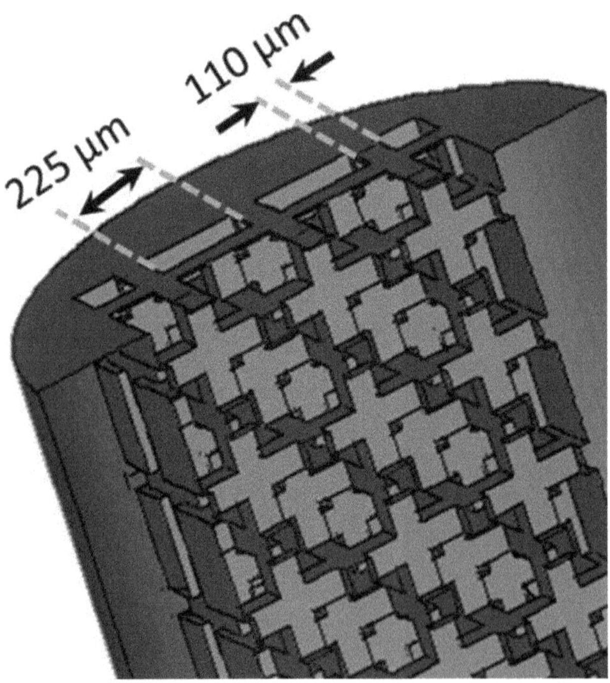


**Figure 4.12** Render of the 3D-printed titanium scaffolds in liquid chromatographic columns. Reproduced from ref. 25 with permission from Elsevier, Copyright 2014.

The printed titanium scaffold based monolithic columns were successfully applied to the reversed-phase liquid chromatographic separations of intact proteins, and similarly to the metal printed columns discussed above, the scaffolds themselves could be regenerated by the pyrolysis of the organic monoliths at 450 °C under $N_2$.

### *4.3.1.2 Printing Chromatographic Stationary Phases*

The above-mentioned printed liquid chromatographic columns were either slurry packed with particles or included an internal organic monolithic phase. Both types of stationary phase are to some extent inherently heterogeneous in structure. Although the trend in liquid chromatography has been continuously moving towards homogeneity using ever smaller particles and denser monoliths, such phases still can't be truly regarded as an ordered structure. Both chromatographic theory and practice show that ordered stationary phases represent an ideal scenario when seeking improved chromatographic efficiency. A stationary phase with random structure extenuates band broadening due to eddy diffusion (multiple path term – loss of efficiency), longitudinal and radial heterogeneity, and significant wall effects, which altogether reduce column chromatographic performance and reproducibility. However, the advent of 3D printing has provided an opportunity to develop customised and regular porous beds that are truly ordered in structure, and which can be used and characterised as liquid chromatography stationary phases. 3D printing can be used to independently control the shape, size, porosity, orientation, and position of stationary phase particles (units) and their surrounding flow-through pores.

The first substantial study into the use of 3D printing to produce ordered stationary phases for liquid chromatographic separations was reported by Fee *et al.* in 2014.[27] The group applied a PolyJet printer to produce stationary phases (porous beds) based upon particle-type and/or monolithic structures, together with internal flow distributors, and external fluid connectors, in an integrated format. Specifically, three porous beds were studied, namely a simple cubic arrangement of 'beads', parallel channel monoliths, and herringbone shaped monoliths, as shown in Figure 4.13. Both radial and fractal flow distributors were studied. The mean bead and channel apothems were observed to be 113.6 ± 3.8 μm (designed as 115 μm) and 148.2 ± 2 μm (designed as 150 μm), respectively. The relatively small error in the observed and designed dimensions shows precise control and reproducibility of the printer used. Measured residence time distributions and porosities were found to be in close agreement with the designed pore densities. These 3D-printed polyacrylate composite columns were reported to be stable up to linear velocities of 594 cm $h^{-1}$. In the same year, they also studied the effect of particle overlap on the chromatographic efficiency.[28] Four different stationary phases were 3D-printed with different overlapping factors: 1.40, 1.45, 1.50, and 1.55. The stationary phases were again printed with a PolyJet printer using acrylonitrile butadiene styrene resin,

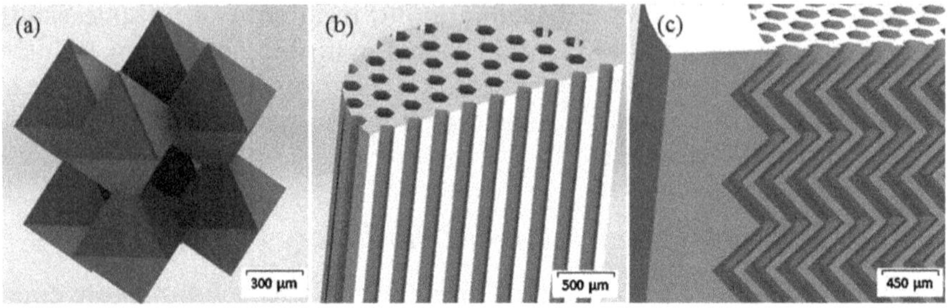


**Figure 4.13** Render of the 3D-printed stationary phases. (a) Simple cubic arrangement of beads, (b) parallel channel monoliths, and (c) herringbone shaped monoliths. Reproduced from ref. 27 with permission from Elsevier, Copyright 2014.

and they were characterised by observing the residence time distribution of a NaCl injection plug. The authors observed an increase in the skewness of the residence time distribution profile with an increase in the overlapping factor. However, the study was found to be inconclusive, because an increase in the particle overlap decreased the pore size and increased the adverse effects resulting from the surface roughness of the 3D-printed particles.

Conventional particle packed beds are currently limited to spherical particles. However, the use of 3D printing to develop beds with particle (units) of customised shape and size enables researchers to develop a better understanding and optimisation of future chromatographic stationary phases. As print resolution continuously improves, some traditional ideas on the 'optimal' particle shape within a chromatographic bed can begin to be re-explored experimentally.

Nawada *et al.*[29] have begun this process, exploring 3D-printed porous beds with different particle shapes, including truncated icosahedrals, tetrahedrals, octahedrals, triangular bipyramidals, and stellar octangulars (Figure 4.14). These particles were arranged in simple cubic, body-centred cubic, and face-centred cubic lattice structures. Consistent results were observed between the designed and printed geometries obtained using a PolyJet printer, with small standard deviations between particles, prints, and print runs. A difference of only 1.3% was observed between the designed and the observed porosities of the resultant beds. Tetrahedral particles and the face-centred cubic arrangement provided the smallest plate height values, which were calculated using the residence time distribution of a sodium chloride injection. These columns were not characterised for liquid chromatographic separations at this early stage, presumably due to the chemical and thermal instability of the print material, as discussed in Chapter 7. Figure 4.15 shows the 3D-printed liquid chromatography column with an integrated porous bed stationary phase.

The above work provides an important starting point for the realisation of 3D-printed liquid chromatographic stationary phases. Taking this forward is now dependent upon several factors, such as the availability of suitable

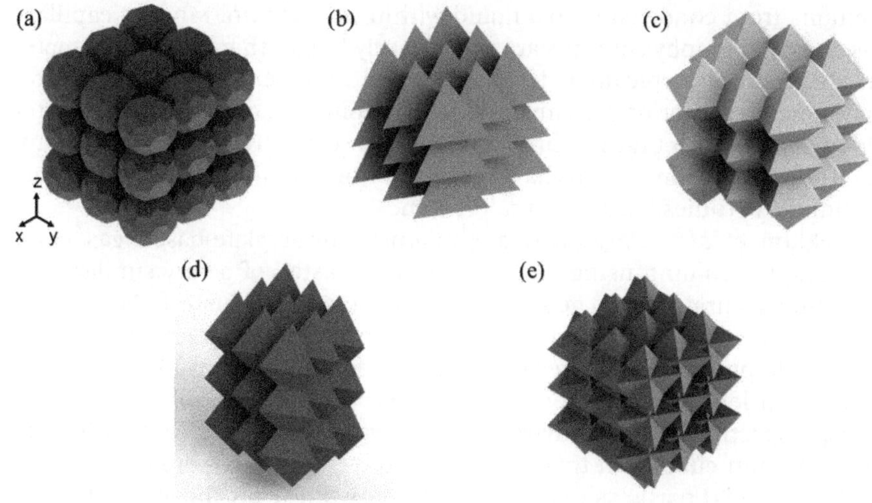


**Figure 4.14** Render of the 3D-printed porous beds with different particle shapes: (a) truncated icosahedral, (b) tetrahedral, (c) octahedral, (d) triangular bipyramidal, and (e) stellar octangular. Reproduced from ref. 29 with permission from Elsevier, Copyright 2017.

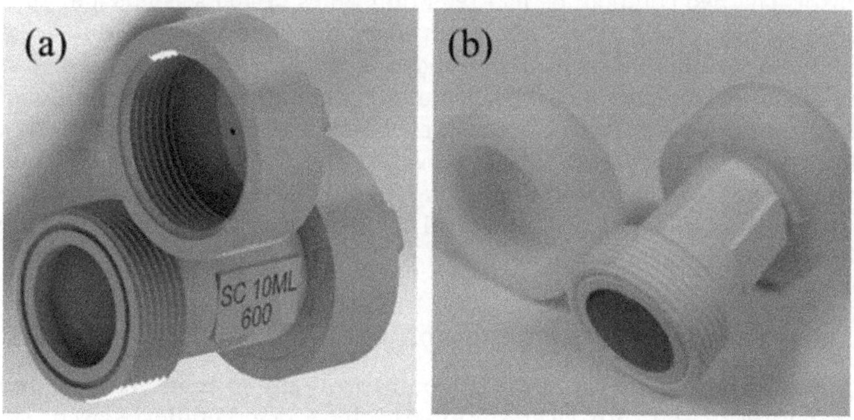


**Figure 4.15** 3D-printed liquid chromatography column: (a) render and (b) photograph. Reproduced from ref. 29 with permission from Elsevier, Copyright 2017.

print and support materials, the available resolution of 3D printers, and the size of the build areas, all of which are improving rapidly.

### 4.3.2 Gas Chromatography

Gas chromatography is used for the separation of gases and volatile species. One of the key parameters in gas chromatography is the column temperature, to keep the sample vaporised and avoid the sample and all that it contains from condensing to a liquid within the column. Modern capillary gas chromatography columns are significantly longer than liquid chromatographic columns, typically 15 to 25 m in length, and hence are coiled. To produce smaller footprint gas chromatography columns, even at much shorter lengths, high aspect ratio column coiling is required. To date, material temperature stability limitations have restricted work in this area, however some preliminary studies have been undertaken.

Lucklum *et al.*[30] 3D-printed a miniature polyacrylate-based gas chromatographic column using SL printing. It consisted of a very similar double-handed spiral column geometry to that shown previously in Figure 4.10, there used for liquid chromatography, and had two integrated M3 fittings in a single printed unit. Internal capillary channels (1 mm I.D.), of 36 cm and 45 cm length, were produced in single and stacked spiral configurations, respectively. The capillary channels were built within a 36 mm × 36 mm × 18 mm cuboid. In this work, the printed columns were packed with Carbosieve®-SII particles to act as a stationary phase for the gas chromatographic retention and separation of ethylene. The size of the packed particles, and the maximum packing pressure that they were able to accomplish with the SL-printed polyacrylate-based column bodies were not disclosed. The target ethylene, which was successfully retained and separated from water, could be detected at an LOD of 35 ppb using a $SnO_2$ metal-oxide gas sensor. The two compounds had retention times of 400 seconds and 1035 seconds, respectively. The same group later modified this 3D-printed gas chromatographic column to accommodate a platinum wire for column heating as shown in Figure 4.16.[31] The column was heated up to a maximum of 45 °C, presumably due to the polyacrylate build material of the column body. However, even this minimal heating lowered the LOD for ethylene detection from 35 ppb to 2.3 ppb. Moreover, an RSD of only 1.2% was reported between different runs using this far-from-ideal column material, demonstrating the potential for future improvements.

### 4.3.3 Thin-layer Chromatography

Thin-layer chromatography (TLC) comes in many formats, based upon the nature of the thin layer stationary phase (*e.g.* cellulose-based, silica-based, carbon-based *etc.*), and operating principles (*e.g.* standard capillary action-based TLC, centrifugal TLC, high-pressure or high-performance TLC, *etc.*). As with liquid chromatography, TLC is typically applied to the separation of non-volatile solutes and is a particularly important laboratory technique in synthetic chemistry and indeed within the field of bio-separations.

Traditional TLC utilises a plate of glass, plastic, or aluminium that is coated with a thin layer of an adsorbent, such as silica, alumina, or cellulose. Investigations into whether it is possible to produce TLC plates, in full or in part, using 3D printing have recently been reported, with the potential to deliver customised designs, and cheaper disposable products.

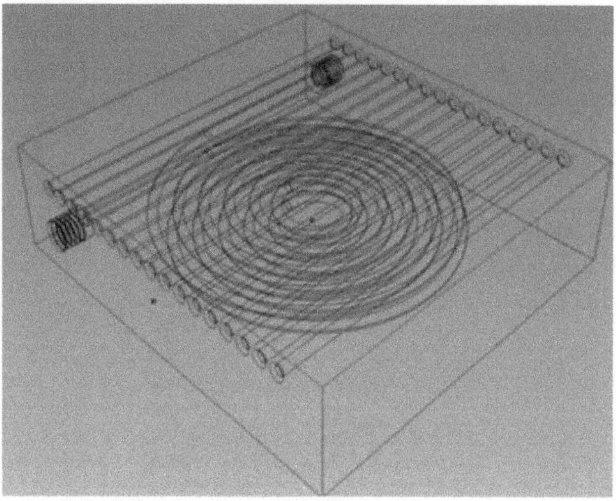

**Figure 4.16** Render of the 3D-printed gas chromatography column with slots for platinum wire. Reproduced from ref. 31 [https://doi.org/10.3390/s17102283] under the terms of a C BY 4.0 license [https://creativecommons.org/licenses/by/4.0/].

Fichou *et al.*[32] have used a direct ink writing technique to deposit complex patterns of silica adsorbent on glass plates in their work to develop printable thin-layer chromatography plates.[32] Direct ink writing is based on the spatially controlled extrusion of semisolid materials in a layer-by-layer arrangement. These materials solidify on a stage or a substrate to develop the desired structures. This relatively simple technique was used to develop different shapes and patterns within the silica layer, and the resulting printed plates were used for the separation of model dyes.[32]

Working towards similar goals, although here in polymer materials, Macdonald *et al.*[33] used a PolyJet printer to produce a complete thin-layer chromatography plate, that showed some limited separation selectivity without any further modification of the printed surface. Incomplete bonding between adjacent layers during PolyJet printing was used to produce a suitably developed surface area, which through capillary action enabled sufficient solvent transport and chromatographic separations. The production time was approx. 1 hour for ten TLC plates (see Figure 4.17), and the cost was *ca.* $3.5 per plate. The plates were 3D-printed with a polyacrylate-based material (VeroClear®), while aligned parallel to the *Z*-axis, as shown in Figure 4.17. The exact composition of VeroClear® and other commercially available PolyJet and SL print resins is mostly proprietary, which in applications such as these limits definitive understanding of the selectivity being exploited upon the print surface. However, in this work, the group attempted to characterise the polymerised VeroClear® surface, which was reported to be rich in carbonyl groups, esters, amides, aliphatic chain groups, and aromatic groups. A net negative charge and a contact angle

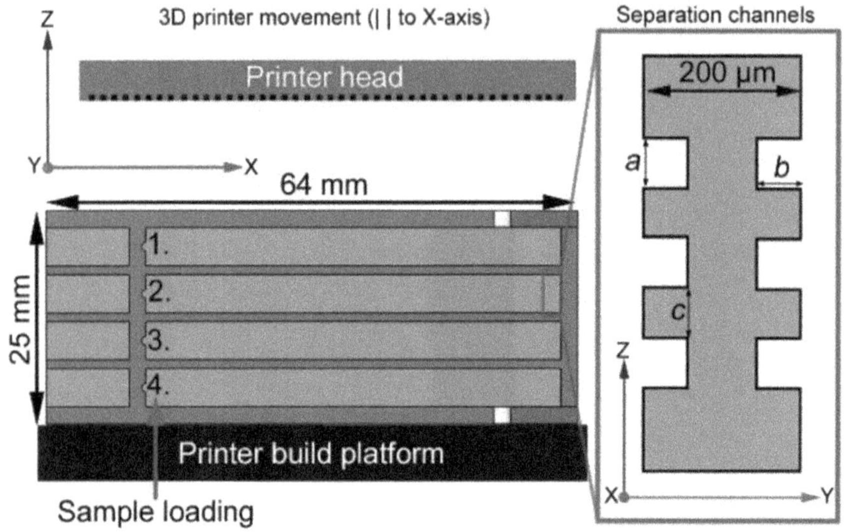


**Figure 4.17** Illustration of the 3D printing process and structure of the 3D-printed thin layer chromatography plates. Reproduced from ref. 33 with permission from American Chemical Society, Copyright 2017.

of 62° was observed. The surface roughness was observed to be 1.34 µm. The 3D-printed VeroClear® plates were successfully used for the separation of water-soluble dyes, fluorescent dyes, and fluorescence tagged proteins (myoglobin and lysozyme). The retardation factors observed for phenol red and bromothymol blue were 0.90 ± 0.05 and 0.34 ± 0.08, respectively. In the applications shown, these 3D-printed plates provided similar and, for some examples, improved separation performance and faster solvent movement than cellulose plates.

## 4.4 Mass Spectrometry

Mass spectrometry is one of the most advanced and widely applied analytical techniques and exists in a huge variety of formats and modes, from hand-held and portable devices, to large sector-field instruments and the most advanced Fourier-transform ion cyclotron resonance mass spectrometers. Application of mass spectrometry is widespread, including for example the field of organic chemistry,[34] proteomics,[35] pharmaceutical chemistry,[36] and environmental analysis,[37] to name a few. This widespread use and applicability has led to a significant level of interest into small-footprint and portable mass spectrometry, where the sensitivity and selectivity of the technique can be utilised at point-of-need.[38] One of the most recognisable examples of portable mass spectrometry is explosives' detection at security within airports, using rapid ion mobility mass spectrometry. However, mass spectrometer components and consumables are typically high cost and highly customised (vendor specific), which gives mass spectrometry a reputation of being a very expensive technique, that is prone to significant downtime. In attempts to address these concerns, and to simultaneously provide new capabilities, 3D printing has recently been applied to produce low-cost components and tailored novel devices for mass spectrometry. Some of the 3D-printed mass spectrometer components/devices that have been recently reported by different research groups are discussed below.

### 4.4.1 Ambient Ionisation Sources

Ambient ionisation sources in mass spectrometry are designed to allow rapid and direct sample introduction *via* ionisation of solutes under atmospheric pressures with minimal preparation steps. As ionisation takes place under atmospheric pressure and not under vacuum, there is more tolerance for polymer material-based components of such sources, than would be possible otherwise. Spray ionisation and surface desorption are two of the most commonly applied ambient ionisation techniques.[39]

Paper spray ionisation is one such ambient ionisation method that has gained significant attention due to its simplicity and low cost. It has been used in a variety of applications, including the quantification of drugs of abuse,[40] pharmaceutical compounds in blood,[41,42] and toxic substances in food.[43] Supporting fast-moving developments in this technique, Salentijn *et al.* have produced three different types of 3D-printed cartridges for paper spray ionisation, for three novel systems, namely fast wetting and continuous solvent supply (FWACSS),[44] controlled and symmetric elution (CASE),[44] and lens and nozzles (LAN).[45] These cartridges were FDM-printed with poly(lactic acid), and they have been individually described below.

The FWACSS cartridge was developed to provide physical stability and protection to the paper tip, and to align the spray with the mass spectrometer, providing fast initial wetting and continuous solvent supply (Figure 4.18). Within the study, it was observed that exposure to organic solvents, such as isopropanol, increased the surface hydrophilicity of the poly(lactic acid) cartridges. Hence, the cartridges were pre-exposed to isopropanol for 5 minutes to facilitate this fast wetting. The 3D-printed FWACSS cartridge provided continuous solvent supply to sustain the spray for tens of minutes.

The 3D-printed CASE cartridge was a modified version of the FWACSS cartridge with two additional lateral wicks, a platform, and minimisation of the contact between the paper tip and the cartridge. The addition of the two wicks led to a symmetrical flow profile and an increased supply of solvent. The platform acted as a solvent guide structure and a secondary reservoir for excess solvent from the initial wetting. Minimised contact between the paper tip and the cartridge prevented unwanted solvent and sample retention around the tip.

The LAN cartridge consisted of four nozzles and an electrostatic lens in addition to the features of the CASE cartridge (Figure 4.19), and it resulted in 15% higher sensitivity and more than 6.5 times lower LOD when compared

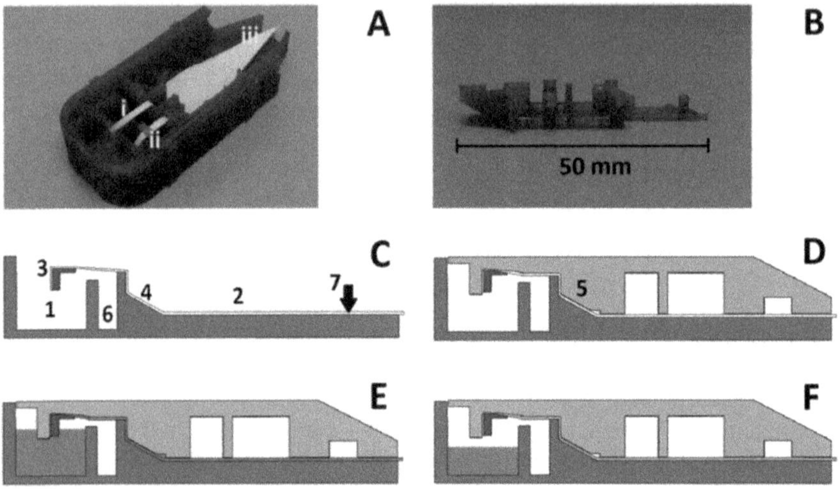


**Figure 4.18** 3D-printed FWACSS cartridge for paper spray ionisation. (A) Photograph of the bottom part with a paper substrate: (i) fast wetting wick, (ii) continuous wetting wick, and (iii) paper tip. (B) Photograph of the top part with a solvent guide structure. (C) Schematic of the bottom part's cross-sectional view (white line represents the paper): (1) solvent reservoir, (2) paper tip chamber, (3) suspended beam, (4) channel for fast wetting, (6) cavity, and (7) paper tip. (D) Schematic of the closed cartridge's cross-sectional view: (5) solvent guide structure. (E) Schematic of the fast wicking process through the fast wetting wick when the reservoir is full of solvent. (F) Schematic of the continuous wicking process through the continuous wetting wick when the solvent level in the reservoir drops. Reproduced from ref. 44 with permission from American Chemical Society, Copyright 2014.

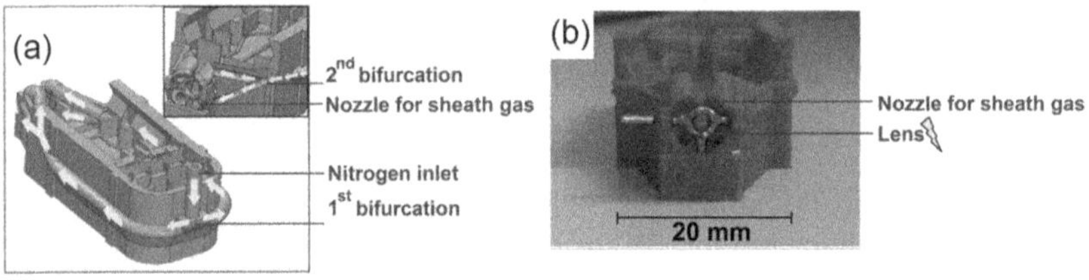


**Figure 4.19** 3D-printed LANS cartridge for paper spray ionisation: (a) render and (b) photograph. Reproduced from ref. 45 with permission from American Chemical Society, Copyright 2017.

with the CASE cartridge. The four nozzles were used around the spray tip for the continuous delivery of sheath gas, and the electrostatic lens was placed immediately in front of the paper tip as a focusing element. The electrostatic lens was neatly and firmly embedded in the printed cartridge by manipulating the FDM print cycle. The cavity for the electrostatic lens was partially printed, then the printing process was paused, and the lens was embedded into the cavity, after which the print was resumed. This process enabled the production of FDM-printed devices with a variety of embedded heterogeneous elements. This approach has also been used to develop reactionware for organic synthesis, which is discussed in Chapter 7. The current range of PolyJet, SL, and SLS/M printers currently lack this capability, because they are programmed to fill any internal cavity with a support material.

The rapid and iterative development of FWACSS, CASE, and LAN cartridges represents a great example of the typical design optimisation process that can now be quickly realised with the use of 3D printing.

In 2012, Baird *et al.* developed a 3D-printed curved plastic electrode system to separate and focus ions generated from spray ionisation for higher sensitivity mass spectrometric detection.[46] The electrode system was printed using an FDM printer and a conductive acrylonitrile butadiene styrene-based material. The conductive material was obtained from MakerGeeks, and it is a proprietary blend of acrylonitrile butadiene styrene, carbon fibre, and carbon black. The printed device consisted of a cylindrical source electrode region (20 mm I.D. and 30 mm length), followed by three curved electrodes (15 mm I.D., 45° swept angle, and 15 mm radius of curvature) (Figure 4.20). The electrodes were separated by 3-mm-thick spacers, which were 3D-printed in non-conductive acrylonitrile butadiene styrene or poly(lactic acid). The electrode system was used for the separation of tetraalkylammonium cations in the open air. The results obtained with the 3D-printed electrode system were found to agree with the ion behaviour simulated by the statistical diffusion simulation algorithm.

Similar FDM-printed conductive and non-conductive components were later used by the same group towards developing a 3D-printed ion mobility spectrometer.[47] An ion mobility spectrometer separates ions based on their interaction with a background gas in combination with electric fields. Ion mobility spectrometers can be used in conjunction with secondary mass spectrometers to provide ultra-high selectivity and sensitivity for the determination of target species.[48] On their own, ion mobility spectrometers are used for a wide variety of applications, including security screening as previously mentioned, for the detection of illicit drugs[49,50] and explosives.[51] The ion

**Figure 4.20** 3D-printed curved plastic electrode system: (a) photograph of the assembly, and (b) render of the electrode assembly and ion trajectories. Reproduced from ref. 46 with permission from The Royal Society of Chemistry.

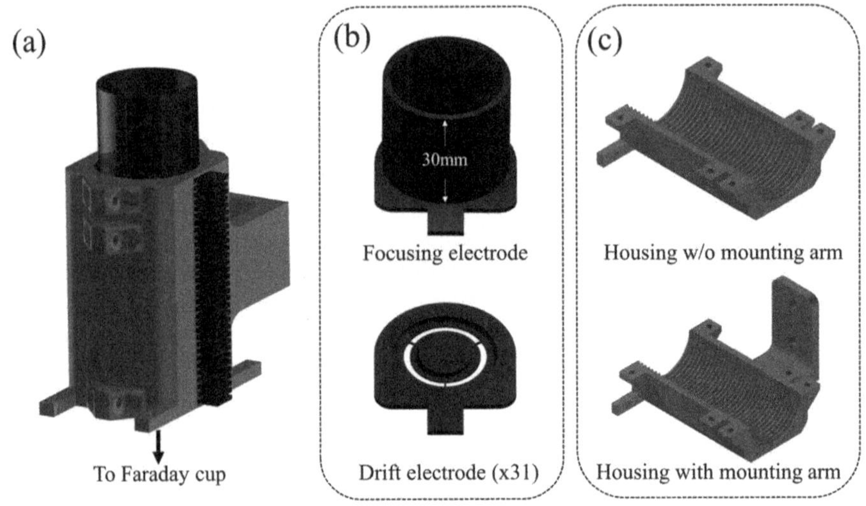


**Figure 4.21** Render of the 3D-printed ion mobility spectrometer: (a) assembly, (b) electrodes, and (c) housing components. Reproduced from ref. 47 with permission from American Chemical Society, Copyright 2017.

separation capability of an ion mobility spectrometer is quantified through its resolving power, which is defined as the drift time of the ion divided by the temporal peak width at half-height of the ion mobility peak. The 3D-printed spectrometer produced consisted of a housing, a focusing electrode, an injection electrode, thirty drift electrodes, housing for a Faraday cup, and housing for a voltage divider circuit (Figure 4.21). The housings and electrodes were 3D-printed with poly(lactic acid) and a conductive poly(ethylene terephthalate glycol)-modified polymer, containing multiwalled carbon nanotubes, respectively. Resolving powers between 24 and 50 for tetraalkylammonium bromide salts, benzylamines, and illicit drugs were obtained in the positive ion mode. Resolving powers between 29 and 42 were observed for sodium alkyl sulphates ($C_8$, $C_{12}$, $C_{16}$, and $C_{18}$) in the negative ion mode. The resolving powers for trichloroethylene, tetrachloroethylene, methyl *tert*-butyl ether, and methyl isobutyl ketone, obtained with the 3D-printed ion mobility spectrometer, were found in accordance with the resolving powers previously predicted from diffusion theory.[52]

Low-temperature plasma ionisation is also gaining significant attention within the field of ambient mass spectrometry. This technique facilitates the direct ionisation of a wide range of solutes without the use of solvents. Low-temperature plasma ionisation mass spectrometry has been used for various applications,[53] such as explosive detection,[54] food analysis,[55] real-time monitoring of chemical reactions,[56] drug abuse testing,[57] bacterial,[58] pesticide[59,60] and crude oil analysis,[61] and micronutrient analysis.[62] Moreover, low power requirements and the simple design of the ion source make it an ideal candidate for use with miniaturised mass spectrometers. In 2016, Martínez-Jarquín *et al.*[53] developed a low-temperature plasma probe using

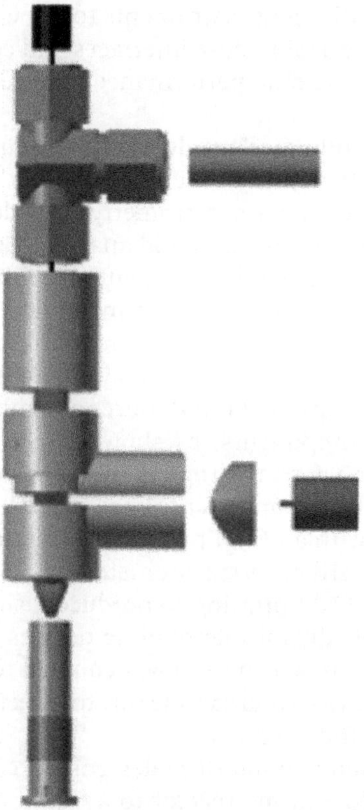

**Figure 4.22** Render of the 3D-printed components for a low-temperature plasma probe. Reproduced from ref. 53 with permission from American Chemical Society, Copyright 2016.

a combination of 3D-printed and off-the-shelf components. The T-connector, top and bottom T fittings, a security ring, and dielectric barrier were all FDM-printed (Figure 4.22) using acrylonitrile butadiene styrene, poly(lactic acid), or polycarbonate materials. Components printed with polycarbonate resulted in lower noise levels, presumably from reduced leaching, and provided a higher resistance to running temperatures than the poly(lactic acid) and acrylonitrile butadiene styrene printed components. The low-temperature plasma probe was coupled with an ion trap analyser, and it was successfully used to detect organic compounds from pure reference standards, biological samples, and pharmaceutical samples.

## 4.4.2 Interfaces

Mass spectrometers are often coupled with different analytical instruments such as gas and liquid chromatographs, and other flow-based methods and sample handling arrangements. This requires a specific interface that can effectively couple these different instruments to the ionisation source of the mass spectrometer. Once again, these interfaces and coupling arrangements are complex, have high material performance requirements, and are generally very expensive.

So-called 'digital microfluidics' involves the manipulation of discrete sample and reagent droplets within a microfluidic chip.[63] On-line coupling of digital microfluidics with mass spectrometry provides an opportunity for automation of sample handling and rapid analysis. Coupling these technologies does however require the development and fabrication of a complex interface, and for this, researchers have again applied 3D printing.

Hu *et al.*[64] have developed a 3D-printed interface for coupling digital microfluidic chips with mass spectrometry. It consisted of a Venturi easy ambient sonic-spray ionisation (V-EASI) source assembled with other (mostly 3D-printed) auxiliary components, as shown in Figure 4.23. The V-EASI source was used for transferring droplets from the microchip to the mass spectrometer. The auxiliary components were made from acrylonitrile butadiene styrene, printed with an FDM printer. The interface was easily clipped onto the inlet of a standard mass spectrometer.

Liu *et al.*[65] also used FDM printing to produce a simple interface, in this instance to couple their digital microfluidic devices with the autosampler of a liquid chromatograph, which itself was coupled to a mass spectrometer (Figure 4.24). It served as a virtual sample tray to directly inject droplets from microfluidic chips into the HPLC.

Jönsson *et al.*[66] studied two interface designs for coupling a microfluidic chip-based immobilised enzyme reactor to a liquid chromatography–mass spectrometry instrument, referred to as brick (two-point bolt clamping) and jewelbox (evenly distributed clamping) interfaces (Figure 4.25). The jewelbox interface provided a more even pressure over the surface of the mounted

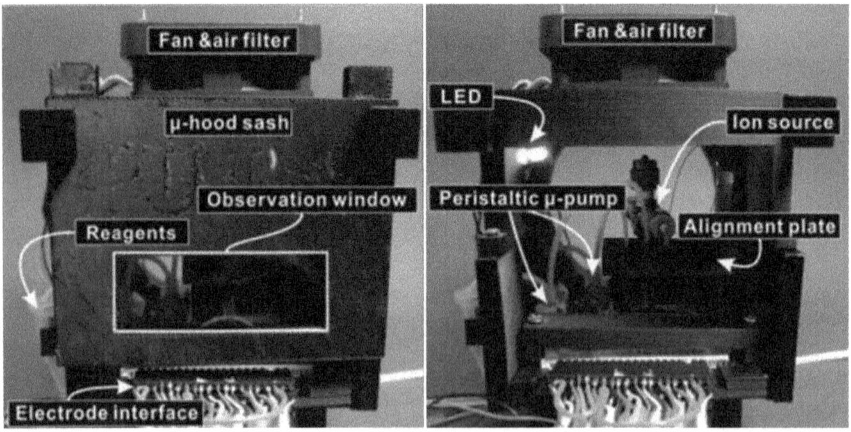


**Figure 4.23** 3D-printed interface to integrate digital microfluidics with mass spectrometers. Reproduced from ref. 64 with permission from The Royal Society of Chemistry.

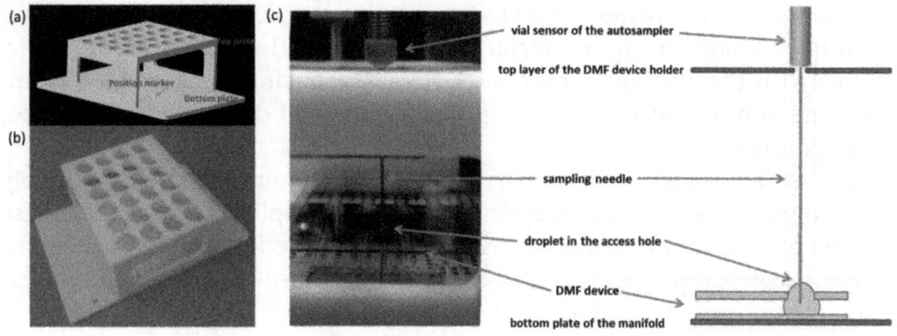


**Figure 4.24** 3D-printed digital microfluidic–autosampler interface: (a) render of the interface, (b) photograph of the interface, and (c) photograph (left) and schematic (right) of the set-up. Reproduced from ref. 65 with permission from American Chemical Society, Copyright 2015.

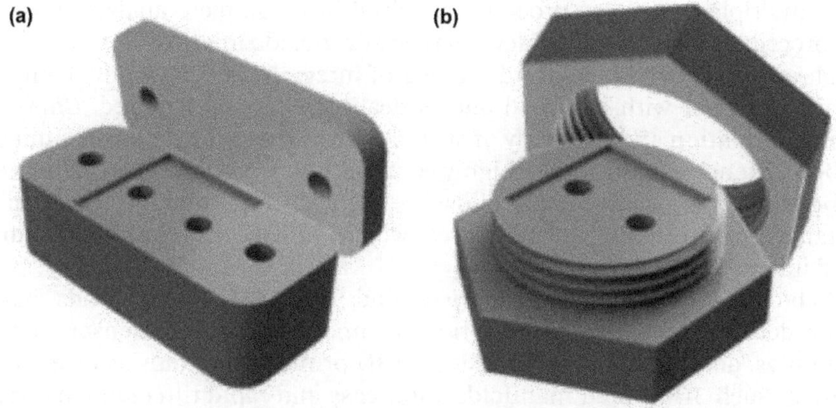


**Figure 4.25** Render of the 3D-printed (a) brick interface and (b) jewelbox interface. Reproduced from ref. 66 with permission from American Chemical Society, Copyright 2017.

chip and more convenient handling than the brick interface. The use of coned fittings with the FDM-printed poly(lactic acid) jewelbox interface provided a pressure tolerance of up to 4.7 Mpa, which is suitably impressive for an FDM-printed device.

The important structures of these brick and jewelbox interfaces, such as the O-ring and chip alignment pockets, were oriented facing up during FDM printing to obtain a smooth surface finish for the pockets.[66] FDM printing lacks the use of dense support structures, hence hanging features, such as downward facing pockets, result in higher surface roughness. The effect of print orientation on the surface roughness of the downward facing and embedded features is more evident during FDM printing as compared to the other 3D printing techniques. However, the same effect can also be observed with SL, PolyJet, and SLS/M printing techniques. These printing techniques result in higher surface roughness of the features that are in contact with the support material. Hence, efforts should be made to minimise the contact between the support material and the printed surface, where possible.

Customised holders have also been 3D-printed to easily and reproducibly couple paper spray ionisation cartridges[45] and multiple reactors[67] with mass spectrometers. FDM printers are usually the preferred method of choice for printing such simple holders and interfaces because they incur the lowest cost and offer easy operation.

## 4.5 Integrated Flow-analysis Manifolds and Lab-on-a-chip Devices

Integrated flow-analysis manifolds and so-called 'lab-on-a-chip' devices are typically macro- and micro-fluidic based assemblies, designed to incorporate multiple laboratory processes involved in a complete analytical assay or procedure. Lab-on-a-chip devices typically include microfluidic channels, high aspect ratio features, and a range of integrated micro-fluidic components, together with electrical and optical elements as required. Conventional formation and assembly of such devices is typically time-consuming, labour intensive, costly and a highly complex process. Moreover, traditional fabrication approaches do not allow the fabrication of three-dimensionally arranged channels and architectures, which limits device miniaturisation and integration.

Lab-on-a-chip microfluidic devices and flow-analysis manifolds have been developed by integrating various components, such as sensors, optical fibres, detectors, and electrodes into 3D-printed manifolds as discussed herein. Such 3D-printed manifolds allow easy and rapid three-dimensional arrangement of different modules.

Mattio *et al.*[68] developed an elaborate 3D-printed multi-pumping flow system for the selective spectrophotometric determination of lead in natural water samples. The flow system consisted of three 3D-printed modules, specifically a resin column, a mixing coil, and a UV-spectroscopy flow-cell (Figure 4.26). All the modules were SL-printed in a methacrylate-based material. The resin column contained TrisKem Pb resin (TrisKem), formed between two layers of glass wool and three inlets for sample and reagent introduction. The mixing coil was a serpentine design with a total length of 50 cm and an I.D. of 1.5 mm. The flow-cell had a 5 cm optical path length, integrated with two optical fibres. Lead was quantified through selective retention onto the crown-ether resin, followed by elution with ammonium oxalate, and spectrophotometric detection following reaction with 4-(2-pyridylazo)-resorcinol (PAR). An LOD of 2.7 μg $L^{-1}$ and a linear range from 3 μg $L^{-1}$ to 120 μg $L^{-1}$ was obtained, consistent with the results obtained using a standard inductively-coupled plasma atomic emission spectroscopy method.

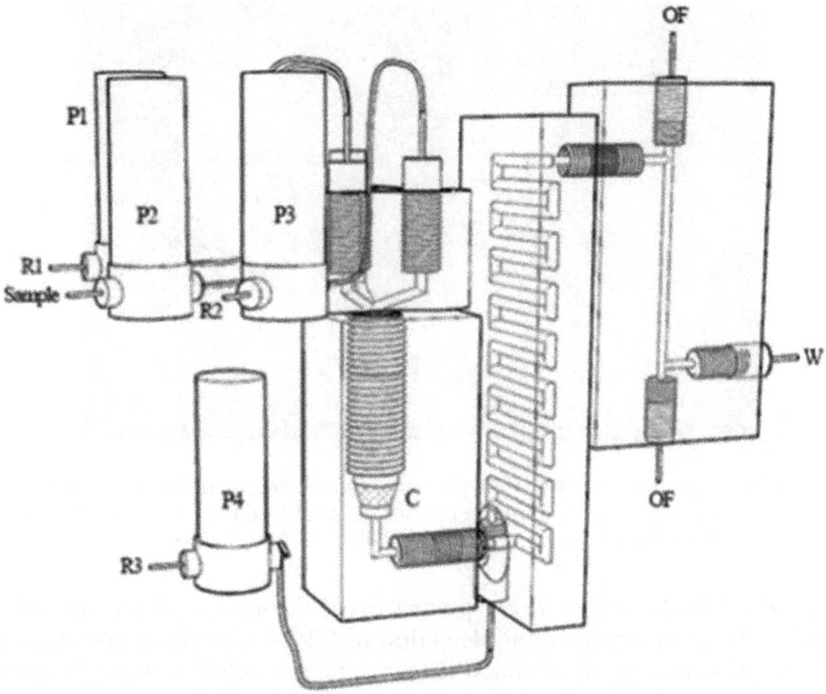


**Figure 4.26** Render of the 3D-printed multi-pumping flow system for the determination of lead in natural water samples. P1–P4: pumps C: resin column, R1: nitric acid, R2: ammonium oxalate, R3: PAR, OF: optical fibre, W: waste. The screws and corresponding screw threads to assemble the units are also shown. Reproduced from ref. 68 with permission from Elsevier, Copyright 2017.

Similarly, Wang *et al.*[69] developed a miniature 3D-printed manifold to integrate various components of a typical flow injection analysis system for the determination of the acid number of oils. The acid number was determined by measuring the change in concentration of the standard potassium hydroxide solution before and after interaction with an oil sample. The system was developed by integrating an oil/base segmentor, a mixing tube, a phase separator, and a pH detector. The manifold was SL-printed with a polyacrylate composite resin. The oil/base segmentor and the mixing tube were printed within the manifold, and dedicated chambers were designed for the phase separator and the pH detector (Figure 4.27). The 3D-printed manifold enabled the accurate determination of acid numbers in the range of 0.20 to 0.64 mg of KOH $g^{-1}$ oil and a sampling frequency of 10 samples per hour. It resulted in an LOD of 0.177 mg of KOH $g^{-1}$ oil, and an RSD of less than 1%. The results obtained with the 3D-printed manifold were in agreement with the results obtained from a conventional flow injection analysis system.

**Figure 4.27** A photograph of the 3D-printed miniaturised manifold for the determination of the acid number of oils. Reproduced from ref. 69 with permission from American Chemical Society, Copyright 2015.

In 2014, Chudobova *et al.*[70] reported the development of an integrated microfluidic platform for rapid detection of MRSA bacteria. It consisted of an FDM-printed acrylonitrile butadiene styrene microfluidic chip, which was used to perform bacterial cultivation, DNA isolation, PCR, and detection of amplified genes. The complex final assembly included a spectrophotometric detector, pumps, a thermoregulatory system, and electromagnets, as shown in Figure 4.28.

In the following year, Gowers *et al.*[71] developed a 3D-printed microfluidic chip to integrate clinical micro-dialysis probes, such as glucose and lactate biosensors, to perform continuous monitoring of human tissue metabolite levels. The microfluidic chip was designed in a unique L-shape, with the horizontal arm used for integrating the biosensors, and the vertical arm used for connections (Figure 4.29). The components were 3D-printed with a PolyJet printer using a polyacrylate-based material. The sensors and electrodes were integrated using 3D-printed modular holders, which were printed with a multi-material PolyJet printer using a rigid and an elastic polyacrylate composite material. A thin layer of elastic material was used to ensure a liquid-tight seal between different components. The holders were precisely aligned with the microfluidic channel that was embedded in the chip using the printed slots. The integrated system was successfully used for monitoring subcutaneous metabolite levels during cycling training.

Similar to the above arrangement of optical fibres and flow-cell for the 3D-printed multi-pumping flow system, Monaghan *et al.*[72] integrated optical fibres with a 3D-printed flow-cell to perform in-line UV-Vis spectroscopy. They used a 50 μm fibre for illumination and 105 μm fibre for detection to minimise the percentage loss in collimation. The set-up was used for in-line monitoring of the carvone semicarbazone synthesis, and it resulted in a wide linear range of detection: 0.5 mM to 25 mM.

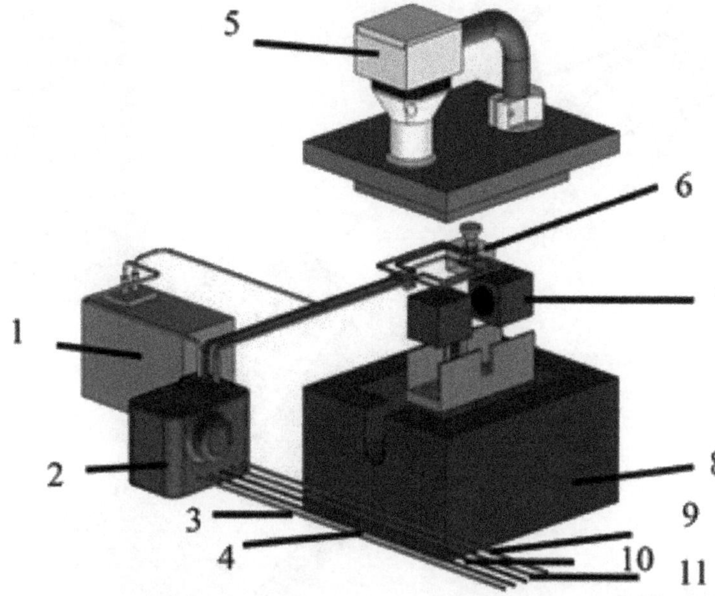


**Figure 4.28** Render of the 3D-printed device for rapid detection of MRSA bacteria. 1: Spectrophotometric detector, 2: pump with the valves, 3: outlet, 4: the first inlet hose, 5: thermoregulatory system, 6: cultivation chip, 7: electromagnet, 8: thermoisolating box, 9: the second inlet hose, 10: the third inlet hose, and 11: the fourth inlet hose. Reproduced from ref. 70 with permission from John Wiley and Sons, Copyright © 2014 WILEY-VCH Verlag GmbH & Co. KGaA, Weinheim.

Low-cost 3D-printed platforms have also been developed to perform electrophoretic separations by integrating electrodes with 3D-printed microfluidic channels. Anciaux *et al.*[73] developed a micro free-flow electrophoresis device for the separation of proteins, such as myoglobin and cytochrome c. This device was fabricated with an FDM printer using acrylonitrile butadiene styrene polymer. The group used acetone vapours to easily bond the different parts of the microfluidic platform, smooth the printed surfaces, and to increase their optical clarity. The smoothed FDM-printed materials resulted in only 14% loss in the fluorescence signal intensity as compared to their glass counterparts. The printed acrylonitrile butadiene styrene surfaces were also found to be resistant to protein adsorption, and the device gave a stable stream profile for micro free-flow electrophoretic separations.

In addition to FDM printing, high-resolution SL printing has also been used to develop similar electrophoretic chips.[74] The use of SL printing allowed the fabrication of closed microfluidic channels, as opposed to the open channels printed with the FDM technique in the above-mentioned case.[73] Shallan *et al.* used an SL-printed polyacrylate composite serpentine channel (10 cm long, 1 mm high, and 0.5 mm wide) for the isotachophoretic separation of three

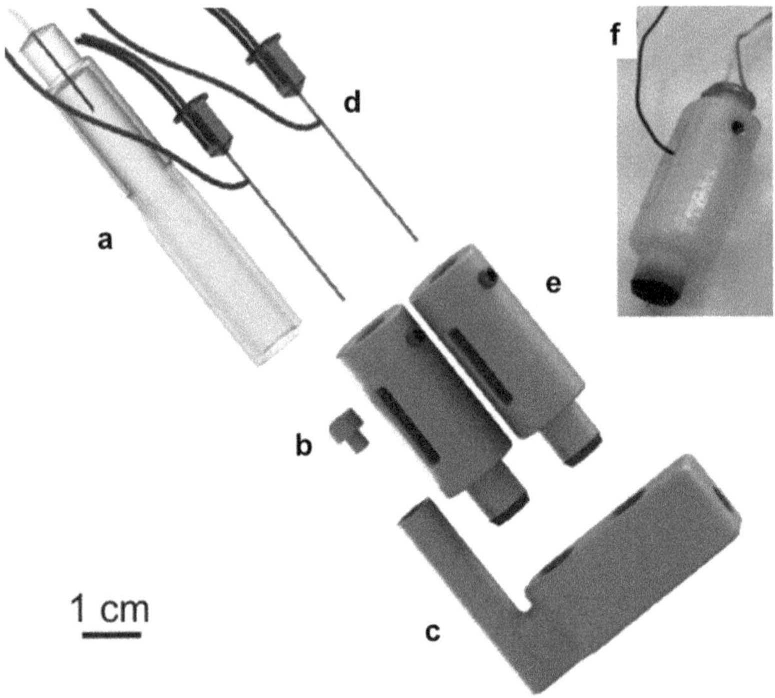


**Figure 4.29** An exploded view of the microfluidic device for continuous monitoring of dialysate: (a) probe outlet holder, (b) microvial rubber insert, (c) microfluidic chip, (d) glucose and lactate needle biosensors, (e) electrode holders, and (f) photograph of an electrode holder containing a needle biosensor. Reproduced from ref. 71 with permission from American Chemical Society, Copyright 2015.

anionic dyes: xylenol orange, carboxynaphthofluorescein, and fluorescein, by applying a voltage of 100 V $cm^{-1}$ (Figure 4.30).[74]

To the best of the authors' knowledge, the first 3D-printed completely integrated 'lab-on-a-chip' device was developed by Comina *et al.*[75] in 2014. The device was designed to perform sample delivery, transport, functionalisation, and readout within a small footprint. It consisted of printed connectors, micromixers, and complex three-dimensional features, facilitating functionalisation, lateral flow, and the introduction of glass microbeads (Figure 4.31). The printed platform could be arranged in several configurations, with both active and passive transport to develop different lab-on-a-chip assays. The device was used for the fluorescence detection of hydrogen peroxide and the colourimetric detection of glucose.

Tang *et al.*[76] incorporated long three-dimensional serpentine mixers in an equally complex 3D-printed lab-on-a-chip device, here to promote on-chip passive turbulent mixing. The three-dimensional serpentine mixer (Figure 4.32) consisted of three times more turns than its two-dimensional

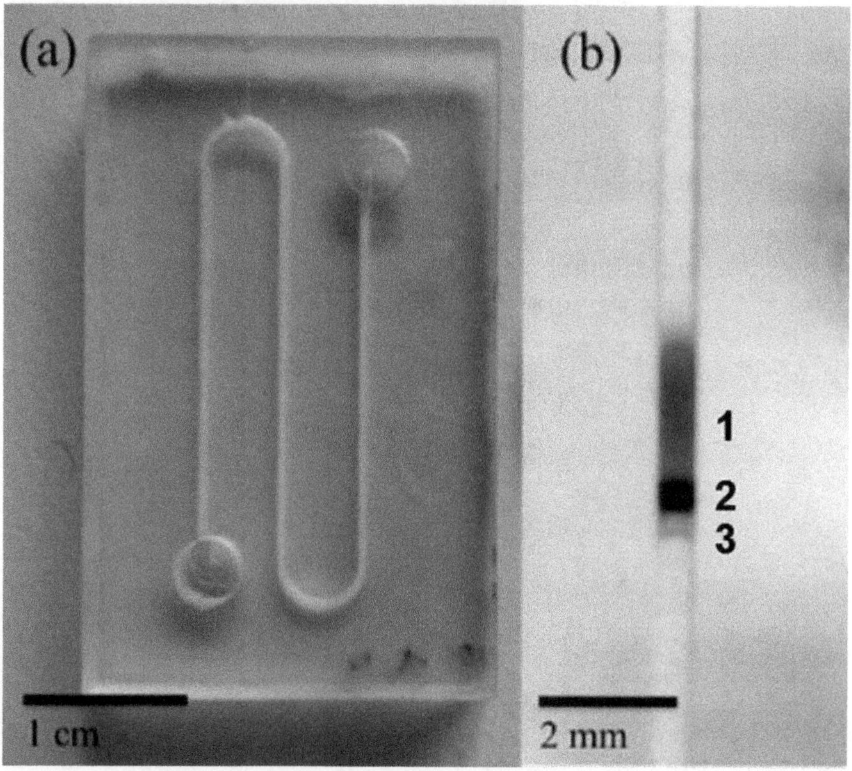


**Figure 4.30** 3D-printed isotachophoretic separation chip: (a) photograph and (b) isotachophoresis of three coloured dyes: xylenol orange (1), carboxynaphthofluorescein (2), and fluorescein (3). Reproduced from ref. 74 with permission from American Chemical Society, Copyright 2014.

counterpart. The cost of 3D printing the device with an SL printer was reported to be approximately \$5. The chip was printed with a methacrylate-based resin, and it was found to be stable up to flow rates of 1 mL $min^{-1}$. The 3D-printed chip was used for the chemiluminescence detection of two proteins: prostate-specific antigen and platelet factor 4. It resulted in an LOD of 0.5 pg $mL^{-1}$ and a linear range from 0.5 pg $mL^{-1}$ to 5 ng $mL^{-1}$ for both the proteins. An RSD of less than 8% was observed for different protein arrays, and the results were found to be in correspondence with the standard ELISA results.

Li *et al.* have developed a fully 3D-printed simple diagnostic microfluidic chip with an integrated porous membrane and liquid reagents to measure nitrate levels in soil.[17] The chip was 3D-printed with a multi-material FDM printer (Figure 4.33). Acrylonitrile butadiene styrene was used to print the chip body, and the porous Lay-Felt material was used to print the membrane. The membrane allowed selective transport of the target anionic solute ($NO_2^-$) from the sample compartment to the reagent compartment. The

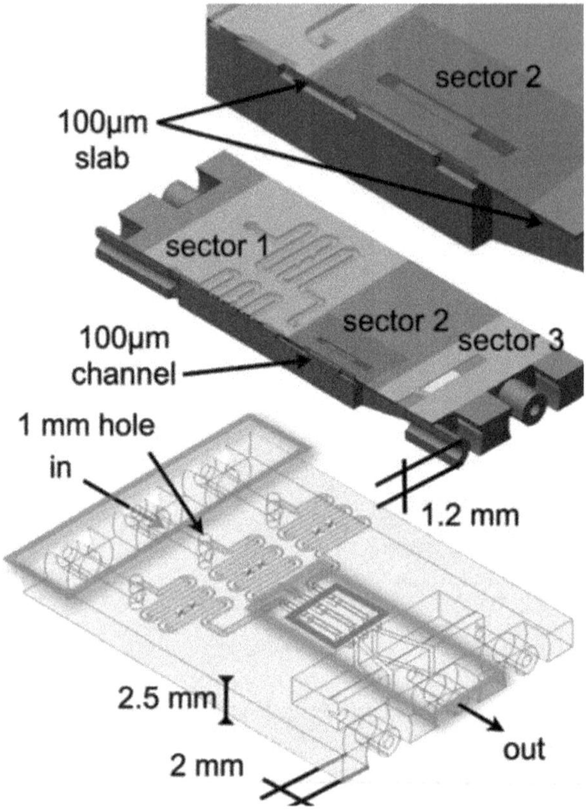


**Figure 4.31** Render of the 3D-printed lab-on-a-chip device for fluorescence detection of hydrogen peroxide and the colourimetric detection of glucose. Reproduced from ref. 75 with permission from The Royal Society of Chemistry.

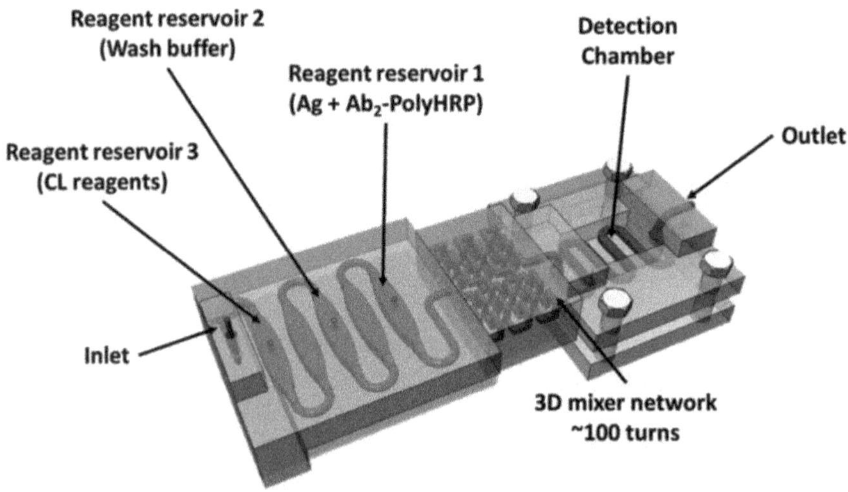


**Figure 4.32** Render of a 3D-printed immunoarray for the detection of cancer biomarker proteins. Reproduced from ref. 76 with permission from The Royal Society of Chemistry.

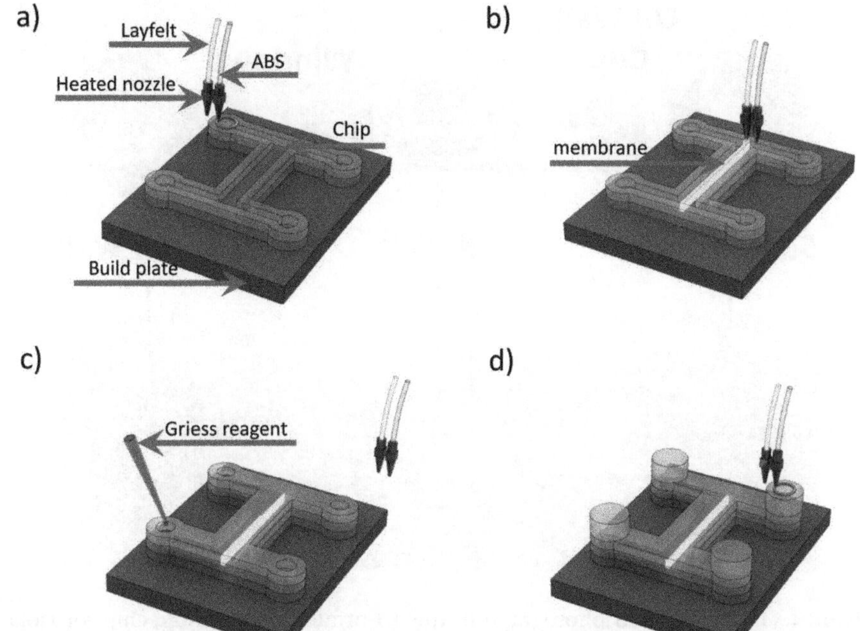


**Figure 4.33** Schematic for the multi-material FDM printing of a microfluidic chip with embedded Griess reagent: (a) ABS printing of the chip, (b) Lay-felt printing of the membrane, (c) introduction of Griess reagent, and (d) sealing the reservoir. Reproduced from ref. 17 with permission from American Chemical Society, Copyright 2017.

reagent compartment was filled with Griess reagent. Liquid Griess reagent was embedded in the chip during FDM printing by pausing and resuming the print cycle (as discussed in Section 4.4). The authors were able to store the Griess reagent in the chip for more than four days, however, prolonged storage decreased the sensitivity of the assay.

Zhang *et al.* have developed 3D-printed microfluidic chips for the adsorption and isolation of nucleic acids without the need for vortex and centrifugation (Figure 4.34).[77] These microfluidic chips utilised a water and oil interface at the junction of a microchannel and a well, to filter lysate contaminants. Cellulose functionalised super magnetic beads were used to transport the nucleic acids across the oil layer. This set-up was used for the extraction of Human Papillomavirus plasmids. A recovery of 61% was observed across the concentration range of $5 \times 10^1$ to $5 \times 10^6$ copies per 100 μL.

3D-printed chips have also been developed to perform controlled actuations of droplets without the need for any additional accessories. The actuations have been performed on superhydrophobic surfaces using gravitational[78] and magnetic forces.[79] Oomen *et al.*[78] coated FDM-printed poly(lactic acid) chips with the commercial, low-friction, and superhydrophobic coating Ultra-Ever Dry to perform gravity actuation of droplets. Gravity actuation was used to perform the discrete transport of a sequence of droplets

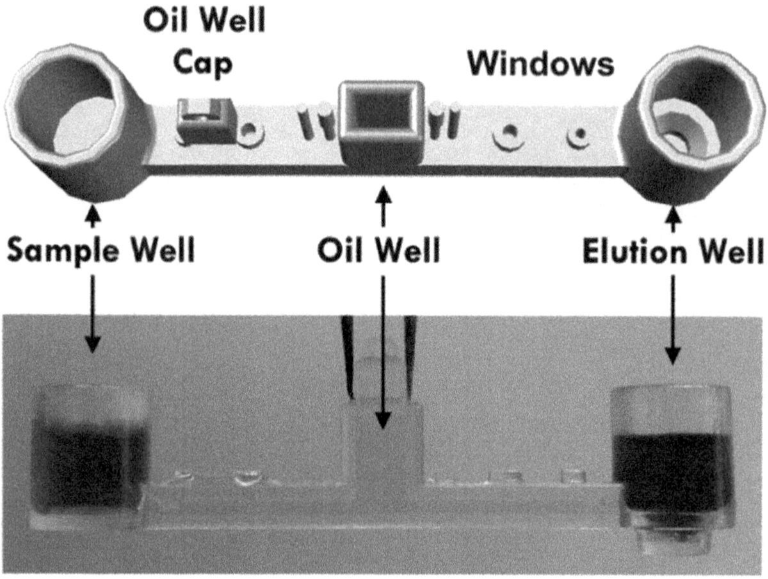


**Figure 4.34** Render and photograph of the 3D-printed microfluidic chip for isolation of nucleic acids. Reproduced from ref. 77 with permission from AIP Publishing, Copyright 2015.

while maintaining their chronological order. Similarly, Hutama *et al.* performed magnetic actuation of droplets for splitting and dispensing them in different ratios.[79] The magnetic actuation chip was composed of individual wells, a weir structure, and differential channel depths to operate different droplet sizes, and five different well sets were arranged in a complementary fashion (Figure 4.35). A chamber was incorporated between different wells to allow the merging and mixing of droplets. The 3D-printed gravity actuation chip was used for the enzyme-coupled amperometric detection of glucose in rat serum, and the magnetic actuation chip for a colourimetric titration determination of water hardness.

## 4.6 Outlook

The reports highlighted throughout this chapter showcase how polymer and metal 3D printing techniques could be used to develop miniaturised, integrated, and geometrically complex analytical devices. These developments have been further summarised in Table 4.1, which lists the different techniques, printers, and print materials that have been used for printing the various analytical devices and technologies that are discussed herein.

It is probably safe to say that each step of standard analytical workflows going forward, ranging from sample collection, identification, and quantification, will be impacted by the continued development of 3D printing.

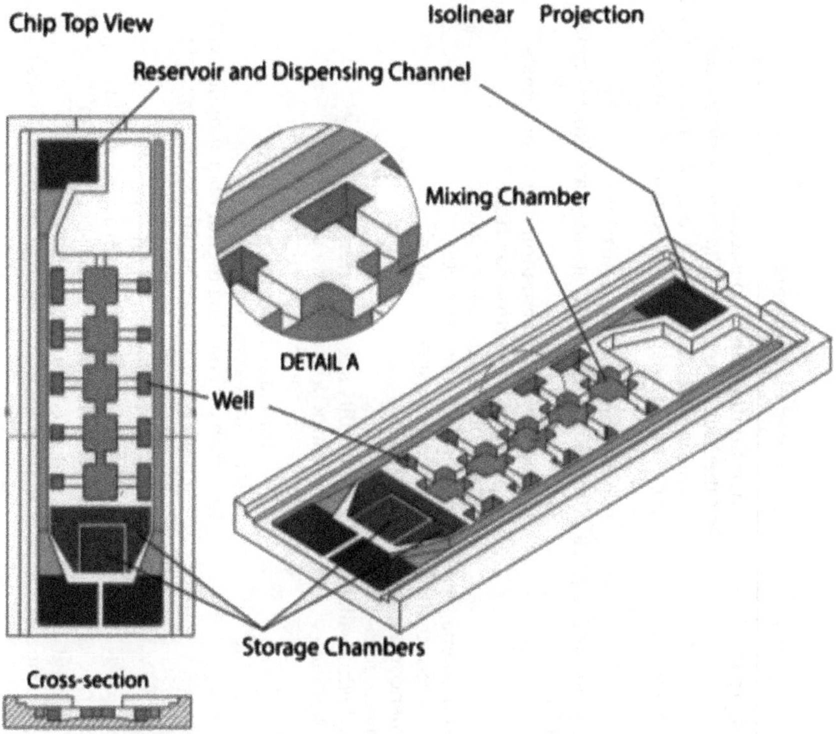


**Figure 4.35** Schematic of the 3D-printed chip for the magnetic actuation of droplets. Reproduced from ref. 79 with permission from The Royal Society of Chemistry.

3D-printed devices have been used, with and without surface functionalisation, for the extraction and separation of target solutes/species from complex matrices. 3D-printed three-dimensionally coiled channel geometries have been used to improve the performance of chromatographic columns and microfluidic mixers, and have been applied to develop bespoke porous beds with a defined shape, size, and orientation of different porous bed elements. Established flow injection analysis manifolds can now be readily 3D-printed into miniaturised integrated devices,[60] and expensive consumables, such as centrifugation vials, flow-cells, liquid–liquid extraction vessels, and micro-reactors can also be 3D-printed to reduce their cost and allow their customisation. In the area of miniaturisation and portable instruments, it is clear how this new production technology is continuing to advance the development of lab-on-a-chip based methods and applications, by minimising the required time, cost, and complexity of production.

The previous chapters have shown how multi-material 3D printing is delivering integration of multiple functionalities within a single printed device, driving important applications and functions such as pumping, extraction,

**Table 4.1** The 3D printing techniques, printers, and print materials that are used for fabricating analytical devices and components discussed within this chapter.

| 3DP technique | Printer: Manufacturer | Printer: Model | Print material used | 3D-printed components[a] |
|---|---|---|---|---|
| FDM | 3Dfactories | Profi3Dmaker | Acrylonitrile butadiene styrene | Microfluidic chip for detection of methicillin-resistant *Staphylococcus aureus*[70] |
| | Aurora | Z-603S | Poly(lactic acid) | Skeleton for coating metal–organic frameworks[7] |
| | Baoyan Automation | HOFI-X1 | Acrylonitrile butadiene styrene | Skeleton for coating metal–organic frameworks[6] |
| | Felix | 1.5 | Poly(lactic acid) | Paper spray ionisation cartridge[44] |
| | | 3 | Acrylonitrile butadiene styrene and Lay-Felt | Microfluidic chip with an integrated membrane[17] |
| | | | Poly(lactic acid) and Lay-Felt | Passive sampler[18] |
| | | | Poly(lactic acid) | Device for gravity actuation of microdroplets[78] |
| | | | | Paper spray ionisation cartridge[45] |
| | GTMax | Graber i3 | Poly(lactic acid) | μ-solid-phase extraction system[3] |
| | Makerfarm | Prusa i3 | Conductive acrylonitrile butadiene styrene | Ion mobility spectrometer[46] |
| | | | Conductive polyethylene terephthalate glycol-modified polymer containing multiwalled carbon nanotubes | Ion mobility spectrometer[47] |
| | | | Poly(lactic acid) with polyhydroxyalkanoate | Ion mobility spectrometer[47] |
| | MakerMex | MM1 Beta | Acrylonitrile butadiene styrene | Low-temperature plasma probe[53] |
| | | | Poly(lactic acid) | Low-temperature plasma probe[53] |
| | | | Polycarbonate | Low-temperature plasma probe[53] |
| | Makers Tool | MendelMax 2 | Poly(lactic acid) | Rotary multi-spray holder[67] |
| | Stratasys | Fortus 400mc | — | Interface for coupling digital microfluidics with a high performance liquid chromatography–mass spectrometer[65] |
| | Ultimaker | 2 | Poly(lactic acid) | Interface for coupling a microreactor with an ultra-high performance liquid chromatography–mass spectrometer[66] |
| | | — | Acrylonitrile butadiene styrene | Micro-free-flow electrophoresis device[73] |
| | UP | Plus | Acrylonitrile butadiene styrene | Interface for coupling digital microfluidics with a mass spectrometer[64] |
| | | Plus 2 | | Enzyme immobilised reactor[5] |
| | Zmorph | Zmorph2.0SX | Lay-Fomm | Porous sorbent[16] |

| | | | | |
|---|---|---|---|---|
| PolyJet printing | 3D Systems | ProJet 3500 HD | Acrylonitrile butadiene styrene | Liquid chromatography stationary phase[27] |
| | | ProJet 3510 HD | Visijet® X | Liquid chromatography stationary phase[29] |
| | Stratasys | Objet Eden 260VS | VeroClear™ | Thin layer chromatography plates[33] |
| | | | | Centerpiece for analytical ultracentrifuge[19] |
| | | Objet260 Connex | VeroWhitePlus™ and TangoBlack™ | Microfluidic chip for continuous monitoring of dialysate[71] |
| SL | 3D Systems | Viper | Accura® 60 | Microfluidic chip for optical fibre based detection[72] |
| | envisionTEC | Perfactory III Mini SXGA+ | HTM140 IV | Distillation column[20] |
| | | Perfactory Micro Hi Res | HTM140 M | Gas chromatography column[30,31] |
| | Formlabs | Form 1+ | Methacrylate-based resin | Microfluidic immunoarray manifold[76] |
| | | | | Multi-pumping flow system[68] |
| | | Form 2 | FLGPCL02 | Microflow injection analysis manifold[2] |
| | Miicraft | — | BV-001 | Preconcentrator[8] |
| | | | | Knotted reactor[12] |
| | Shanghai Aowei Digital Technology Limited | SPS 250E | Somos® 14120 | Flow injection analysis manifold[69] |
| SLM | Realizer | SLM 50 | Stainless Steel (316L) | Liquid chromatography column body[22] |
| | | | Titanium alloy (Ti-6Al-4V) | Liquid chromatography column body[22,24,26] |

[a]The devices have been named in this table as per the names proposed in their respective reports. In the case of the absence of any clearly proposed name, generic names based on their applications have been used.

and optical and electrochemical detection. Novel 3D-printable materials, such as porous[17] or conductive[33,40] thermoplastic filaments, and conductive inks[72,73] have been developed to advance this cause, with new functional print materials emerging constantly.

## References


1. C. Basheer, A. A. Alnedhary, B. M. Rao, S. Valliyaveettil and H. K. Lee, *Anal. Chem.*, 2006, **78**, 2853–2858.
2. H. Wang, D. J. Cocovi-Solberg, B. Hu and M. Miró, *Anal. Chem.*, 2017, **89**, 12541–12549.
3. E. M. Kataoka, R. C. Murer, J. M. Santos, R. M. Carvalho, M. N. Eberlin, F. Augusto, R. J. Poppi, A. L. Gobbi and L. W. Hantao, *Anal. Chem.*, 2017, **89**, 3460–3467.
4. C. Calderilla, F. Maya, V. Cerdà and L. O. Leal, *Talanta*, 2017, **175**, 463–469.
5. C.-K. Su, S.-C. Yen, T.-W. Li and Y.-C. Sun, *Anal. Chem.*, 2016, **88**, 6265–6273.
6. Z. Wang, J. Wang, M. Li, K. Sun and C.-j. Liu, *Sci. Rep.*, 2014, **4**, 5939.
7. Z. Shi, C. Xu, F. Chen, Y. Wang, L. Li, Q. Meng and R. Zhang, *RSC Adv.*, 2017, **7**, 49947–49952.
8. C.-K. Su, P.-J. Peng and Y.-C. Sun, *Anal. Chem.*, 2015, **87**, 6945–6950.
9. V. Camel, *Spectrochim. Acta, Part B*, 2003, **58**, 1177–1233.
10. P. H. McCluskey, R. L. Snyder and R. A. Condrate, *J. Solid State Chem.*, 1989, **83**, 332–339.
11. S. Blain, P. Appriou and H. Handel, *Analyst*, 1991, **116**, 815–820.
12. C.-K. Su, M.-H. Hsieh and Y.-C. Sun, *Anal. Chim. Acta*, 2016, **914**, 110–116.
13. Z. Fang, S. Xu, L. Dong and W. Li, *Talanta*, 1994, **41**, 2165–2172.
14. X.-P. Yan and Y. Jiang, *TrAC, Trends Anal. Chem.*, 2001, **20**, 552–562.
15. S. Cerutti, R. G. Wuilloud and L. D. Martinez, *Appl. Spectrosc. Rev.*, 2005, **40**, 71–101.
16. M. Belka, S. Ulenberg and T. Bączek, *Anal. Chem.*, 2017, **89**, 4373–4376.
17. F. Li, P. Smejkal, N. P. Macdonald, R. M. Guijt and M. C. Breadmore, *Anal. Chem.*, 2017, **89**, 4701–4707.
18. U. Kalsoom, C. K. Hasan, L. Tedone, C. Desire, F. Li, M. C. Breadmore, P. N. Nesterenko and B. Paull, *Anal. Chem.*, 2018, **90**, 12081–12089.
19. A. Desai, J. Krynitsky, T. J. Pohida, H. Zhao and P. Schuck, *PLoS One*, 2016, **11**, e0155201.
20. S. Mardani, L. S. Ojala, P. Uusi-Kyyny and V. Alopaeus, *Chem. Eng. Process.*, 2016, **109**, 136–148.
21. B. Paull, *LCGC*, 2017, **30**, 611–612.
22. S. Sandron, B. Heery, V. Gupta, D. Collins, E. Nesterenko, P. N. Nesterenko, M. Talebi, S. Beirne, F. Thompson and G. G. Wallace, *Analyst*, 2014, **139**, 6343–6347.
23. E. P. Nesterenko, P. N. Nesterenko, D. Connolly, F. Lacroix and B. Paull, *J. Chromatogr. A*, 2010, **1217**, 2138–2146.

24. V. Gupta, M. Talebi, J. Deverell, S. Sandron, P. N. Nesterenko, B. Heery, F. Thompson, S. Beirne, G. G. Wallace and B. Paull, *Anal. Chim. Acta*, 2016, **910**, 84–94.
25. R. J. Vonk, A. Vaast, S. Eeltink and P. J. Schoenmakers, *J. Chromatogr. A*, 2014, **1359**, 162–169.
26. V. Gupta, S. Beirne, P. N. Nesterenko and B. Paull, *Anal. Chem.*, 2018, **90**, 1186–1194.
27. C. Fee, S. Nawada and S. Dimartino, *J. Chromatogr. A*, 2014, **1333**, 18–24.
28. S. Nawada, C. Fee and S. Dimartino, *Chromatogr. Today*, 2014, **362**, 18–20.
29. S. Nawada, S. Dimartino and C. Fee, *Chem. Eng. Sci.*, 2017, **164**, 90–98.
30. F. Lucklum, S. Janssen, W. Lang and M. Vellekoop, *Procedia Eng.*, 2015, **120**, 703–706.
31. N. A. Zaidi, M. W. Tahir, M. J. Vellekoop and W. Lang, *Sensors*, 2017, **17**, 2283.
32. D. Fichou and G. E. Morlock, *Anal. Chem.*, 2017, **89**, 2116–2122.
33. N. P. Macdonald, S. A. Currivan, L. Tedone and B. Paull, *Anal. Chem.*, 2017, **89**, 2457–2463.
34. F. W. McLafferty, *Mass Spectrometry of Organic Ions*, Elsevier, 2012.
35. R. Aebersold and M. Mann, *Nature*, 2003, **422**, 198.
36. R. N. Rao and M. K. Talluri, *J. Pharm. Biomed. Anal.*, 2007, **43**, 1–13.
37. S. D. Richardson, *Anal. Chem.*, 2008, **80**, 4373–4402.
38. Z. Ouyang and R. G. Cooks, *Annu. Rev. Anal. Chem.*, 2009, **2**, 187–214.
39. Z. Takats, J. M. Wiseman and R. G. Cooks, *J. Mass Spectrom.*, 2005, **40**, 1261–1275.
40. R. D. Espy, S. F. Teunissen, N. E. Manicke, Y. Ren, Z. Ouyang, A. van Asten and R. G. Cooks, *Anal. Chem.*, 2014, **86**, 7712–7718.
41. N. E. Manicke, P. Abu-Rabie, N. Spooner, Z. Ouyang and R. G. Cooks, *J. Am. Soc. Mass Spectrom.*, 2011, **22**, 1501–1507.
42. N. E. Manicke, Q. Yang, H. Wang, S. Oradu, Z. Ouyang and R. G. Cooks, *Int. J. Mass Spectrom.*, 2011, **300**, 123–129.
43. Z. Zhang, R. G. Cooks and Z. Ouyang, *Analyst*, 2012, **137**, 2556–2558.
44. G. I. Salentijn, H. P. Permentier and E. Verpoorte, *Anal. Chem.*, 2014, **86**, 11657–11665.
45. G. I. Salentijn, R. Oleschuk and E. Verpoorte, *Anal. Chem.*, 2017, **89**, 11419–11426.
46. Z. Baird, P. Wei and R. G. Cooks, *Analyst*, 2015, **140**, 696–700.
47. A. Hollerbach, Z. Baird and R. G. Cooks, *Anal. Chem.*, 2017, **89**, 5058–5065.
48. A. B. Kanu, P. Dwivedi, M. Tam, L. Matz and H. H. Hill, *J. Mass Spectrom.*, 2008, **43**, 1–22.
49. T. P. Forbes and M. Najarro, *Analyst*, 2016, **141**, 4438–4446.
50. A. Sergio, G. Salvador, G. Miguel, B. Judit, A. Manel, B. Marcelo, P.-A. Clara and G. Nieves, *Drug Test. Anal.*, 2015, **7**, 280–289.
51. G. A. Harris, M. Kwasnik and F. M. Fernández, *Anal. Chem.*, 2011, **83**, 1908–1915.
52. A. B. Kanu, M. M. Gribb and H. H. Hill, *Anal. Chem.*, 2008, **80**, 6610–6619.

53. S. Martínez-Jarquín, A. Moreno-Pedraza, H. Guillén-Alonso and R. Winkler, *Anal. Chem.*, 2016, **88**, 6976–6980.
54. J. F. Garcia-Reyes, J. D. Harper, G. A. Salazar, N. A. Charipar, Z. Ouyang and R. G. Cooks, *Anal. Chem.*, 2010, **83**, 1084–1092.
55. J. F. García-Reyes, F. Mazzoti, J. D. Harper, N. A. Charipar, S. Oradu, Z. Ouyang, G. Sindona and R. G. Cooks, *Rapid Commun. Mass Spectrom.*, 2009, **23**, 3057–3062.
56. X. Ma, S. Zhang, Z. Lin, Y. Liu, Z. Xing, C. Yang and X. Zhang, *Analyst*, 2009, **134**, 1863–1867.
57. S. Soparawalla, F. K. Tadjimukhamedov, J. S. Wiley, Z. Ouyang and R. G. Cooks, *Analyst*, 2011, **136**, 4392–4396.
58. J. I. Zhang, A. B. Costa, W. A. Tao and R. G. Cooks, *Analyst*, 2011, **136**, 3091–3097.
59. A. Albert, A. Kramer, S. Scheeren and C. Engelhard, *Anal. Methods*, 2014, **6**, 5463–5471.
60. J. S. Wiley, J. F. García-Reyes, J. D. Harper, N. A. Charipar, Z. Ouyang and R. G. Cooks, *Analyst*, 2010, **135**, 971–979.
61. M. Benassi, A. Berisha, W. Romão, E. Babayev, A. Römpp and B. Spengler, *Rapid Commun. Mass Spectrom.*, 2013, **27**, 825–834.
62. M. J. Stein, E. Lo, D. G. Castner and B. D. Ratner, *Anal. Chem.*, 2012, **84**, 1572–1578.
63. M. J. Jebrail, M. S. Bartsch and K. D. Patel, *Lab Chip*, 2012, **12**, 2452–2463.
64. J.-B. Hu, T.-R. Chen, C.-H. Chang, J.-Y. Cheng, Y.-C. Chen and P. L. Urban, *Analyst*, 2015, **140**, 1495–1501.
65. C. Liu, K. Choi, Y. Kang, J. Kim, C. Fobel, B. Seale, J. L. Campbell, T. R. Covey and A. R. Wheeler, *Anal. Chem.*, 2015, **87**, 11967–11972.
66. A. Jönsson, R. R. Svejdal, N. Bøgelund, T. T. Nguyen, H. Flindt, J. P. Kutter, K. D. Rand and J. P. Lafleur, *Anal. Chem.*, 2017, **89**, 4573–4580.
67. C. J. Pulliam, R. M. Bain, H. L. Osswald, D. T. Snyder, P. W. Fedick, S. T. Ayrton, T. G. Flick and R. G. Cooks, *Anal. Chem.*, 2017, **89**, 6969–6975.
68. E. Mattio, F. Robert-Peillard, C. Branger, K. Puzio, A. Margaillan, C. Brach-Papa, J. Knoery, J.-L. Boudenne and B. Coulomb, *Talanta*, 2017, **168**, 298–302.
69. J. Wang, C. Xu and Q. Song, *Energy Fuels*, 2015, **29**, 1040–1044.
70. D. Chudobova, K. Cihalova, S. Skalickova, J. Zitka, M. A. M. Rodrigo, V. Milosavljevic, D. Hynek, P. Kopel, R. Vesely and V. Adam, *Electrophoresis*, 2015, **36**, 457–466.
71. S. A. Gowers, V. F. Curto, C. A. Seneci, C. Wang, S. Anastasova, P. Vadgama, G.-Z. Yang and M. G. Boutelle, *Anal. Chem.*, 2015, **87**, 7763–7770.
72. T. Monaghan, M. Harding, R. A. Harris, R. J. Friel and S. D. Christie, *Lab Chip*, 2016, **16**, 3362–3373.
73. S. K. Anciaux, M. Geiger and M. T. Bowser, *Anal. Chem.*, 2016, **88**, 7675–7682.
74. A. I. Shallan, P. Smejkal, M. Corban, R. M. Guijt and M. C. Breadmore, *Anal. Chem.*, 2014, **86**, 3124–3130.

75. G. Comina, A. Suska and D. Filippini, *Lab Chip*, 2014, **14**, 2978–2982.
76. C. Tang, A. Vaze and J. Rusling, *Lab Chip*, 2017, **17**, 484–489.
77. L. Zhang, R. N. Deraney and A. Tripathi, *Biomicrofluidics*, 2015, **9**, 064118.
78. P. Oomen, J. Mulder, E. Verpoorte and R. Oleschuk, *Anal. Chim. Acta*, 2017, **988**, 50–57.
79. T. Hutama and R. Oleschuk, *Lab Chip*, 2017, **17**, 2640–2649.

CHAPTER 5

# 3D Printing in Pharmaceutical Chemistry

## 5.1 Introduction

Conventional pharmaceutical manufacturing, formulation, and production techniques involve expensive equipment and facilities, require highly trained personnel, and are typically designed around high production volumes. In addition, both suppliers and users of pharmaceuticals are increasing concerned with accurate and reproducible control of product quality, and more specifically with issues such as the spatial distribution of drugs and excipients, microstructure of solid dosage forms, drug release profiles, and the move towards patient-specific formulation.

In the area of pharmaceutical formulation, production and delivery, printing techniques have the potential to reduce time and cost of traditional manufacturing techniques, *e.g.* the powder-fed tablet press for example, and help move pharmaceutical manufacturing toward a more versatile and responsive position. 3D printing techniques have already gained significant interest in the pharmaceutical industry after the U.S. Food and Drug Administration (USFDA) approval of a 3D-printed tablet, namely Spritam® in 2015. Spritam® is an orodispersible tablet manufactured by Aprecia Pharmaceuticals, which contains levetiracetam for the treatment of epilepsy.[1] The binder jetting technique was utilised to produce a high drug content (up to 1000 mg) and fast dissolving tablet (10 seconds in 15 mL water) for paediatric and geriatric patients who have difficulty swallowing.

More recently, we have seen a range of 3D printing techniques applied in a variety of ways within the pharmaceutical sector, including the

3D Printing in Chemical Sciences: Applications Across Chemistry
By Vipul Gupta, Pavel Nesterenko and Brett Paull

Published by the Royal Society of Chemistry, www.rsc.org

production of various oral, implantable, and transdermal drug delivery systems. These investigations and developments are described below in this chapter.

## 5.2 Development of Pharmaceutical 3D Printing

### 5.2.1 Binder Jetting

Binder jetting was the first printing technique that was used to print a drug delivery dosage form as early as 1996.[2] It was reported by Cima's group, who utilised the binder jetting technique to develop several unique dosage forms with controlled drug release profiles. The drug release profile was controlled by regulating the spatial distribution of the drugs and modifying the local composition and microstructure of the dosage forms, as shown schematically in Figure 5.1. Soluble poly(ethylene oxide) powder was used as the bulk material, insoluble poly(ε-caprolactone) powder as the top and bottom wall material, and 20% acid-modified poly(ε-caprolactone) and 5% poly(ε-caprolactone) in chloroform as the binding solution. Methylene blue and alizarine yellow dyes were manually deposited as model drugs in the desired locations before the final layer was printed. This work laid the foundation for the use of more advanced 3D printing to fabricate customised formulations and previously unobtainable solid dosage arrangements and forms.

Following this report, the group applied the same procedure for manufacturing oral dosage forms with active pharmaceutical ingredients (API) and standard pharmaceutical excipients.[3] Three types of oral tablets were prepared using chlorpheniramine maleate as the API, cellulose powder and spray-dried lactose as the matrices, and acrylates/dimethylaminoethyl methacrylate copolymer (Eudragit® E-100, Eudragit® RL PO), and poly(vinylpyrrolidone) as the binders.[3] In this study, the API solutions were introduced through the print heads as opposed to the previously described manual introduction.[2] The line spacings and the number of passes were

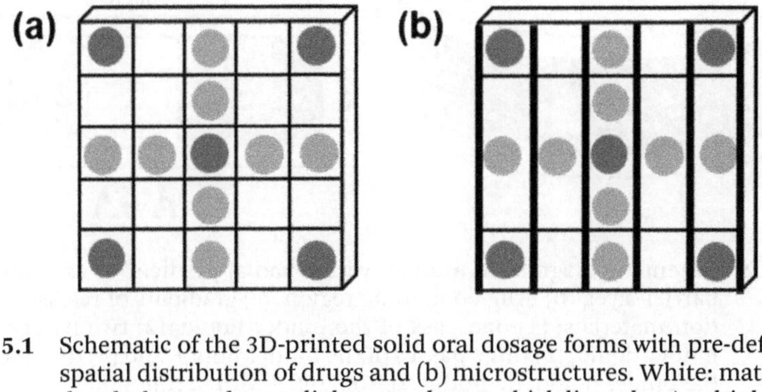


**Figure 5.1** Schematic of the 3D-printed solid oral dosage forms with pre-defined (a) spatial distribution of drugs and (b) microstructures. White: matrix powder, dark grey: drug 1, light grey: drug 2, thick line: slow/multiple passes of the binder, and thin line: fast/single pass of the binder.

varied to regulate the polymer and the API concentrations. Line spacings of the binder solutions were successfully able to control the lag times and drug release profiles of the 3D-printed tablets. Incorporating APIs in the print heads of 3D-printed tablets allowed excellent content uniformity and dosage control.

Incorporation of APIs in the print heads with the binder solutions usually limits the percentage of drug loading. In an attempt to increase drug loading percentages, Yu *et al.*[4] pre-mixed APIs with the matrix powder. Pre-mixing of the drug and the matrix powder allowed at least a 68% drug loading, as demonstrated by the 3D printing of a model drug, namely acetaminophen, in a hydroxypropyl methylcellulose matrix.

### 5.2.1.1 *Effect of Excipients*

Yu *et al.* have studied the effect of various pharmaceutical excipients, such as ethyl cellulose, Eudragit® RS-100, stearic acid, and sodium lauryl sulphate on the properties of binder jetting printed tablets.[4] Eudragit RS-100 is a copolymer of ethyl acrylate, methyl methacrylate, and a low content of methacrylic acid ester containing quaternary ammonium groups. All four excipients resulted in a decrease in the rate of drug release from hydroxypropyl methylcellulose printed tablets. The degree of retardation at pH 6.8 was observed in the order of sodium lauryl sulphate < stearic acid < ethyl cellulose < Eudragit® RS-100. Radial gradients of these excipients were further studied to formulate complex drug release profiles by varying the number of binder solution passes as shown in Figure 5.2. Incorporation of ethyl cellulose in the printed tablets allowed (1) a tablet hardness value of 73.82 ± 5.47 N $cm^{-2}$, (2) minimisation of the mass loss to 0.27 ± 0.06% during the friability testing, and (3) the formation of smooth tablet walls.[5]

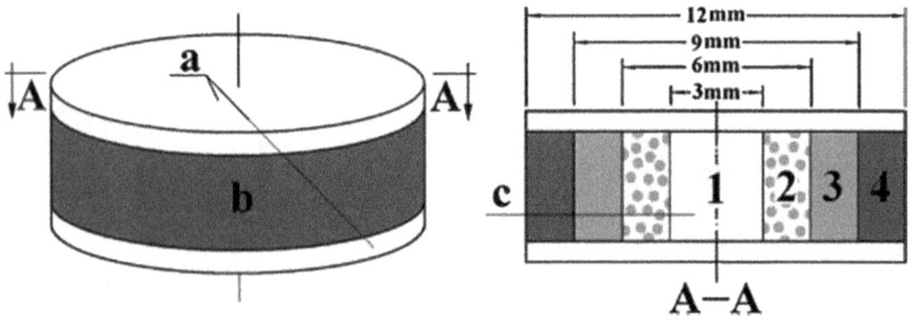


**Figure 5.2** Schematic diagram of a tablet with a radial gradient of excipients: (a) barrier layer, (b) drug-containing region, (c) gradients of release-retardation materials; (1) one pass of the binder liquid, (2) two passes of the binder liquid, (3) three passes of the binder liquid, and (4) four passes of the binder liquid. Reproduced from ref. 4 with permission from Elsevier, Copyright 2007.

Eudragit® E-100 is a cationic/dimethylaminoethyl methacrylate copolymer that dissolves below pH 5 and is permeable above pH 5. Eudragit® RL PO is an ammonio-methacrylic acid copolymer that does not dissolve, but is permeable throughout the pH range. Accordingly, for tablets printed with the E-100 binder the polymer matrix dissolved at pH 2, resulting in an erosion mediated drug release.[3] However, tablets printed with the RL PO binder resulted in a diffusion-mediated drug release due to the permeability of the polymer matrix.[3] Table 5.1 lists different pharmaceutical polymers (with their form, composition, and properties) that are used in the various 3D-printed dosage forms.

These results demonstrated the potential of adding different pharmaceutical excipients during the binder jetting 3D printing process to improve and vary the properties of 3D-printed tablets.

### 5.2.2 Fused Deposition Modelling

As opposed to the binder jetting technique, FDM printing can potentially deliver a simpler route to the production of complex structured solid dosage forms with lower cost, higher resolution, and higher mechanical strength. Indeed, the potential use of FDM printing for the production of drug-loaded solid dosage forms was first investigated and reported by Gaisford's group in 2014.[6] Commercial poly(vinyl alcohol) filaments were loaded with fluorescein as a model drug by incubating poly(vinyl alcohol) filaments in a 2% w/v fluorescein solution in ethanol for 24 hours (Figure 5.3). This procedure resulted in a drug loading of only 0.29 ± 0.01% in the poly(vinyl alcohol) filaments, and most of the drug was present on the filament surface due to the slow rate of drug diffusion. However, fluorescein was found to be uniformly distributed inside the printed tablets, which suggests uniform re-distribution of drugs in the softened filament during FDM printing. Similar results were obtained for loading mesalazine (5-aminosalicylic acid) and PAS (4-aminosalicylic acid) drugs into poly(vinyl alcohol) filaments through incubation.[7]

The percentage of drug loading in FDM filaments is improved by using a hot-melt extrusion process to actively incorporate drugs into polymers as opposed to relying on the passive drug diffusion during incubation. The hot-melt extrusion process has been widely used in the pharmaceutical industry and has resulted in more than ten commercial pharmaceutical products and over 300 patents.[8] A hot-melt extrusion process, as the name suggests, involves actively pumping raw materials at high temperatures through a die. Rotating screws are used to pump and mix the raw materials during the extrusion process. The hot-melt extrusion process was successfully used by Gaisford's group to load poly(vinyl alcohol) filaments with 3.95% w/w paracetamol (mixing ratio was 4% w/w) (Figure 5.4).[8] The drug was found to be homogeneously dispersed within the polymer matrix as a solid dispersion.

**Table 5.1** A list of branded pharmaceutical polymers (and form, composition, and properties) that are used in the various 3D-printed dosage forms.

| Trade name, manufacturer | Form and composition | Properties |
|---|---|---|
| Eudragit® E-100, Evonik | Granules, poly(butyl methacrylate-*co*-(2-dimethylaminoethyl) methacrylate-*co*-methyl methacrylate) 1 : 2 : 1 | Cationic copolymer, soluble in gastric fluid up to pH 5, swellable and permeable above pH 5.0, low viscosity, high pigment binding capacity, excellent adhesion, and high effectiveness in thin coatings |
| Eudragit® EPO, Evonik | Powder, poly(butyl methacrylate-*co*-(2-dimethylaminoethyl) methacrylate-*co*-methyl methacrylate) 1 : 2 : 1 | Cationic copolymer, soluble in gastric fluid up to pH 5, swellable and permeable above pH 5.0, low viscosity, high pigment binding capacity, excellent adhesion, and high effectiveness in thin coatings |
| Eudragit® RL PO, Evonik | Powder, poly(ethyl acrylate-*co*-methyl methacrylate-*co*-trimethylammonioethyl methacrylate chloride) 1 : 2 : 0.2 | Cationic copolymer, insoluble, high permeability, and pH-independent swelling. |
| Eudragit® RL, Evonik | Granules, poly(ethyl acrylate-*co*-methyl methacrylate-*co*-trimethylammonioethyl methacrylate chloride) 1 : 2 : 0.2 | Cationic copolymer, insoluble, high permeability, and pH-independent swelling |
| Eudragit RS 100, Evonik | Granules, poly(ethyl acrylate-*co*-methyl methacrylate-*co*-trimethylammonioethyl methacrylate chloride) 1 : 2 : 0.1 | Cationic polymer, insoluble, low permeability, and pH-independent swelling |
| Eudragit® L, Evonik | Powder, poly(methacylic acid-*co*-methyl methacrylate) 1 : 1 | Anionic polymer, dissolution above pH 6.0, effective and stable enteric coatings with a fast dissolution in the upper bowel, and granulation of drug substances in powder form for controlled release |
| Kollicoat® IR, BASF | Powder, copolymer of poly(vinyl alcohol) and poly(ethylene glycol), 3 : 1, with 0.3% colloidal silica, Mw 45 000 | Instant release coatings and quick dissolving formulations, soluble in water, and low viscosity |
| Soluplus®, BASF | poly(ethylene glycol)/poly(vinyl acetate)/poly(vinylcaprolactame), Mw 90 000–140 000 | Polymeric solubiliser with amphiphilic chemical structure, and improves bioavailability of poorly soluble drugs |
| Kollidon® SR, BASF | Copolymer of 80% poly(vinylacetate) and 19% poly(vinylpyrrolidone) containing 0.8% sodium laurylsulfate and 0.2% silica | Water soluble, pH-independent sustained release matrix, and inert to drugs |

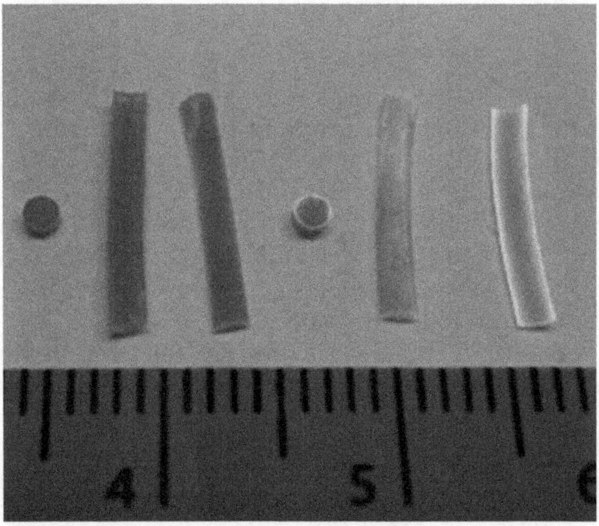


**Figure 5.3** Photographs of the poly(vinyl alcohol) filament (left) as received and (right) loaded with fluorescein. Reproduced from ref. 6 with permission from Elsevier, Copyright 2014.

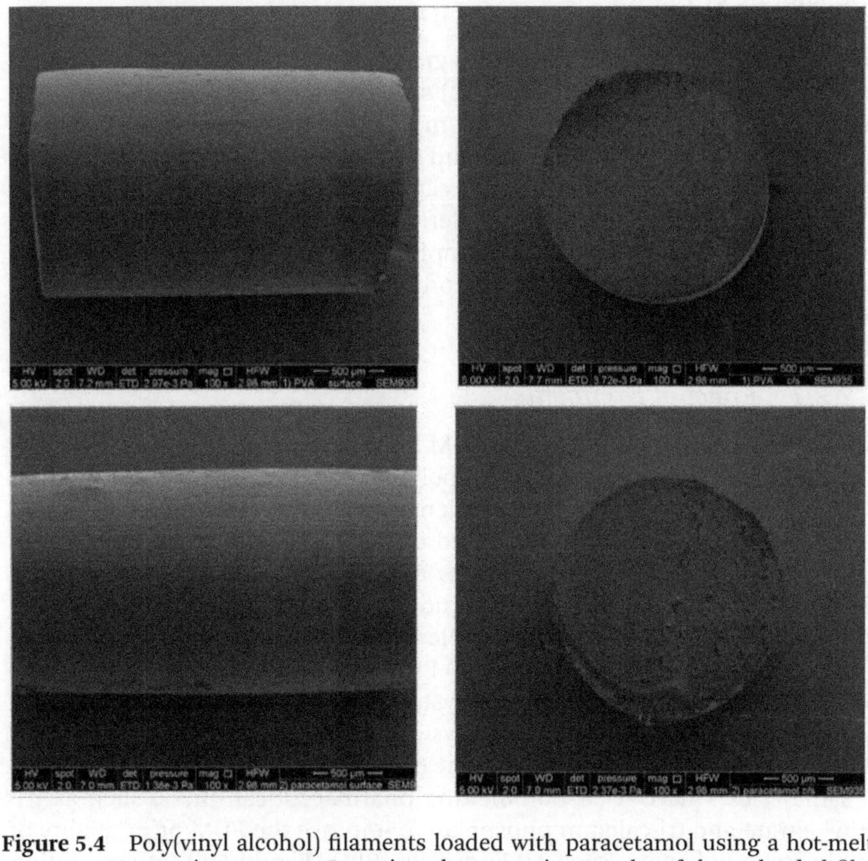

**Figure 5.4** Poly(vinyl alcohol) filaments loaded with paracetamol using a hot-melt extrusion process. Scanning electron micrographs of the unloaded filament (top) and loaded filament (bottom). Reproduced from ref. 8 with permission from Elsevier, Copyright 2015.

The parameters of FDM filaments, such as the melt flow index, flexural modulus, and surface roughness determine their 3D printing performance. The melt flow index indicates the force required to extrude a molten filament. The flexural modulus indicates the ability of a polymeric filament to resist bending. Accordingly, higher viscosity (lower melt flow index) polymers require stiffer (higher flexural modulus) filaments.[9] The acceptable surface roughness of a filament depends on the groove depth and the surface of the FDM printer's pinch rollers.[10]

Based on the above-mentioned factors, Genina *et al.*[9] investigated the ability to use different poly(ethylene-vinyl acetate) compositions for FDM printing of dosage forms. A decrease in the vinyl acetate content resulted in a decrease in the melt flow index and an increase in the flexural modulus. Poly(ethylene-vinyl acetate) with a vinyl acetate content of 16%, a melt flow index of 28 g per 10 min, and a flexural modulus of 60 MPa provided the best print quality out of the 12 different studied grades.[9] The poly(ethylene-vinyl acetate) filament was successfully loaded with up to 15% indomethacin using the hot-melt extrusion process.[9] Both the drug-free and drug-loaded poly(ethylene-vinyl acetate) were successfully printed at a temperature of 165 °C.

Taking this further, Melocchi *et al.*[11] have also developed FDM filaments from several different pharmaceutical grade polymers: (1) insoluble polymers (ethylcellulose and Eudragit® RL), (2) soluble polymers (Kollicoat® IR and poly(ethylene oxide)), (3) enteric polymers (Eudragit® L and hydroxypropyl methylcellulose acetate succinate), and (4) erodible polymers (poly(ethylene glycol)/poly(vinyl acetate)/poly(vinyl caprolactame) (Soluplus®), poly(vinyl alcohol), and hydrophilic cellulose derivatives). These filaments have been successfully used to 3D-print disk samples, which demonstrated low content variability and had properties in accordance with their respective pharmaceutical polymers.

### 5.2.2.1 *Effect of Excipients*

Alhijjaj *et al.*[12] have improved the FDM printing compatibility of Soluplus® and Eudragit® EPO pharmaceutical polymers by premixing these polymers with poly(ethylene glycol), poly(ethylene oxide), and Tween® 80. Incorporation of these constituents decreased the melt viscosity and increased the plasticity of Soluplus® and Eudragit® EPO polymers, and hence improved their print quality. These polymer blends were also successfully loaded with 10% felodipine. A lower rate of drug release was observed from the Soluplus® system as compared to the Eudragit® EPO system at pH 1.2. The lower rate of drug release from the Soluplus® system was presumably due to its lower solubility at pH 1.2, and the higher crystalline content due to the presence of poly(ethylene glycol) and poly(ethylene oxide).

Sadia *et al.*[13] have used non-melting pharmaceutical fillers, such as triethyl citrate and tri-calcium phosphate to improve the FDM print quality of methacrylic polymers, such as Eudragit® EPO (Figure 5.5). Incorporation

**Figure 5.5** Scanning electron micrograph of the methacrylate filament loaded with 50% tri-calcium phosphate. Reproduced from ref. 13 with permission from Elsevier, Copyright 2016.

of non-melting fillers results in uniform flow and higher reproducibility for FDM printing of methacrylic polymers. The optimum polymer formulation for FDM printing of Eudragit EPO was found to be 46.75 : 3.25 : 50 Eudragit® EPO: triethyl citrate: tri-calcium phosphate. This polymer composition was successfully loaded with acidic mesalazine and captopril (1-[(2S)-3-mercapto-2-methylpropionyl]-L-proline), basic theophylline (1,3-dimethylxanthine), and neutral prednisolone active pharmaceutical ingredients. These drug-loaded filaments were used to 3D-print caplets that resulted in 0% mass loss during friability testing.

Okwuosa *et al.*[14] have used thermostable fillers, such as talc, to allow low-temperature FDM printing. Exposure to high temperatures during FDM printing and hot-melt extrusion processes often results in the degradation or phase change of thermolabile active pharmaceutical ingredients, such as 4-aminosalicylic acid[7] and captopril.[13] Accordingly, FDM printing at lower temperatures, such as 110 °C, has been studied by incorporating talc as a thermostable filler within polyvinylpyrrolidone. An optimised formulation for low-temperature FDM printing was found to be 50% polymer (polyvinylpyrrolidone), 12.5% plasticiser (triethyl citrate), 27.5% filler (talc), and 10% active pharmaceutical ingredients (theophylline or dipyridamole). This formulation was successfully used to 3D-print tablets with acceptable mechanical strength, low content variability, and immediate drug release profiles. Moreover, it maintained the presence of 101.18 ± 3.91% and 99.56 ± 0.48% dipyridamole and theophylline, respectively, in the printed tablets.

### 5.2.3 Stereolithography

The SL technique can enable production of solid dosage forms with higher structural resolution, greater design flexibility, and lower cost as compared to the binder jetting techniques, and can be applied to deliver higher drug content with lower thermal degradation as compared to FDM printing. The first successful use of an SL printer for the production of a solid dosage form was reported in 2016 by Wang *et al.*[15] Poly(ethylene glycol) diacrylate was used as the monomer, diphenyl(2,4,6-trimethylbenzoyl)phosphine oxide as the photoinitiator, and 4-aminosalicylic acid and paracetamol as the model

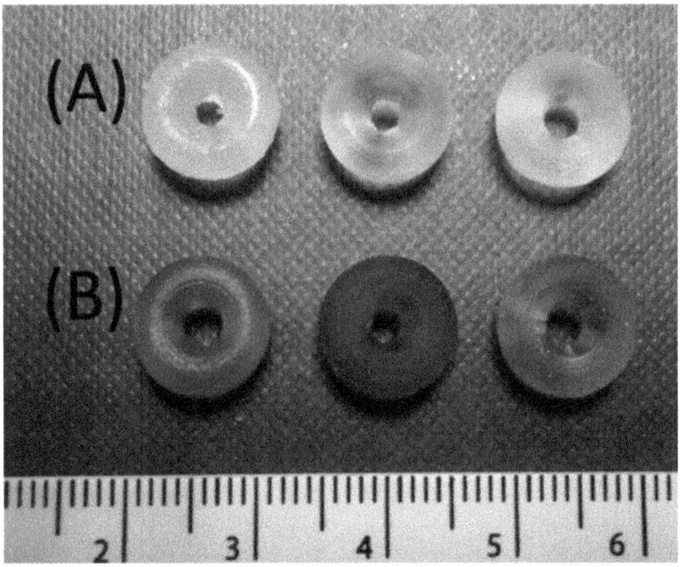


**Figure 5.6** 3D-printed tablets loaded with (A) paracetamol, and (B) 4-aminosalicylic acid. Left to right: ratios of poly(ethylene glycol) diacrylate and poly(ethylene glycol) 300 as 35:65, 65:35, and 90:10. Reproduced from ref. 15 with permission from Elsevier, Copyright 2016.

active pharmaceutical ingredients. The polymer resin with a 5.9% drug loading was successfully used to 3D-print torus-shaped tablets as shown in Figure 5.6. The resulting tablets had a drug content of 5.40% and 5.69%, for 4-aminosalicylic acid and paracetamol, respectively. They further varied the ratios of poly(ethylene glycol) diacrylate and poly(ethylene glycol) 300 to control the rate of drug release.[15] An increase in the poly(ethylene glycol) diacrylate content led to a decrease in the rate of drug release, due to the higher cross-linking of poly(ethylene glycol) diacrylate as compared to poly(ethylene glycol) 300.

### 5.2.4 PolyJet Printing

Similar to SL printing, PolyJet printing can also be applied to the production of solid dosage forms with a high degree of resolution and design flexibility. In addition, PolyJet printers can deliver significant time savings when compared to the binder jetting and FDM techniques, and offer benefits of scale-up as compared to most SL printers.

The successful use of PolyJet printing for the fabrication of a solid dosage form (Figure 5.7) was recently reported by Clark *et al.*[16] They used poly(ethylene glycol) diacrylate as the monomer, 1-[4-(2-hydroxyethoxy)-phenyl]-2-hydroxy-2-methyl-1-propane-1-one (Irgacure® 2959) as the photoinitiator, and ropinirole hydrochloride as the model active pharmaceutical ingredient. The

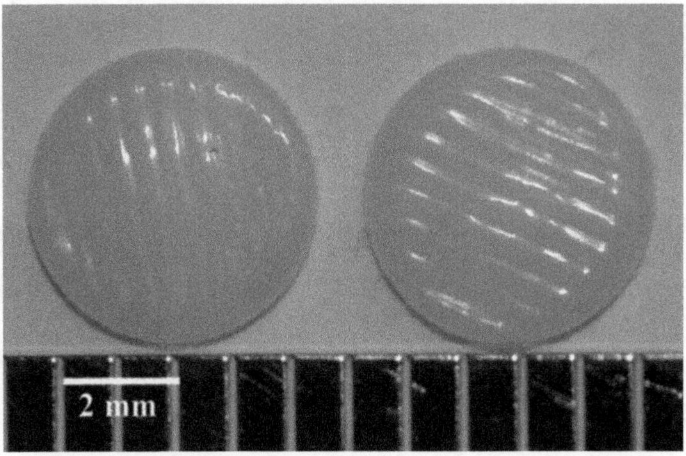


**Figure 5.7** Photographs of the PolyJet-printed solid oral dosage forms. Reproduced from ref. 16 with permission from Elsevier, Copyright 2017.

formulation with 0.41 mg drug loading resulted in a drug content of 0.39 ± 0.01 mg in the printed tablets. The percentage recovery of 97 ± 0.4% indicates minimal drug degradation during the PolyJet printing process. The drug was found to be in its amorphous form, and its release kinetics was observed to be as per the Korsmeyer–Peppas model with Fickian-type diffusion.

### 5.2.5 Selective Laser Sintering

Both binder jetting and FDM techniques suffer from low print resolution. PolyJet printing and SL techniques solve this issue, as shown above, but currently are also hindered by the limited number of biocompatible UV polymerisable monomers and photoinitiators. However, the use of SLS-based printers to produce solid dosage forms can circumvent each of these limitations, as described in Chapter 1.

The successful use of the SLS technique to 3D-print a solid dosage form (Figure 5.8) was first reported in 2017 by Fina *et al.*[17] Kollicoat® IR and Eudragit® L-100 were used as the matrices for printing both immediate, and modified release tablets. Paracetamol was used as the model active pharmaceutical ingredient and 3% Candurin® gold sheen was used to aid laser sintering, because of its absorbance at the utilised 445 nm wavelength. The formulations had 5, 20, and 35% paracetamol loadings, and no signs of degradation were observed in the SLS-printed tablets. Kollicoat® IR and Eudragit® L-100 3D-printed tablets demonstrated pH-independent (dose-dependent) and pH-dependent (dose-independent) drug release profiles, respectively. An increase in the drug content resulted in a decrease in the porosity, and hence a decrease in the rate of drug release from the SLS-printed tablets.

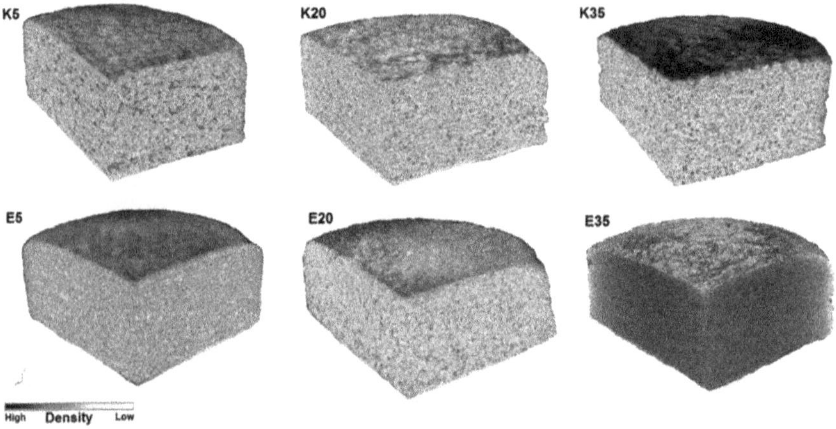


**Figure 5.8** X-ray micro-CT images of the SLS-printed tablets. Reproduced from ref. 17 with permission from Elsevier, Copyright 2017.

The SLS-printed tablets successfully resisted high crushing strengths (284 N to more than 485 N) and resulted in less than 1% mass loss during friability testing.[17] Hence, the mechanical properties of the SLS-printed tablets were found to be in accordance with the US Pharmacopoeia requirements for uncoated tablets.

## 5.3 Solid Oral Drug Delivery Systems

Solid oral dosage forms are the most commonly used drug delivery systems, because of their simple and independent administration by patients. To date, 3D printing techniques have been used to develop various solid oral dosage forms, including personalised medicines, fast disintegrating dosage forms, orodispersible films, floating tablets, nanocapsule delivery devices, and hydrogels as described below.

### 5.3.1 Personalised Medicines

Recently, there has been an increasing demand for customised production of oral dosage forms for application in the area of personalised medicines. Personalised medicine takes into account each patient's pharmacogenomics, anatomical, and physiological particulars to provide them with a more effective treatment. However, personalised medicine presents various manufacturing challenges, such as the customisation of dosages, drug release profiles, and active pharmaceutical ingredients combinations for each prescription.

The dimensions of a 3D-printed dosage form can be used to control its dosage to meet personalised requirements. Alhnan's group has observed a linear correlation between the volume and the dosage of FDM-printed tablets.[18,19]

They studied different tablet volumes of prednisolone-loaded poly(vinyl alcohol)[19] and theophylline-loaded Eudragit® RL.[18] The concentration of prednisolone was varied from 2–10 mg, and the concentration of theophylline was varied from 60–300 mg. Prednisolone was loaded into poly(vinyl alcohol) filaments through incubation (1.9% loading), whereas theophylline was loaded into Eudragit® RL filaments through hot-melt extrusion (50% loading). A good correlation ($R^2 > 0.99$) was observed between the programmed and the observed drug content in the FDM-printed tablets.[18,19]

In addition to the size, Gaisford's group has also investigated different shapes of printed dosage forms to modify the rate of drug release and the drug content.[8] FDM-printed tablets in five different shapes: cube, pyramid, cylinder, sphere, and torus, were studied for their effect on the rate of drug release. The drug release rate from these tablets was found to be dependent on the surface-area-to-volume ratio. Different tablets with an equivalent surface area resulted in a drug release rate in the order of pyramid > torus > cube > sphere and cylinder. Hence, just varying the shape of the 3D-printed tablets allowed the modification of the time of 90% drug release, from 2 hours for the pyramid, to 12 hours for the sphere and the cylinder. Owing to the dependence of the drug release rate on the surface-area-to-volume ratio, smaller FDM-printed tablets result in a higher rate of drug release as compared to their bigger counterparts.[18]

The shape of a dosage form is known to impact patient compliance.[20] Hence, 3D printing can be used to develop new and personalised shapes of dosage forms to increase a patient's willingness to take medicine. Gaisford's group has studied ten different tablet shapes for their end-user acceptability regarding grasping and swallowing. The shapes were the disc, torus, sphere, tilted diamond, capsule, pentagon, heart, diamond, triangle, and cube, as shown in Figure 5.9.[21] The torus was found to be the most preferred shape by patients for the ease of both grasping and swallowing. The sphere was

**Figure 5.9** Different shapes of 3D-printed tablets corresponding to different weights of 3D-printed (A) size 0, (B) size 1, (C) size 2, and (D) size 3 capsules. Shapes, from left to right: disc, torus, sphere, tilted diamond, capsule, pentagon, heart, diamond, triangle, and cube. (scale is in cm). Reproduced from ref. 21 with permission from Elsevier, Copyright 2017.

found to be the least preferred tablet shape. Smaller tablet sizes were also preferred over larger sizes; however, the preference of size was dependent on the tablet shape. A total of 350 tablets were 3D-printed, and the RSD values of their tablet mass ranged between 2% to 4% for all shapes. Good reproducibility between a large number of tablets supports the potential of 3D printing for large-scale customised production.

As mentioned above, binder jetting techniques have been applied to control the drug release profile and drug content of a 3D-printed dosage form by varying the three-dimensional position of the drugs, the composition of the binder liquids, and the thickness and microstructure of the printed walls.[2] The rate of drug release is inversely proportional to the distance between the drug molecules and the external walls, and hence varying the spatial distribution of the drug within the dosage forms results in different release profiles.[2]

Different binders result in different degrees of polymer binding. Hence, binder compositions can also be modulated to control the mechanism and rate of drug release. Wu *et al.*[2] studied poly(ε-caprolactone) and acid-modified poly(ε-caprolactone) binders for controlling the rate of drug release from poly(ethylene oxide) matrix tablets. The use of a poly(ε-caprolactone) binder resulted in a combination of diffusion and erosion type release mechanisms. Poly(ε-caprolactone) binder printed dosage forms resulted in the dissolution of unbound poly(ethylene oxide) particles and erosion of bound poly(ethylene oxide) layers; whereas the use of an acid-modified poly(ε-caprolactone) binder predominantly resulted in an erosion-type release mechanism through the erosion of bound poly(ethylene oxide) layers.

The architecture of the printed walls determines the rate of erosion and the path length of diffusion. Hence, the thickness and microstructure of the printed walls can be modulated to control the drug release rate. As expected, thinner and more porous walls have resulted in higher release rates as compared to their thicker and less porous counterparts.[2]

FDM print parameters, such as filament composition, drug loading, and infill percentage can be used to customise drug release profiles and dosage. An increase in the drug loading of the FDM-printed dosage forms increases the rate of drug release due to a decrease in the polymer content, and hence a reduction in the erosion and diffusion resistance.[22,23] Higher drug loading can also result in a higher percentage of the drug being present on the dosage form's surface, which would further increase the rate of drug release.

Pharmaceutical enteric polymers are routinely used to control the temporal and spatial release of conventional dosage forms in the gastrointestinal tract. Gaisford's group has successfully replicated this effect in 3D-printed dosage forms.[23] They used three grades of hypromellose acetate succinate (LG, MG, and HG) to control the drug release kinetics of FDM-printed dosage forms. The drug release rate (LG > MG > HG) was found to be in accordance with the succinyl group content (14–18% for LG, 10–14% for MG and 4–8% for HG grade) and the pH-threshold of the polymers. All three polymer grades successfully resisted drug release at the gastric pH (pH = 1.2).

The same group has further coated FDM-printed tablets with enteric polymers such as Eudragit® L-100 in a fluidised bed coater.[24] The enteric-coated tablets were found to be resistant to gastric pH conditions, and they provided a sustained release throughout the distal intestine and colon section of the *in vitro* human gastrointestinal tract dissolution model.

FDM printing can be performed at different infill percentages to control the print density. Hence, infill percentage can be used to modulate various physical properties of a printed dosage form, such as weight, mechanical strength, dissolution profile, and drug content. An increase in the infill percentage increases the density of the printed dosage forms, as shown in Figure 5.10, and hence increases their resistance towards drug release.[6,7,23]

Cima's group has used the binder jetting technique to fabricate complex shapes and configurations of oral dosage forms as described below.[25] Immediate-extended release tablets were fabricated by incorporating a fast and a slow eroding section in the same dosage form as shown in Figure 5.11(a).[25] Eudragit® E-100 and RLPO binder solutions were used to fabricate the fast and slow eroding sections, respectively. The fast eroding section resulted in

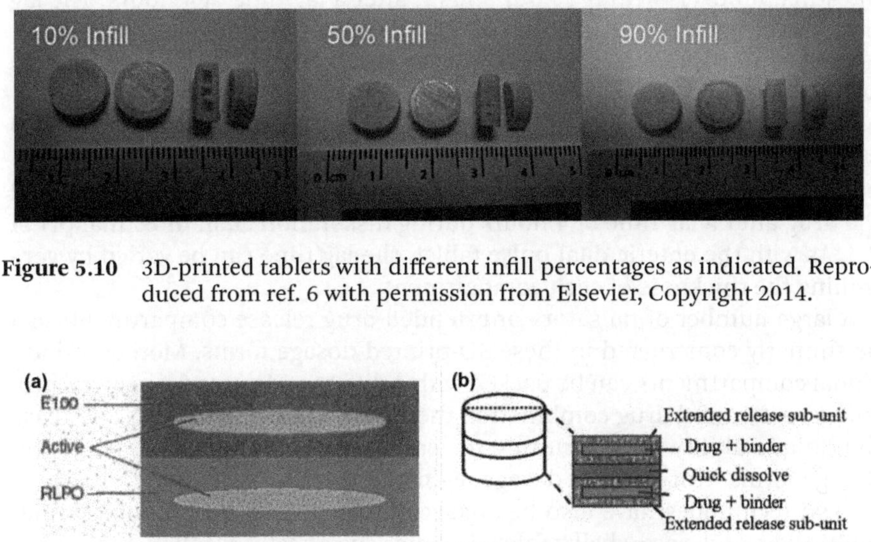


**Figure 5.10** 3D-printed tablets with different infill percentages as indicated. Reproduced from ref. 6 with permission from Elsevier, Copyright 2014.

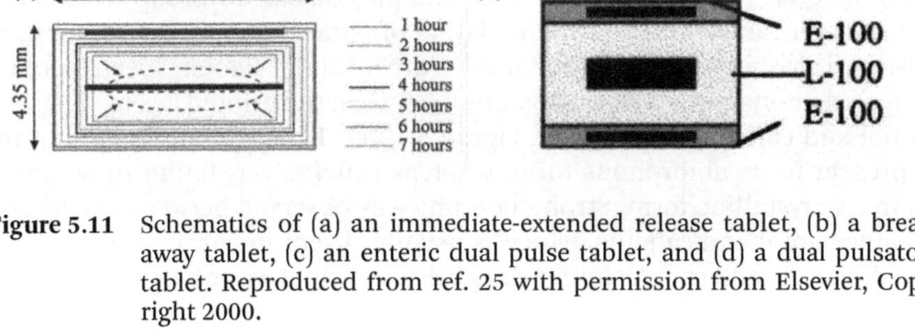


**Figure 5.11** Schematics of (a) an immediate-extended release tablet, (b) a breakaway tablet, (c) an enteric dual pulse tablet, and (d) a dual pulsatory tablet. Reproduced from ref. 25 with permission from Elsevier, Copyright 2000.

an immediate drug release (0.5 hours), and the slow eroding section resulted in an extended drug release (6 hours). Microcrystalline cellulose was used as the matrix powder, and chlorpheniramine maleate was used as the model active pharmaceutical ingredient.

Break-away tablets were fabricated with multiple discrete sub-units that were separated by quick dissolve sections for easy breakage, as shown in Figure 5.11(b).[25] The quick dissolve sections were formulated with 22% w/w poly(vinylpyrrolidone) in water as a binder and a 30 : 30 : 40 mixture of microcrystalline cellulose, spray-dried lactose, and Eudragit® L-100 as a matrix. The break-away tablets were designed to split into individual units within 15–40 minutes during dissolution. The release profiles of individual sub-units can be pre-defined. These tablets can be used to distribute low pharmacokinetic transport drugs along the gastrointestinal tract to achieve higher bioavailability.[25]

Enteric dual pulse tablets were fabricated by printing a drug in two separate compartments as shown in Figure 5.11(c).[25] A Eudragit® L-100 binder was used to resist the drug release in the gastric pH. These tablets released the drug in two individual pulses upon coming in contact with a simulated intestinal fluid: (1) within 1 hour and (2) after a lag time of 4 hours. The lag time can be varied by controlling the thickness of the inner compartment.

Dual pulsatory tablets were fabricated by printing drug compartments with different polymers as shown in Figure 5.11(d).[25] The outer drug compartment was printed with cationic polymer Eudragit® E-100, which eroded within 1 hour during dissolution at a gastric pH of 1.2. The inner drug compartment was printed with anionic polymer Eudragit® L-100, which released the drug after a lag time of 4 hours during dissolution at an intestinal pH of 7.5. As with the enteric dual pulse tablet, the lag time can be varied by controlling the thickness of each compartment.

A large number of pulsatory or extended drug release compartments can be similarly constructed in these 3D-printed dosage forms. Moreover, individual compartments can be packed with a different drug and dosage to provide a customised drug combination therapy. These dosage forms showcase the unique ability of 3D printing to readily design and develop new solid dosage forms with complex dosage regimes.

FDM techniques have also been successfully applied to produce similar multi-layer and core–shell tablet designs, specifically through the use of multi-material FDM printers.[26,27] For example, Gaisford's group developed a multi-layer tablet with alternating layers of paracetamol and caffeine in a poly(vinyl alcohol) matrix as shown in Figure 5.12(a).[26] Raman spectroscopic analysis demonstrated a clear separation between the printed layers of paracetamol and caffeine as shown in Figure 5.12(b). Paracetamol was found to be present in its amorphous form, whereas caffeine was found to be present in its crystalline form. Strong bonding was observed between the layers of paracetamol and caffeine, which is usually absent in conventionally produced bilayer tablets.[28] A similar drug release profile was observed for each

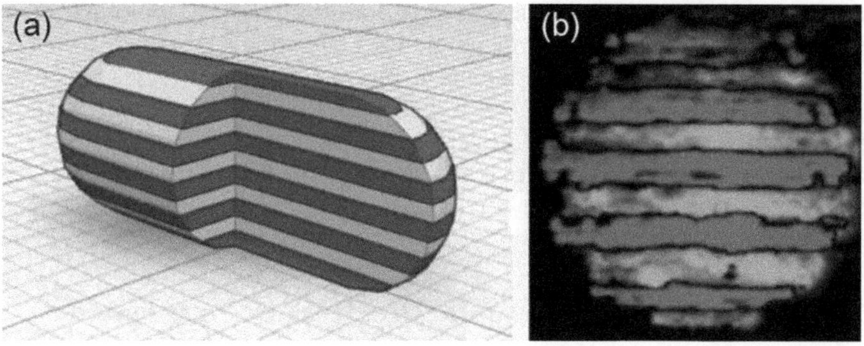


**Figure 5.12** 3D-printed multi-layer tablet with paracetamol and caffeine: (a) render and (b) 2D Raman mapping spectroscopy of the printed tablets. Reproduced from ref. 26 with permission from American Chemical Society, Copyright 2015.

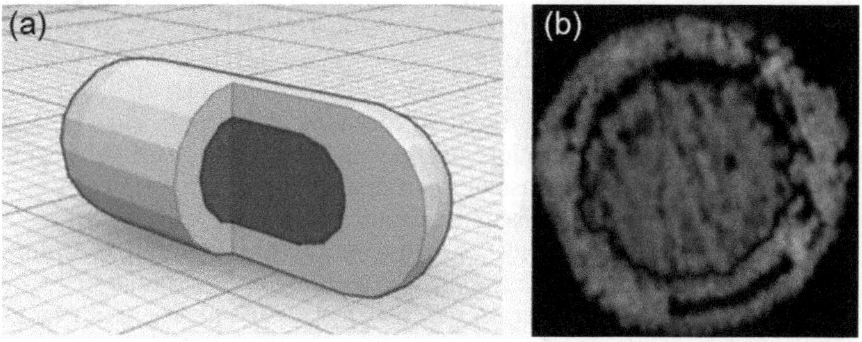


**Figure 5.13** 3D-printed core–shell tablet with paracetamol and caffeine: (a) render and (b) 2D Raman mapping spectroscopy of the printed tablets. Reproduced from ref. 26 with permission from American Chemical Society, Copyright 2015.

drug. The rate of drug release increased with an increase in the drug loading, as described earlier.

Goyanes *et al.*[26] and Li *et al.*[27] have 3D-printed core–shell dosage forms to incorporate multiple drugs and multiple concentrations of the same drug in a single dosage form, respectively. Drugs and their dosage were modulated by individually regulating the core and shell layers. A core–shell dosage form of paracetamol and caffeine in a poly(vinyl alcohol) matrix, which was 3D-printed with a multi-material FDM printer, is shown in Figure 5.13.[26] The drug present in the shell layers was released first, and that present in the core layers was released after a lag time. The lag time is dependent on the properties of the shell layer, which can be varied to control the drug release profile. Shell layers with paracetamol and a lower drug loading resulted in a slower erosion-dissolution rate, and hence a longer lag time as compared to the shell layers with caffeine and a higher drug loading.[26]

Li *et al.* achieved two different concentrations of glipizide (2.2% and 4.8% in poly(vinyl alcohol)) in the core and shell layers to produce controlled or delayed drug release dosage forms.[27] As expected, the drug present in the shell layers was released first followed by a lag time before the release of the drug present in the core layers. This core–shell tablet produced sustained drug delivery over a 5 hour period.

## 5.3.2 Fast Disintegrating Solid Oral Drug Delivery Systems

Fast disintegrating tablets allow easy swallowing, rapid drug absorption, and higher patient convenience and compliance.[29,30] These capabilities have rendered fast disintegrating solid dosage forms as an active area of interest. In conventional direct compression tablet manufacturing methods, tablet porosity is inversely proportional to the compression pressure. Usually, high compression pressure is exerted to ensure adequate tablet strength and robustness. This process often renders compressed tablets with low porosity, and hence a slow rate of disintegration. Thus, conventional tablet manufacturing techniques often make it difficult to fabricate fast disintegrating tablets while maintaining the required mechanical strength. However, 3D printing can be applied to produce mechanically robust and fast disintegrating solid dosage forms, because of its unique ability to independently control the density of active and excipient components.

Yang and co-workers have successfully used the binder jetting technique to fabricate such tablets.[5,31] They applied a combination of solid outer print layers and porous inner print layers to deliver fast disintegrating paracetamol tablets, as shown in Figure 5.14.[5] The matrix was composed of paracetamol, lactose, poly(vinylpyrrolidone) K30, mannitol, and colloidal silicon dioxide

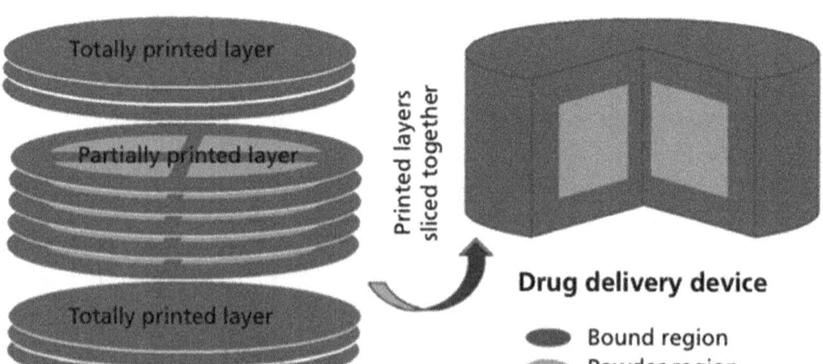


**Figure 5.14** Structure of a fast disintegrating paracetamol tablet. Reproduced from ref. 5 with permission from John Wiley and Sons, Copyright 2009 Royal Pharmaceutical Society of Great Britain.

in the ratio of 40 : 20 : 9.5 : 30 : 0.5 by weight. The binder solution was composed of methylene blue (0.5%, w/v) and poly(vinylpyrrolidone) K30 (5.0%, w/v) in 75% (v/v) of ethanol in water. Methylene blue was incorporated to easily visualise the bound components. The average disintegration and wetting time of the printed tablets was found to be 23.4 seconds and 67.6 seconds, respectively. Dissolution tests demonstrated a release of 98.5% drug content within the first 2 minutes. A good content uniformity was observed, with the tablet weight and drug content quoted as 326.5 ± 4.2 mg ($n$ = 20) and 131.5 ± 2.2 mg ($n$ = 12), respectively. An acceptable tablet hardness value of 63.4 ± 5.4 N $cm^{-2}$ was observed. The high mechanical strength of these 3D-printed tablets was presumably due to the incorporation of poly(vinylpyrrolidone) in both the tablet matrix and the binder solution. Poly(vinylpyrrolidone) is known to act as an adhesive and solid binder in conventional pharmaceutical preparations.[32]

Similarly, Wang *et al.*[33] have also used the binder jetting technique, here to develop a new zero-order controlled release dosage form. Conventional manufacturing methods used for the fabrication of zero-order dosage forms are either complicated or are linked to a specific interaction between the drug and the matrix.[33] The 3D-printed zero-order dosage forms were designed in a core–shell fashion with an immediate release core and a release rate regulating shell, as shown in Figure 5.15. The rate of drug (pseudoephedrine

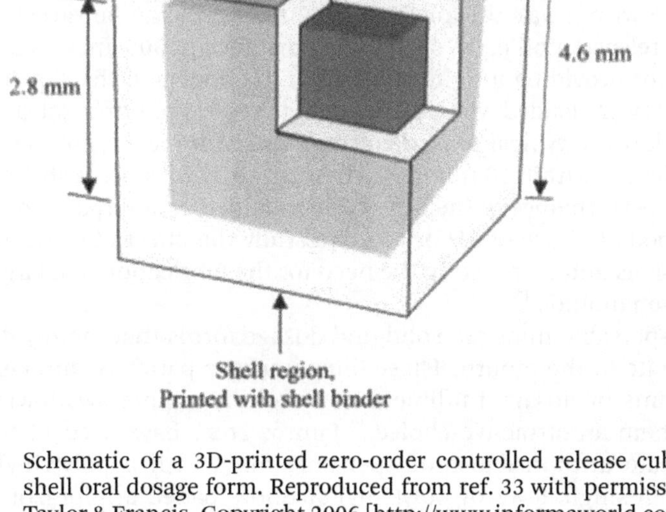


**Figure 5.15** Schematic of a 3D-printed zero-order controlled release cubic core–shell oral dosage form. Reproduced from ref. 33 with permission from Taylor & Francis, Copyright 2006 [http://www.informaworld.com].

hydrochloride) release was controlled by varying the ratio of Kollidon® SR and hydroxypropyl methylcellulose in the powder bed. The release rates were observed to increase with an increase in the ratio of hydroxypropyl methylcellulose. Kollidon® SR is a poly(vinyl acetate)/poly(vinylpyrrolidone)-based copolymer (see Table 5.1), which provides an impermeable network, whereas the hydrated hydroxypropyl methylcellulose matrix allows drug diffusion. The rate of drug release from the printed dosage forms was not dependant on either the pH or the hydrodynamic stress of the dissolution medium. The chemical and dissolution properties of the printed dosage forms were found to be stable even after their exposure to 25 °C/60% relative humidity and 40 °C/75% relative humidity for one month.

FDM printing is limited in the production of such fast disintegrating formulations, because of the high mechanical strength and high density of the resulting solid dosage forms.[34] However, other 3D printing techniques, such as SL, SLS, and PolyJet printing should allow fabrication of similar fast disintegrating and zero-order sustained release formulations, because of their ability to fabricate similar core–shell formulations with rapidly dissolving cores.

### 5.3.3 Miscellaneous Solid Oral Drug Delivery Systems

In addition to the above-mentioned controlled and fast disintegrating solid oral drug delivery systems, 3D printing has also been successfully used to fabricate other solid oral dosage forms, such as capsules,[11] orodispersable films,[35] floating tablets,[36] nanocapsule delivery devices,[37] and hydrogels,[38] as described below.

Melocchi *et al.* have successfully used FDM printers to fabricate hydroxypropyl cellulose capsule shells to provide oral pulsatile release of a drug similar to the ChronoCap® pulsatile delivery system.[11] The capsule shells were fabricated with a wall thickness of 600 μm, which can be varied to control the drug release profile. An overlapping distance of 300 μm was found to be optimal for providing a tight fit between the body and the cap. These capsule shells were loaded with paracetamol to study their drug release profiles. The result was a typical pulsatile drug release profile with 90% of the drug being released within 10 minutes. These 3D-printed capsule shells delivered a similar performance to their injection moulded counterparts, demonstrating the potential use of 3D printing to study the effect of different capsule shell designs and eliminating the need for the fabrication of a large number of injection moulds.[11]

Orodispersable films are solid oral dosage forms that rapidly dissolve or disintegrate in the mouth. These films improve patient compliance, avoid false claims of dosage fulfilment, and do not require swallowing, which makes them an attractive choice.[39] Jamróz *et al.* have used FDM printing to fabricate aripiprazole orodispersable films with a poly(vinyl alcohol) matrix.[35] A single layer of poly(vinyl alcohol)-loaded aripiprazole filament was deposited, which resulted in a film thickness of ~198 μm. These

3D-printed films were found to be soft, flexible, and durable. However, the durability of the 3D-printed films was less than that of the cast films, because of the perforated nature of the former and the continuous nature of the latter. Although the 3D-printed films were twice the thickness of the cast films, the disintegration time was only 5 seconds longer. The 3D-printed orodispersible films resulted in a faster rate of drug release as compared to the cast films, due to the amorphous form of aripiprazole in the 3D-printed films and the perforated structure of the 3D-printed layers.

Floating tablets are low-density systems with enough buoyancy to float over gastric contents. They result in prolonged and predictable residence times in the stomach.[40] Increased residence time in the stomach can increase the solubility of weak basic drugs, reduce dosing frequency, and increase the drug bioavailability.[41] Chai *et al.* have applied FDM printing to produce domperidone intragastric floating sustained release tablets with a hydroxypropyl cellulose matrix.[36] The tablets were printed with 10% domperidone, 2 wall layers and 0% infill. The majority of domperidone was found to be in its amorphous form. The 3D-printed floating tablets stayed in a rabbit's stomach for 8 hours, as shown in Figure 5.16. These 3D-printed

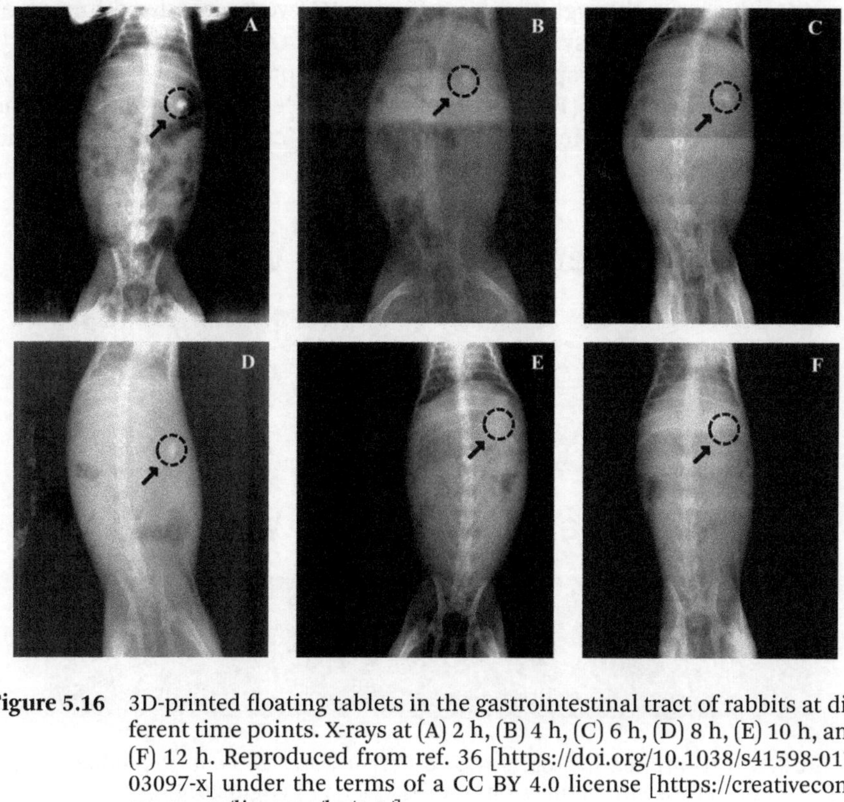


**Figure 5.16** 3D-printed floating tablets in the gastrointestinal tract of rabbits at different time points. X-rays at (A) 2 h, (B) 4 h, (C) 6 h, (D) 8 h, (E) 10 h, and (F) 12 h. Reproduced from ref. 36 [https://doi.org/10.1038/s41598-017-03097-x] under the terms of a CC BY 4.0 license [https://creativecommons.org/licenses/by/4.0/].

floating tablets resulted in a relative bioavailability of 222.5 ± 62.9% as compared to the commercial reference tablets.

Multi-functional drug delivery systems, such as nanocapsule-loaded solid dosage forms, pose significant manufacturing challenges.[37] However, Gaisford's group have successfully used FDM printing to produce such a drug delivery system, specifically poly(ε-caprolactone) and Eudragit® RL 100 tablets loaded with deflazacort nanocapsules.[37] FDM-printed tablets were incubated in a solution of deflazacort nanocapsules (138 nm). Incorporation of a channelling agent such as mannitol in the tablet matrix increased the percentage of drug loading. Also, hollow core (50% infill) tablets resulted in a higher percentage of drug loading and a higher rate of drug release as compared to their solid core counterparts. Tablets printed with the Eudragit® RL 100 matrix generally resulted in a higher drug content and a higher rate of drug release as compared to the tablets printed with poly(ε-caprolactone). The differences in the drug content and release rate were presumably due to the higher swelling index of Eudragit® RL 100 as compared to poly(ε-caprolactone). The 3D-printed tablets containing nanocapsules demonstrated higher drug loading percentages as compared to similar systems prepared conventionally.

Hydrogels can be used as controlled drug delivery devices, and are typically loaded with a drug through passive diffusion. However, passive diffusion of a drug does not allow easy determination and control of the drug content.[38] Hence, Gaisford's group applied an SL printer to produce poly(ethylene glycol) diacrylate hydrogels (Figure 5.17).[38] Poly(ethylene glycol) diacrylate was used as the photopolymerisable substrate, poly(ethylene glycol) 300 and

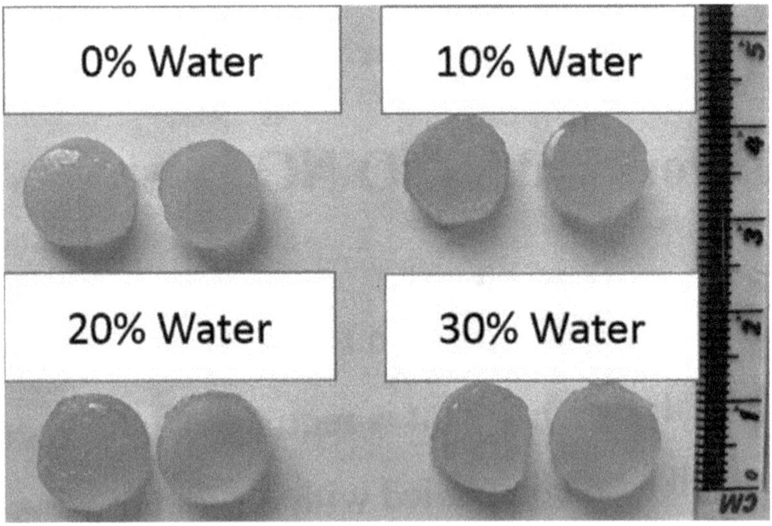


**Figure 5.17** 3D-printed hydrogels with different water content. Reproduced from ref. 38 with permission from Elsevier, Copyright 2017.

water were used to control the cross-linking density, diphenyl(2,4,6-trimethylbenzoyl)phosphine oxide and riboflavin with triethanolamine were used as the photoinitiators, and ibuprofen was used as a model drug. The rate of drug release from the 3D-printed hydrogels increased with an increase in the water content. Hydrogels with up to 30% w/w water and 10% w/w ibuprofen were successfully printed. 3D-printed hydrogels were able to resist the drug release at gastric pH as per the USFDA regulations for a delayed release dosage form.

## 5.4 Implantable Drug Delivery Systems

In addition to the oral drug delivery systems, drug implants are now routinely used to treat clinical conditions. Drug implants can offer significant advantages by allowing local drug delivery instead of systemic drug delivery. Drugs delivered through systemic routes can suffer from dose-limited efficacy and side effects.[42,43] Whereas, drugs delivered through sustained release implants that are present in the proximity of the required sites can provide high therapeutic efficacy and low systemic toxicity.[44] Conventional methods of fabricating drug implants are again rather complex, involving the need for complex shapes, the control of internal architecture, and control of complex drug release profiles. As discussed above, 3D printing can potentially help overcome many of these difficulties; hence it has been investigated for the production of drug implants with controlled drug release profiles, by pre-defining the spatial position of the drugs within the implants.

Lin *et al.* used the binder jetting technique to develop subdermal implants for the controlled delivery of ethinylestradiol.[45] The effects of different drug distribution strategies and implant matrices on the drug release profiles of the 3D-printed implants were studied. Three subdermal implants were developed, with (1) a single channel distribution of the drug within a poly(ε-caprolactone) matrix (implant I, Figure 5.18), (2) an homogeneous distribution of the drug within a poly(ε-caprolactone) matrix (implant II), and (3) a concentration gradient distribution of the drug within a 50 : 50 poly(ε-caprolactone) and poly(D,L-lactide-*co*-glycolide) matrix (implant III). The rate of drug release was found to be in the order of implant III > implant II > implant I. The poly(ε-caprolactone) matrix experiences a slow rate of erosion. Hence, the implants I and II experienced a diffusion mediated drug release. As expected, a higher rate of drug release and a shorter drug delivery period was observed for implant II as compared to implant I. Poly(D,L-lactide-*co*-glycolide) has a faster rate of erosion as compared to poly(ε-caprolactone), therefore implant III demonstrated the highest rate of drug release and the shortest drug delivery period out of all the three studied implants.

Binder jetting has also been used to develop other complex drug delivery implants, such as a compartmentalised implant to incorporate multiple drugs for combination therapy (Figure 5.19).[44,46] The formulation was

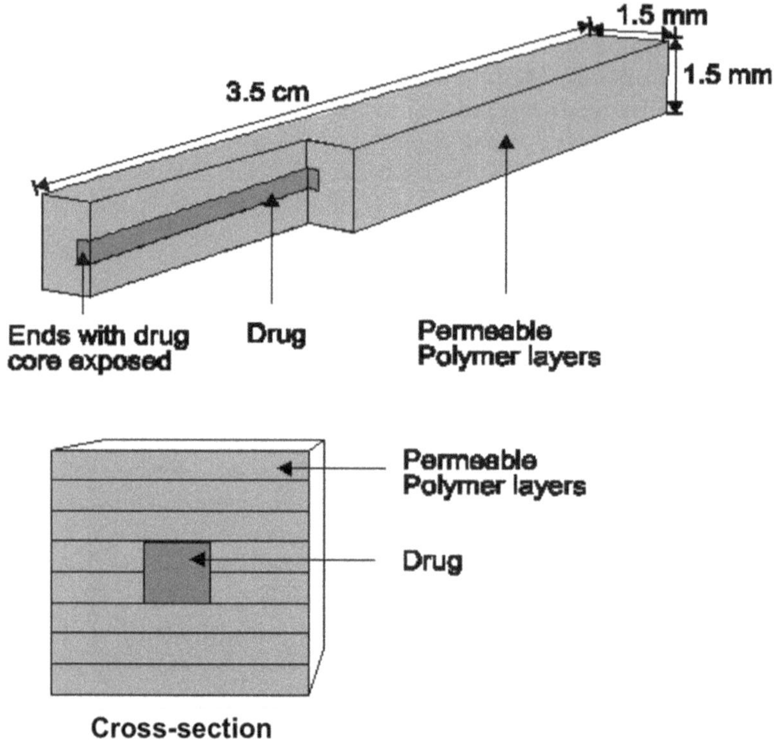


**Figure 5.18** Schematic of a 3D-printed implant with a pre-defined drug distribution channel. Reproduced from ref. 45 with permission from Springer Nature, Copyright 2001 American Association of Pharmaceutical Scientists.

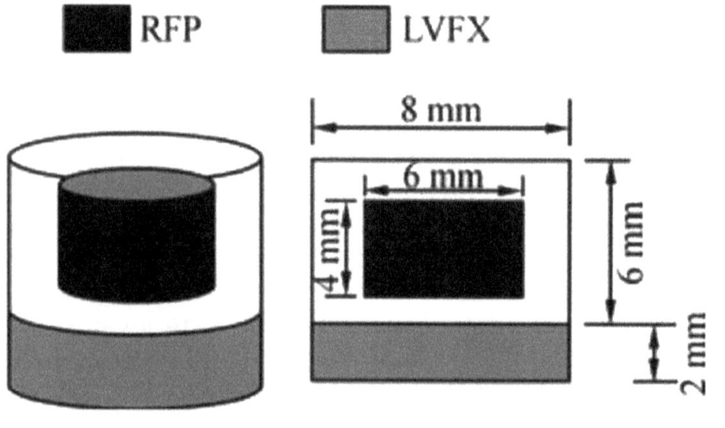


**Figure 5.19** Schematic of a 3D-printed implant with multiple drugs: rifampicin (RFP) and levofloxacin (LVFX). Reproduced from ref. 46 with permission from Springer Nature, Copyright 2009 Wuhan University of Technology and Springer Berlin Heidelberg.

composed of poly(lactic acid) powder as the matrix, 20 : 80 ethanol and acetone mixture as the binder, and levofloxacin and rifampicin as the model active pharmaceutical ingredients. Incorporation of levofloxacin in both compartments resulted in a bi-modal release profile.[44] Incorporation of rifampicin in the upper compartment and levofloxacin in the lower compartment resulted in an immediate release of levofloxacin and a delayed release of rifampicin, as expected.[46] The 3D-printed implants exhibited a more porous and uniform structure as compared to the implants fabricated by powder compression.[44] Interconnected and uniformly distributed micropores of 50–100 μm diameter were observed in the 3D-printed implants, resulting in a porosity of 61.3%.[44]

Drug implants such as intrauterine devices and subcutaneous rods have been widely used for obstetric and gynaecological purposes. They are conventionally fabricated through methods such as injection moulding, compression moulding, freeze-drying, extrusion, and solvent casting.[47–51] Such manufacturing techniques limit the production of intrauterine devices to standard shapes, sizes, and doses, and these generic designs do not conform to individual requirements. Improper shapes, sizes, and doses can result in ectopic pregnancies, pelvic inflammatory diseases, and uterine perforations.[52,53] 3D-printed intrauterine devices can be easily personalised to avoid these complications and increase patient compliance.

To the best of the authors' knowledge, Genina *et al.*[9] reported the first 3D-printed intrauterine device and subcutaneous rod in 2015. The FDM technique was used to fabricate customised T-shaped intrauterine devices and subcutaneous rods. Ethylene-vinyl acetate copolymer filaments were loaded with 5% and 15% indomethacin by hot-melt extrusion. Hot-melt extrusion was performed below the melting point of the drug, and FDM printing was performed above the melting point. Hence, indomethacin was present only in its crystalline state in the filaments, whereas it was found in both crystalline and amorphous state in the printed prototypes. The rate of drug release from the 5% loaded intrauterine devices and subcutaneous rods was found to be higher than from their 15% loaded counterparts, which is contrary to the trends observed by Gaisford's group for FDM-printed tablets.[22,23] 3D-printed intrauterine devices and subcutaneous rods resulted in an initial burst release of the drug followed by a slow release. The drug release followed the Korsmeyer–Peppas model with the $n$ value of ≤0.45 for all samples except the 15% drug-loaded subcutaneous rods. This indicates that all samples, except for the 15% drug-loaded subcutaneous rods, demonstrated the Fickian diffusion mechanism. The same group further developed similar intrauterine devices with a poly(ε-caprolactone) matrix, because it offers high drug permeability and high biocompatibility.[54] In this case, the drug release was found to follow the Higuchi square root model.

The poly(ε-caprolactone) matrix was further used by Tappa *et al.*[55] to develop hormone eluting intrauterine devices. Estrogen and progesterone were loaded onto poly(ε-caprolactone) pellets before hot-melt extrusion. Oil coating was used to distribute the hormones on the pellets evenly. Silicon oil was used for coating, because of its high thermal stability. The hormone-loaded poly(ε-caprolactone) fibres were successfully used to print intrauterine devices, pessaries, subdermal rods, and meshes (Figure 5.20).[55]

Ding *et al.* 3D-printed novel drug release implants using a thermo-sensitive hydrogel (1-tetradecanol).[56] The architecture of these implants was inspired by the lotus root and was intended to achieve automatic local-inflammation-triggered drug release patterns. Three different internal architectures of the implants were studied: (1) a cylindrical cavity of 3.3 mm diameter (Cy3.3), (2) a cylindrical cavity of 3.6 mm diameter (Cy3.6), and (3) a spherical cavity with two cylindrical cavities (Sp5&Cy3) as shown in Figure 5.21. The implants were loaded with fluorescein isothiocyanate and silver nanoparticles. These implants successfully retained drugs up to 37 °C and released the drug when exposed to 39 °C for 25 minutes. The mechanical properties and the drug release profiles of these implants were found to be dependent on their internal architectures.

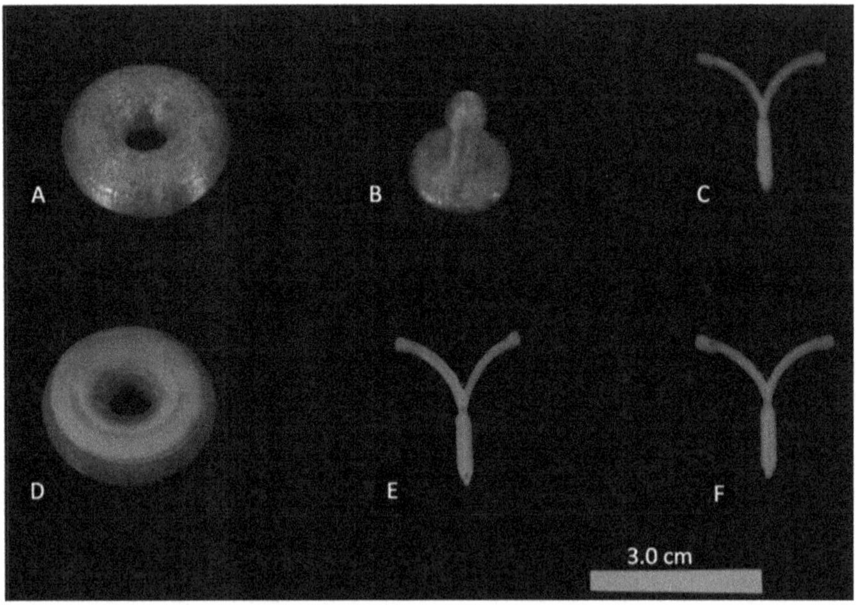


**Figure 5.20** 3D-printed (A) doughnut-shaped pessary, (B) Gellhorn-shaped pessary, (C) intrauterine device, (D) pessary printed with a combination of poly(lactic acid) and estradiol-loaded poly(ε-caprolactone), (E) estrone-loaded poly(ε-caprolactone) intrauterine system, and (F) estradiol-loaded poly(ε-caprolactone) intrauterine system. Reproduced from ref. 55 [https://doi.org/10.1371/journal.pone.0182929] under the terms of a CC BY 4.0 license [https://creativecommons.org/licenses/by/4.0/].

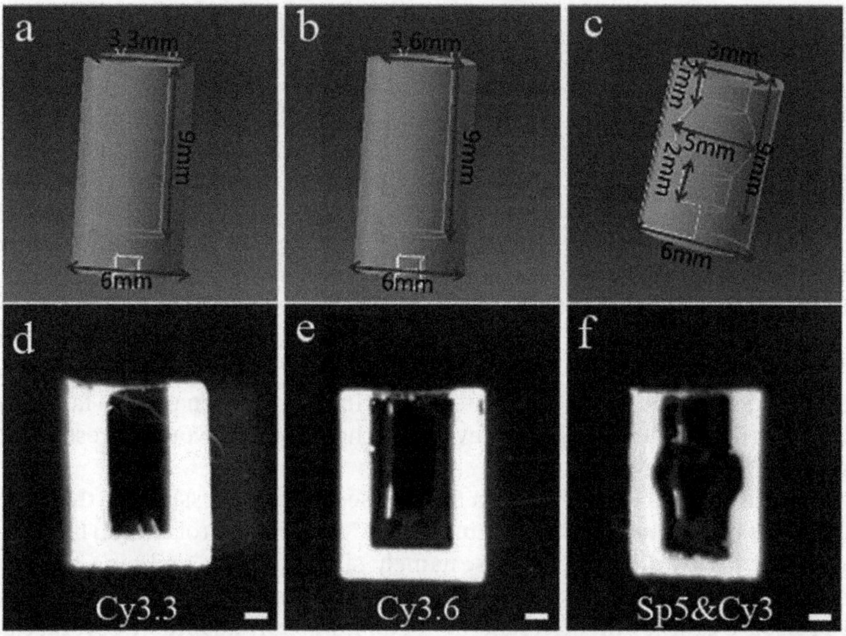


**Figure 5.21** 3D-printed automatic local-inflammation-triggered drug release implants: (a) schematic of Cy3.3, (b) schematic of Cy3.6, (c) schematic of Sp5Cy3, (d) 3D-printed Cy3.3, (e) 3D-printed Cy3.6, and (f) 3D-printed Sp5&Cy3. Reproduced from ref. 56 with permission from The Royal Society of Chemistry.

3D printing has also been used to fabricate antibiotic and chemotherapeutic catheters. Generally, simple and low-cost FDM printers are used to print such catheters. Weisman *et al.*[52] used poly(lactic acid) filaments loaded with 1% w/w gentamicin sulphate and 2.5% w/w methotrexate to print antibiotic and chemotherapeutic catheters, respectively.[52] Gentamicin-loaded catheters demonstrated an average zone of inhibition of 857.6 $mm^2$ for *E. coli*, and methotrexate loaded filaments successfully inhibited osteosarcoma cell growth.

In addition to the antibiotic catheters, poly(lactic acid) filaments loaded with gentamicin were also used by Ballard *et al.* to 3D-print antibiotic hernia meshes.[57] These 3D-printed hernia meshes demonstrated an average zone of inhibition of 110 $mm^2$ for *E. coli* and 120 $mm^2$ for *S. aureus*. Similar poly(lactic acid) filaments loaded with nitrofurantoin have also previously resulted in more than 85% biofilm inhibition; hence they can also be used to fabricate catheters and other antibacterial medical devices.[58,59]

3D printing has been extensively used for the design and development of various types of non-drug eluting implants such as orthopaedic implants,[60,61] dental implants,[62] and radiology implants,[63] to name a few. The discussion of these types of implants is beyond the scope of this book, and their discussion can be found in several review papers.[64–66]

## 5.5 Transdermal Drug Delivery Systems

Similar to the implants, transdermal systems can also provide local drug delivery and avoid first-pass metabolism and the adverse effects of systemic delivery. Transdermal drug delivery further offers ease of administration for chronic diseases such as diabetes.[67] The advent of 3D scanning and 3D printing techniques is revolutionising the field of transdermal drug delivery by allowing the development of personalised and novel patches. 3D scanning is used to scan the desired body part to model the required transdermal patch, which can be easily 3D-printed. 3D printing has been used to fabricate customised transdermal drug delivery patches, such as wound dressings,[68] patches,[69] and microneedles.[70]

Gaisford's group FDM-printed a customised wound dressing for the nose, which was based on a 3D scan (Figure 5.22).[68] A poly(ε-caprolactone) filament was loaded with different metals, namely silver, copper, and zinc to print antibacterial dressings. These 3D-printed dressings resulted in a sustained drug release profile and offered a higher anatomically adaptable coverage as compared to conventional flat dressings.

They further compared FDM and SL printing for the production of customised transdermal drug delivery patches.[69] FDM-printed patches were fabricated with Flex EcoPLA™ filaments (a copolymer of poly(lactic acid)), and SL-printed patches were fabricated with a resin mixture of poly(ethylene glycol) diacrylate and poly(ethylene glycol). Both patches were designed to achieve 2% w/w salicylic acid in the printed product. The SL-printed patches resulted in a drug content of 1.9% w/w, whereas the FDM-printed patches resulted in a drug content of only 0.4% w/w. Significantly, low drug content

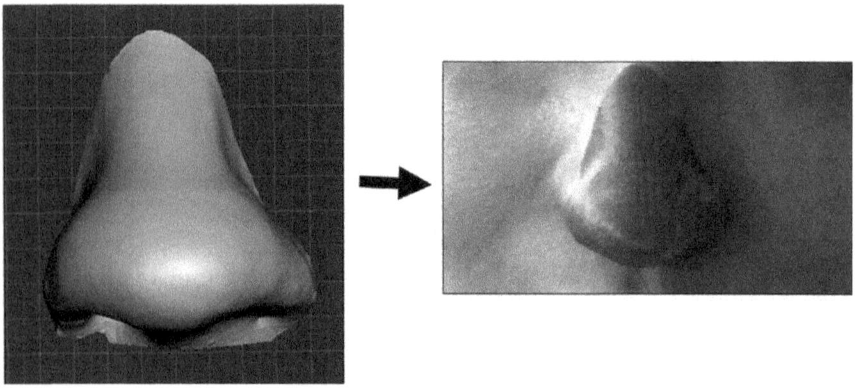

**Figure 5.22** 3D scanned model of a nose (left) and 3D-printed copper loaded nose dressing (right). Reproduced from ref. 68 with permission from Elsevier, Copyright 2017.

was observed in the FDM-printed patches because of the thermal degradation of the drug during FDM printing as discussed above. The SL-printed patches further resulted in a higher print resolution and a higher drug diffusion coefficient as compared to their FDM-printed counterparts.

A recent advance in the field of transdermal drug delivery is the use of microneedles.[71] Microneedles penetrate the outermost and hardest layer of the skin (stratum corneum) to enable successful drug delivery without any pain or bleeding. Microneedles are usually less than 500 μm in height and are grouped in large numbers.[72] Design parameters such as shape, size, composition, and spacing can be tailored to modify performance.[70]

Johnson *et al.* used the CLIP 3D printing technique to produce microneedles with customised shape, size, spacing, and aspect ratios, as shown in Figure 5.23.[70] As discussed in Chapter 1, CLIP 3D printing is similar to the SL 3D printing technique, except that the former allows continuous printing as opposed to the discontinuous printing in the latter. The use of conventional SL printing could damage the microneedles while mechanically separating them from the window. Hence, the CLIP approach was used to print microneedles with poly(ethylene glycol), poly(ε-caprolactone), and poly(acrylic acid) build materials. A consistent morphology was observed between all the materials and less than 10% variation was observed from the programmed dimensions. The 3D-printed microneedles were able to penetrate the murine skin. The drug release profile was successfully controlled by varying the composition of the microneedles.

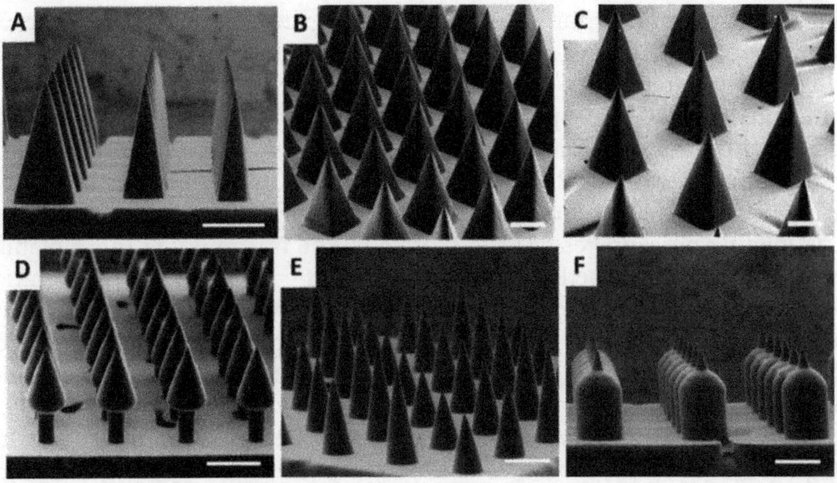


**Figure 5.23** 3D-printed microneedles for transdermal drug delivery: (A) microneedles of aspect ratio 2, 3, and 4 (left to right), (B) 1000 μm tall and 0.5 base width spacing microneedles, (C) 1000 μm tall and 1.5 base width spacing microneedles, (D) arrowhead microneedles, (E) tiered microneedles, and (F) turret microneedles. Scale bar is 500 μm. Reproduced from ref. 70 [https://doi.org/10.1371/journal.pone.0162518] under the terms of a CC BY 4.0 license [https://creativecommons.org/licenses/by/4.0/].

Multi-material CLIP-printed microneedles were also fabricated by manually exchanging the poly(ε-caprolactone) and poly(acrylic acid) resins as and when required.

## 5.6 Outlook

As described above, 3D printing has already shown great potential in pharmaceutical manufacturing, particularly in the area of drug delivery dosage forms. Table 5.2 lists various 3D-printed drug delivery systems along with the utilised printing techniques and print materials that have been highlighted above. 3D printing can facilitate easy customisation of the drug content, shape, and release profile of a dosage form, and has the clear potential to eventually be used at a local pharmacy by local pharmacists for tailored/personalised patient subscriptions. This could facilitate the successful adoption of personalised medicines on a whole new level. 3D printing can be used to independently customise active and non-active components of a dosage form, allowing rapid design and development of new and personalised formulations, such as hormonal replacement devices with physiological cyclic rhythms.[2]

Each 3D printing technique offers a unique capability for fabricating different drug delivery systems. Binder jetting and FDM techniques produce systems with higher biocompatibility as compared to the PolyJet and SL techniques. The greater biocompatibility of binder jetting and FDM-printed dosage forms has seen the use of these techniques preferred over other techniques for the production of 3D-printed drug delivery systems. However, PolyJet and SL techniques offer higher print resolution and more design flexibility as compared to other 3D printing techniques.

Despite their obvious potential, 3D printing techniques currently also suffer from serious limitations with respect to their use for pharmaceutical manufacturing, which will need to be addressed in the near future. These include the low resolution of FDM and binder jetting techniques, the low drug loading percentages of FDM filaments, the instability of thermolabile drugs in the FDM-printed dosage forms, the toxic nature of the photoinitiators used in SL and PolyJet techniques, and a limited number of biocompatible 3D-printable materials. Each of the above is currently being researched. The resolution of each 3D printing technique is continuously improving, and new biocompatible 3D-printable materials are being added every year. The drug loading percentage of an FDM filament can be further improved by adding plasticisers.[26] Incorporation of different pharmaceutical excipients can allow the modulation of different material and dosage form properties. An increase in 3D printing resolution would allow the fabrication of drug delivery dosage forms with better appearance and would offer a higher level of control over drug release profiles. The adoption of 3D printing for commercial pharmaceutical manufacturing would further require strict regulation of material and process quality parameters to comply with pharmaceutical good manufacturing practices.

**Table 5.2** The 3D printing techniques, printers, and print materials that are used for fabricating different drug delivery systems.

| 3DP technique | Printer: Manufacturer | Printer: Model | Print material used | 3D-printed dosage forms[a] |
|---|---|---|---|---|
| Binder jetting | Assembled at Shanghai Folichif Co. Ltd | | Matrix: hydroxypropyl methylcellulose, poly(vinylpyrrolidone), and silicon dioxide<br>Binder: sodium lauryl sulphate/stearic acid/ethyl cellulose/Eudragit® RS-100 in ethanol<br>Model drug: paracetamol<br>Drug incorporated in the matrix | Tablets[4] |
| | | | Matrix: lactose, poly(vinylpyrrolidone), mannitol, and silicon dioxide<br>Binder: poly(vinylpyrrolidone) in ethanol<br>Model drug: paracetamol<br>Drug incorporated in the matrix | Fast disintegrating tablet[5,31] |
| | | | Matrix: poly(lactic acid)<br>Binder: ethanol and acetone<br>Model drug: levofloxacin<br>Drug incorporated in the matrix | Implants[44] |
| | | | Matrix: poly(lactic acid)<br>Binder: ethanol and acetone<br>Model drug: levofloxacin and rifampicin<br>Drug incorporated in the binder | Implants[46] |
| | In-house | | Matrix: cellulose<br>Binder: Eudragit® E-100 in ethanol/Eudragit® RLPO in acetone<br>Model drug: chlorpheniramine maleate | Immediate-extended release tablets[25] |
| | | | Matrix: cellulose/lactose<br>Binder: Eudragit® E-100 in ethanol/Eudragit® RLPO in acetone/poly(vinylpyrrolidone) in water<br>Model drug: chlorpheniramine maleate | Tablets[3] |

*continued*

**Table 5.2** (*continued*)

| 3DP technique | Printer: Manufacturer | Printer: Model | Print material used | 3D-printed dosage forms[a] |
|---|---|---|---|---|
| Binder jetting | In-house | | Matrix: lactose and Eudragit® L-100<br>Binder: Eudragit® E-100 in acetone and Eudragit® L-100 in ethanol<br>Model drug: diclofenac<br>Drug incorporated with the binder | Dual pulsatory tablets[25] |
| | | | Matrix: lactose and Eudragit® L-100<br>Binder for the active unit: Eudragit® RLPO in acetone<br>Binder for the quick dissolve unit: poly(vinylpyrrolidone) in water<br>Model drug: diclofenac<br>Drug incorporated with the binder | Break-away tablets[25] |
| | | | Matrix: lactose and Eudragit® L-100<br>Binder: Eudragit® L-100 in ethanol<br>Model drug: diclofenac<br>Drug incorporated with the binder | Enteric dual pulse tablets[25] |
| | | | Matrix: poly(vinylpyrrolidone) and hydroxypropyl methylcellulose<br>Binder for immediate release: pseudoephedrine hydrochloride in water<br>Binder for controlled release: triethyl citrate in ethanol<br>Model drug: pseudoephedrine hydrochloride<br>Drug incorporated with the binder | Zero-order controlled release dosage form[33] |
| | | | Matrix: polyethylene oxide<br>Binder: poly(ϵ-caprolactone)/acid-modified poly(ϵ-caprolactone) in chloroform | Tablets[2] |
| | Therics | Theriform | Matrix: poly(ϵ-caprolactone) and poly(lactide-*co*-glycolide)<br>Model drug: ethinyl estradiol | Implants[45] |

| | | | | |
|---|---|---|---|---|
| FDM | Cluovo Technologies | Delta-MK2 | Filament: poly(vinyl alcohol)<br>Model drug: glipizide<br>Drug incorporated by hot-melt extrusion | DuoTablet[27] |
| | ExtrusionBot | | Filament: poly(lactic acid)<br>Model drug: gentamicin<br>Drug incorporated by hot-melt extrusion | Hernia meshes[57] |
| | MakerBot | Replicator 2X | Eudragit® RL-100/poly(ϵ-caprolactone) | Tablets loaded with nanocapsules[37] |
| | | | Filament: ethylene vinyl acetate<br>Model drug: ϒ-indomethacin<br>Drug incorporated by hot-melt extrusion | Intrauterine systems and subcutaneous rods[9] |
| | | | Filament: Eudragit® (E PO)<br>Model drug: theophylline/5-aminosalicylic acid/captopril/prednisolone<br>Drug incorporated by hot-melt extrusion | Immediate release tablets[13] |
| | | | Filament: Eudragit® (RL/RS/E)/hydroxypropyl cellulose<br>Model drug: theophylline<br>Drug incorporated by hot-melt extrusion | Immediate and extended release tables[18] |
| | | | Filament: Flex EcoPLA™<br>Model drug: salicylic acid<br>Drug incorporated by hot-melt extrusion | Personalised topical drug delivery system[69] |
| | | | Filament: hydroxypropyl cellulose<br>Model drug: domperidone<br>Drug incorporated by hot-melt extrusion | Intragastric floating tablets[36] |
| | | | Filament: hydroxypropyl methylcellulose acetate succinate (LG/MG/HG)<br>Model drug: paracetamol<br>Drug incorporated by hot-melt extrusion | Modified release tablets[23] |
| | | | Filament: poly(vinyl alcohol)<br>Model drug: 5-aminosalicylic acid/4-aminosalicylic acid<br>Drug incorporated by incubation in the ethanolic drug solution | Modified release tablets[7] |
| | | | Filament: poly(vinyl alcohol)<br>Model drug: budesonide<br>Drug incorporated by hot-melt extrusion | Controlled release tablet[24] |

*continued*

**Table 5.2** (*continued*)

| 3DP technique | Printer: Manufacturer | Printer: Model | Print material used | 3D-printed dosage forms[a] |
|---|---|---|---|---|
| FDM | MakerBot | Replicator 2X | Filament: poly(vinyl alcohol)<br>Model drug: fluorescein<br>Drug incorporated by incubation in the ethanolic drug solution | Tablets[6] |
| | | | Filament: poly(vinyl alcohol)<br>Model drug: paracetamol<br>Drug incorporated by hot-melt extrusion | Tablets[8] |
| | | | Filament: poly(vinyl alcohol)<br>Model drug: paracetamol/caffeine<br>Drug incorporated by hot-melt extrusion | Tablets[26] |
| | | | Filament: poly(vinyl alcohol)<br>Model drug: paracetamol/caffeine<br>Drug incorporated by hot-melt extrusion | Caplet[22] |
| | | | Filament: poly(vinyl alcohol)<br>Model drug: prednisolone<br>Drug incorporated by incubation in the methanolic drug solution | Extended release tablets[19] |
| | | | Filament: poly(ϵ-caprolactone)<br>Model drug: indomethacin<br>Drug incorporated by hot-melt extrusion | Intrauterine systems[54] |
| | | | Filament: poly(ϵ-caprolactone)<br>Model drug: silver, copper, zinc.<br>Drug incorporated by hot-melt extrusion | Antimicrobial wound dressings[68] |
| | | | Filament: polyvinylpyrrolidone<br>Model drug: theophylline<br>Drug incorporated by hot-melt extrusion | Immediate release tablets[14] |
| | | | Hydroxypropyl cellulose | Tablets[21] |
| | | Replicator 5th Gen | Filament: poly(ϵ-caprolactone)<br>Model drug: estrone, estradiol, estriol, and progesterone<br>Drug incorporated by hot-melt extrusion | Intrauterine systems[55] |

| | | | | |
|---|---|---|---|---|
| FDM | Makerfarm | Prusa i3 | Filament: Hydroxypropyl cellulose (LF/EF)/hydroxypropyl methylcellulose/ethyl cellulose N14/Soluplus®/Eudragit® L100<br>Model drug: paracetamol<br>Drug incorporated by hot-melt extrusion | Controlled release tablet[34] |
| | UP! | Plus | Filament: poly(lactic acid)<br>Model drug: nitrofurantoin<br>Drug incorporated by hot-melt extrusion | Discs for biofilm prevention[58] |
| | Zmorph | Zmorph2.0SX | Filament: poly(vinyl alcohol)<br>Model drug: aripiprazole<br>Drug incorporated by hot-melt extrusion | Orodispersable films[35] |
| SL | Formlabs | Form 1+ | Resin: poly(ethylene glycol) diacrylate<br>Model drug: paracetamol/ibuprofen/4-aminosalicylic acid/salicylic acid | Modified release dosage forms[15] |
| | | | Resin: poly(ethylene glycol) diacrylate and polyethylene glycol<br>Drug: ibuprofen | Drug-loaded hydrogels[38] |
| | | | Resin: poly(ethylene glycol) diacrylate and polyethylene glycol<br>Model drug: salicylic acid | Personalised topical drug delivery system[69] |
| CLIP | Carbon | — | Resin: trimethylolpropane triacrylate, poly(ethylene glycol) dimethacrylate, polycaprolactone trimethacrylate, acrylic acid<br>Model drug: rhodamine | Microneedles[70] |
| PolyJet printing | Dimatix materials | DMP-2830 Fujifilm | Resin: poly(ethylene glycol) diacrylate<br>Model drug: ropinirole hydrochloride | Tablets[16] |
| SLS | Sintratec | Sintratec Kit | Matrix: Kollicoat® IR and Eudragit® L100-55<br>Model drug: paracetamol | Tablets[17] |

[a]The dosage forms have been named in this table as per the names proposed in their respective reports. In the case of the absence of any clearly proposed name, generic names have been used.

## References

1. J. Jacob, N. Coyle, T. G. West, D. C. Monkhouse, H. L. Surprenant and N. B. Jain, *US Pat.*, 9339489 B2, 2014.
2. B. M. Wu, S. W. Borland, R. A. Giordano, L. G. Cima, E. M. Sachs and M. J. Cima, *J. Controlled Release*, 1996, **40**, 77–87.
3. W. Katstra, R. Palazzolo, C. Rowe, B. Giritlioglu, P. Teung and M. Cima, *J. Controlled Release*, 2000, **66**, 1–9.
4. D. G. Yu, X. L. Yang, W. D. Huang, J. Liu, Y. G. Wang and H. Xu, *J. Pharm. Sci.*, 2007, **96**, 2446–2456.
5. D. G. Yu, X. X. Shen, C. Branford-White, L. M. Zhu, K. White and X. L. Yang, *J. Pharm. Pharmacol.*, 2009, **61**, 323–329.
6. A. Goyanes, A. B. Buanz, A. W. Basit and S. Gaisford, *Int. J. Pharm.*, 2014, **476**, 88–92.
7. A. Goyanes, A. B. Buanz, G. B. Hatton, S. Gaisford and A. W. Basit, *Eur. J. Pharm. Biopharm.*, 2015, **89**, 157–162.
8. A. Goyanes, P. R. Martinez, A. Buanz, A. W. Basit and S. Gaisford, *Int. J. Pharm.*, 2015, **494**, 657–663.
9. N. Genina, J. Holländer, H. Jukarainen, E. Mäkilä, J. Salonen and N. Sandler, *Eur. J. Pharm. Sci.*, 2016, **90**, 53–63.
10. M. K. Agarwala, V. R. Jamalabad, N. A. Langrana, A. Safari, P. J. Whalen and S. C. Danforth, *Rapid Prototyping J.*, 1996, **2**, 4–19.
11. A. Melocchi, F. Parietti, A. Maroni, A. Foppoli, A. Gazzaniga and L. Zema, *Int. J. Pharm.*, 2016, **509**, 255–263.
12. M. Alhijjaj, P. Belton and S. Qi, *Eur. J. Pharm. Biopharm.*, 2016, **108**, 111–125.
13. M. Sadia, A. Sośnicka, B. Arafat, A. Isreb, W. Ahmed, A. Kelarakis and M. A. Alhnan, *Int. J. Pharm.*, 2016, **513**, 659–668.
14. T. C. Okwuosa, D. Stefaniak, B. Arafat, A. Isreb, K.-W. Wan and M. A. Alhnan, *Pharm. Res.*, 2016, **33**, 2704–2712.
15. J. Wang, A. Goyanes, S. Gaisford and A. W. Basit, *Int. J. Pharm.*, 2016, **503**, 207–212.
16. E. A. Clark, M. R. Alexander, D. J. Irvine, C. J. Roberts, M. J. Wallace, S. Sharpe, J. Yoo, R. J. Hague, C. J. Tuck and R. D. Wildman, *Int. J. Pharm.*, 2017, **529**, 523–530.
17. F. Fina, A. Goyanes, S. Gaisford and A. W. Basit, *Int. J. Pharm.*, 2017, **529**, 285–293.
18. K. Pietrzak, A. Isreb and M. A. Alhnan, *Eur. J. Pharm. Biopharm.*, 2015, **96**, 380–387.
19. J. Skowyra, K. Pietrzak and M. A. Alhnan, *Eur. J. Pharm. Sci.*, 2015, **68**, 11–17.
20. *Guidance for Industry: Size, Shape, and Other Physical Attributes of Generic Tablets and Capsules*, P. Quality/CMC, USFDA, Maryland, USA, 2015.
21. A. Goyanes, M. Scarpa, M. Kamlow, S. Gaisford, A. W. Basit and M. Orlu, *Int. J. Pharm.*, 2017, **530**, 71–78.

22. A. Goyanes, M. Kobayashi, R. Martínez-Pacheco, S. Gaisford and A. W. Basit, *Int. J. Pharm.*, 2016, **514**, 290–295.
23. A. Goyanes, F. Fina, A. Martorana, D. Sedough, S. Gaisford and A. W. Basit, *Int. J. Pharm.*, 2017, **527**, 21–30.
24. A. Goyanes, H. Chang, D. Sedough, G. B. Hatton, J. Wang, A. Buanz, S. Gaisford and A. W. Basit, *Int. J. Pharm.*, 2015, **496**, 414–420.
25. C. Rowe, W. Katstra, R. Palazzolo, B. Giritlioglu, P. Teung and M. Cima, *J. Controlled Release*, 2000, **66**, 11–17.
26. A. Goyanes, J. Wang, A. Buanz, R. Martínez-Pacheco, R. Telford, S. Gaisford and A. W. Basit, *Mol. Pharm.*, 2015, **12**, 4077–4084.
27. Q. Li, H. Wen, D. Jia, X. Guan, H. Pan, Y. Yang, S. Yu, Z. Zhu, R. Xiang and W. Pan, *Int. J. Pharm.*, 2017, **525**, 5–11.
28. A. Abebe, I. Akseli, O. Sprockel, N. Kottala and A. M. Cuitiño, *Int. J. Pharm.*, 2014, **461**, 549–558.
29. S. V. Sastry, J. R. Nyshadham and J. A. Fix, *Pharm. Sci. Technol. Today*, 2000, **3**, 138–145.
30. W. Habib, R. Khankari and J. Hontz, *Crit. Rev. Ther. Drug Carrier Syst.*, 2000, **17**, 12.
31. D.-G. Yu, C. Branford-White, Y.-C. Yang, L.-M. Zhu, E. W. Welbeck and X.-L. Yang, *Drug Dev. Ind. Pharm.*, 2009, **35**, 1530–1536.
32. V. Bühler, *Polyvinylpyrrolidone for the Pharmaceutical Industry*, BASF Aktiengesellschaft, 2nd edn, 1992.
33. C.-C. Wang, M. R. Tejwani, W. J. Roach, J. L. Kay, J. Yoo, H. L. Surprenant, D. C. Monkhouse and T. J. Pryor, *Drug Dev. Ind. Pharm.*, 2006, **32**, 367–376.
34. J. Zhang, X. Feng, H. Patil, R. V. Tiwari and M. A. Repka, *Int. J. Pharm.*, 2017, **519**, 186–197.
35. W. Jamróz, M. Kurek, E. Łyszczarz, J. Szafraniec, J. Knapik-Kowalczuk, K. Syrek, M. Paluch and R. Jachowicz, *Int. J. Pharm.*, 2017, **533**, 413–420.
36. X. Chai, H. Chai, X. Wang, J. Yang, J. Li, Y. Zhao, W. Cai, T. Tao and X. Xiang, *Sci. Rep.*, 2017, **7**, 2829.
37. R. Beck, P. Chaves, A. Goyanes, B. Vukosavljevic, A. Buanz, M. Windbergs, A. Basit and S. Gaisford, *Int. J. Pharm.*, 2017, **528**, 268–279.
38. P. R. Martinez, A. Goyanes, A. W. Basit and S. Gaisford, *Int. J. Pharm.*, 2017, **532**, 313–317.
39. S. Stegemann, M. Gosch and J. Breitkreutz, *Int. J. Pharm.*, 2012, **430**, 197–206.
40. V. K. Pawar, S. Kansal, G. Garg, R. Awasthi, D. Singodia and G. T. Kulkarni, *Drug Delivery*, 2011, **18**, 97–110.
41. C. M. Lopes, C. Bettencourt, A. Rossi, F. Buttini and P. Barata, *Int. J. Pharm.*, 2016, **510**, 144–158.
42. A. Dash and G. Cudworth II, *J. Pharmacol. Toxicol. Methods*, 1998, **40**, 1–12.
43. A. Göpferich, *J. Controlled Release*, 1997, **44**, 271–281.
44. W. Huang, Q. Zheng, W. Sun, H. Xu and X. Yang, *Int. J. Pharm.*, 2007, **339**, 33–38.

45. S. Lin, P.-Y. Chao, Y. W. Chien, A. Sayani, S. Kumar, M. Mason, T. West, A. Yang and D. Monkhouse, *AAPS PharmSciTech*, 2001, **2**, 55–65.
46. W. Wu, Q. Zheng, X. Guo and W. Huang, *J. Wuhan Univ. Technol., Mater. Sci. Ed.*, 2009, **24**, 977.
47. M. M. Baum, I. Butkyavichene, J. Gilman, S. Kennedy, E. Kopin, A. M. Malone, C. Nguyen, T. J. Smith, D. R. Friend and M. R. Clark, *J. Pharm. Sci.*, 2012, **101**, 2833–2843.
48. S. L. Fialho, F. Behar-Cohen and A. Silva-Cunha, *Eur. J. Pharm. Biopharm.*, 2008, **68**, 637–646.
49. Y. A. Han, M. Singh and B. B. Saxena, *Contraception*, 2007, **76**, 132–138.
50. R. K. Malcolm, K.-L. Edwards, P. Kiser, J. Romano and T. J. Smith, *Antiviral Res.*, 2010, **88**, S30–S39.
51. N. Zheng, M. Zhou, C. Du, S. Wang and W. Lu, *Arch. Pharmacal Res.*, 2013, **36**, 1487–1493.
52. J. A. Weisman, J. C. Nicholson, K. Tappa, U. Jammalamadaka, C. G. Wilson and D. K. Mills, *Int. J. Nanomed.*, 2015, **10**, 357.
53. E. Espey and T. Ogburn, *Obstet. Gynecol.*, 2011, **117**, 705–719.
54. J. Holländer, N. Genina, H. Jukarainen, M. Khajeheian, A. Rosling, E. Mäkilä and N. Sandler, *J. Pharm. Sci.*, 2016, **105**, 2665–2676.
55. K. Tappa, U. Jammalamadaka, D. H. Ballard, T. Bruno, M. R. Israel, H. Vemula, J. M. Meacham, D. K. Mills, P. K. Woodard and J. A. Weisman, *PLoS One*, 2017, **12**, e0182929.
56. X. Ding, X. He, C. Xue, C. Wu, L. Xie, T. Chen, J. Wei, X. Cheng and X. Wang, *RSC Adv.*, 2016, **6**, 76785–76788.
57. D. H. Ballard, J. A. Weisman, U. Jammalamadaka, K. Tappa, J. S. Alexander and F. D. Griffen, *Surgery*, 2017, **161**, 1479–1481.
58. N. Sandler, I. Salmela, A. Fallarero, A. Rosling, M. Khajeheian, R. Kolakovic, N. Genina, J. Nyman and P. Vuorela, *Int. J. Pharm.*, 2014, **459**, 62–64.
59. J. J. Water, A. Bohr, J. Boetker, J. Aho, N. Sandler, H. M. Nielsen and J. Rantanen, *J. Pharm. Sci.*, 2015, **104**, 1099–1107.
60. Z. A. Starosolski, J. H. Kan, S. D. Rosenfeld, R. Krishnamurthy and A. Annapragada, *Pediatr. Radiol.*, 2014, **44**, 216–221.
61. N. Xu, F. Wei, X. Liu, L. Jiang, H. Cai, Z. Li, M. Yu, F. Wu and Z. Liu, *Spine*, 2016, **41**, E50–E54.
62. R. Van Noort, *Dent. Mater.*, 2012, **28**, 3–12.
63. R. Ricotti, A. Vavassori, A. Bazani, D. Ciardo, F. Pansini, R. Spoto, V. Sammarco, F. Cattani, G. Baroni and R. Orecchia, *Phys. Medica*, 2016, **32**, 1698–1706.
64. C. L. Ventola, *Pharmacol. Ther.*, 2014, **39**, 704.
65. H. H. Malik, A. R. J. Darwood, S. Shaunak, P. Kulatilake, A. A. El-Hilly, O. Mulki and A. Baskaradas, *J. Surg. Res.*, 2015, **199**, 512–522.
66. C. M. B. Ho, S. H. Ng and Y.-J. Yoon, *Int. J. Precis. Eng. Manuf.*, 2015, **16**, 1035–1046.
67. L. K. Prasad and H. Smyth, *Drug Dev. Ind. Pharm.*, 2016, **42**, 1019–1031.
68. Z. Muwaffak, A. Goyanes, V. Clark, A. W. Basit, S. T. Hilton and S. Gaisford, *Int. J. Pharm.*, 2017, **527**, 161–170.

69. A. Goyanes, U. Det-Amornrat, J. Wang, A. W. Basit and S. Gaisford, *J. Controlled Release*, 2016, **234**, 41–48.
70. A. R. Johnson, C. L. Caudill, J. R. Tumbleston, C. J. Bloomquist, K. A. Moga, A. Ermoshkin, D. Shirvanyants, S. J. Mecham, J. C. Luft and J. M. DeSimone, *PLoS One*, 2016, **11**, e0162518.
71. Y.-C. Kim, J.-H. Park and M. R. Prausnitz, *Adv. Drug Delivery Rev.*, 2012, **64**, 1547–1568.
72. S. Henry, D. V. McAllister, M. G. Allen and M. R. Prausnitz, *J. Pharm. Sci.*, 1998, **87**, 922–925.

CHAPTER 6

# 3D Printing in Biochemistry

## 6.1 Introduction

Biochemical assays play an important part in understanding the biochemical processes and mechanisms in all living organisms. Today, these assays can be found in new point-of-care diagnostic devices, aiding drug discovery, helping to develop 'organ-on-a-chip' and tissue constructs, and are used in the fields of genomics, proteomics, and metabolomics. With increasing human populations leading to the increased threat of epidemics and disease, together with a growing demand for more effective drugs, transplants, medical implants, and personalised medicines, the need to develop simple and rapid biochemical assays is only set to continue at pace. High sensitivity and high throughput biochemical assays are a major focus of the microfluidic device and integrated lab-on-a-chip research communities.[1] Within this field, new devices and platforms can automate complex assays, work with low sample volumes, and provide access to the cellular micro-environment. As discussed in previous chapters, 3D printing offers numerous advantages for the production of microfluidic platforms, and hence it is gaining increasing application in biochemistry.

Within this chapter, the focus is on the use of 3D printing techniques to aid greater understanding of complex biological and biochemical systems, including cellular analysis, biomolecular analysis, and biological drug screening. This chapter also provides a brief discussion on the toxicity of different polymers and resins that are used for 3D printing and the ways to improve their biocompatibility. Discussion of bioprinting is not included herein, as this is covered elsewhere in significant detail in

3D Printing in Chemical Sciences: Applications Across Chemistry
By Vipul Gupta, Pavel Nesterenko and Brett Paull

Published by the Royal Society of Chemistry, www.rsc.org

numerous recent reviews and books.[2–4] Moreover, the discussion has been focused on biochemical applications rather than biotechnological applications (such as prosthetics), which again can be found covered elsewhere in recent reviews.[5–7]

## 6.2 Cell Biology

The cell is the basic structural and functional unit of living organisms. Hence, cell biology plays a crucial role in medical diagnosis and biomedical engineering.[8] Microfluidic platforms have been used for cellular assays,[9,10] cell–cell communication studies,[11,12] and organ-on-a-chip development.[13] However, the expensive and two-dimensionally restrictive nature of traditionally used soft-lithography has limited the capabilities and applications of these biochemical microfluidic devices. As discussed in chapter 2, 3D printing provides a unique opportunity to produce multi-functional and three-dimensional biocompatible microfluidic platforms. These platforms have been used to study cell biology through cellular assays, cell sorting, and electroporation procedures, as discussed below.

### 6.2.1 Cellular Assays

Cellular assays are used to identify and quantify cellular responses such as metabolism, migration, proliferation, and cell death in the presence of different external stimuli. These assays can provide extensive information about the physiological state of a group of cultured cells under various conditions. However, they often require the use of microfluidic platforms, which are either impractical or very expensive to produce with conventional fabrication techniques. Recently, 3D-printed microfluidic platforms have been developed to provide low-cost and simple arrangements for such assays.

2D cell cultures produced in a petri dish are not well suited for culture sampling or many downstream applications that require chemical or enzymatic detachment. Moreover, they do not adequately mimic the 3D architecture and the physiological functions of the living tissues.[8] Different techniques have been studied to deliver 3D cell culture systems, such as the hanging drop method, gyratory rotation, and liquid overlay cultures.[14] An alternative technology is to cultivate cells around degradable scaffolds[15] or biocompatible matrices,[16] which allow a better diffusion of oxygen and nutrients.[17] Microfluidic devices, often made from glass, such as co-flow and co-extrusion micro-devices, have been used to generate such hydrogel cell-enclosing spheres[18] and core–shell structure capsules.[19] However, these devices require the three-dimensional arrangement of multiple co-axial capillaries for sample injection and subsequent capsule formation. Recently, Alessandri *et al.*[20] used SL printing to produce a new co-extrusion device (Figure 6.1) for generating cell-encapsulated hollow

**Figure 6.1** 3D-printed co-extrusion device for generating cell-encapsulated hollow hydrogel spheres (scale bar: 1 mm). Reproduced from ref. 20 with permission from The Royal Society of Chemistry.

hydrogel spheres and 3D cell cultures. It consisted of a narrow 200 µm channel and was used to encapsulate human neural stem cells derived from human-induced pluripotent stem cells. The 3D-printed co-extrusion device had 20 times smaller dead volume than the hand-made glass analogue, and a higher percentage of the capsule was available for the cell culture.[21]

Microwells allow *in vitro* cellular assays of cell cultures ranging from an individual cell to 3D functional cultures. They require small sample volume, perform high-throughput analysis, and are used for low-cost assays for drug discovery processes.[22–24] Several microfabrication techniques such as soft lithography,[25–28] biochemical patterning,[29,30] direct printing,[31] and laser processing[32,33] have been used for the production of microwells for cell culture assays. However, these approaches are typically rather time-consuming, labour-intensive, expensive, and inefficient.[8] Hence, Yang *et al.*[8] investigated the use of SL printing to produce miniaturised poly(ethylene glycol) diacrylate microwell chips for the culturing and multi-dimensional analysis of single cells (1D), cell clusters (2D), and cell spheroids (3D). Circular, triangular, square, and star-shaped microwells were developed, as shown in Figure 6.2. Each well was 15 µm in diameter and 10 µm in depth. These dimensions were selected to match single-cell volumes and to control the shape of the cell culture (by modulating the microwell shape). The 3D-printed microwell chip successfully facilitated the study of cell behaviour and drug response with 1D, 2D, and 3D cell cultures.

**Figure 6.2** Graphical representation of the different shaped 3D-printed microwells. Reproduced from ref. 8 with permission from The Royal Society of Chemistry.

Spence's group has developed a 3D-printed fluidic device to incorporate Transwell membrane cell culture inserts for cellular assays and cell–cell communication studies (Figure 6.3).[34] The device was 3D-printed with a PolyJet printer using a polyacrylate-based material and consisted of six parallel channels with small openings on the top surface. The openings were used to incorporate the cell culture inserts, such that the semi-permeable membrane of the inserts provides an interface between the fluidic channel and the cell culture. This device was used to determine the release of adenosine triphosphate (ATP) from stored blood erythrocytes.[35] It utilised polystyrene membrane based inserts with a pore size of 0.4 μm. The inserts were sealed in the 3D-printed device by wrapping polytetrafluoroethylene tape on the insert side-walls. Its successful use in a week-long study compared well to single-use poly(dimethylsiloxane) devices. The multiple use 3D-printed devices lowered the cost of experiments and reduced the variability in measurements over time. The device was further used to study tissue–tissue communication between rat INS-1 cells and the endothelium.[36] This 3D-printed device allowed the investigation of a complex physiological process such as a cellular pathway in a controlled manner.

Huang *et al.*[37] have studied the migration and physical behaviour of cancer cells using a 3D-printed replica of a rat's vascular morphology. Rat capillary structures were replicated using a biocompatible poly(ethylene glycol) diacrylate hydrogel, printed using an SL printer (Figure 6.4). The

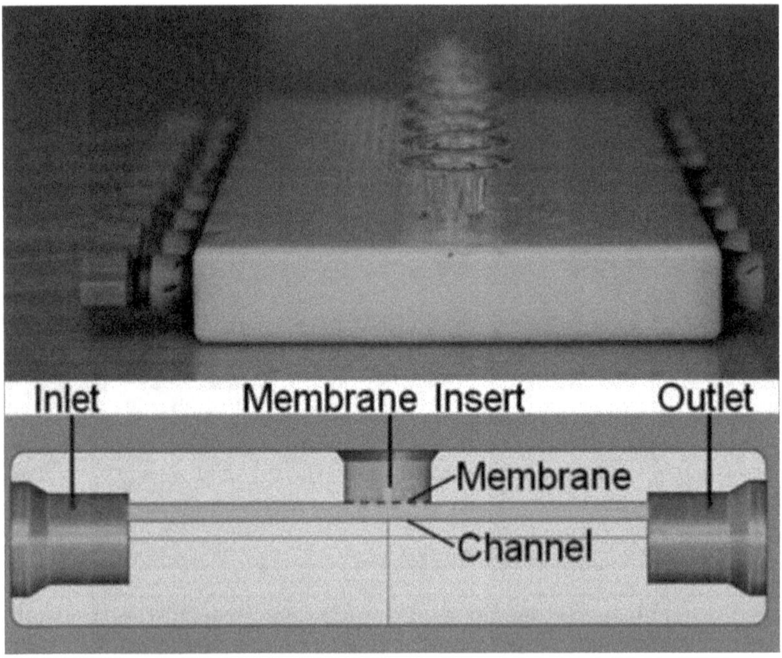


**Figure 6.3** 3D-printed fluidic device to incorporate Transwell membrane inserts. Reproduced from ref. 34 with permission from American Chemical Society, Copyright 2013.

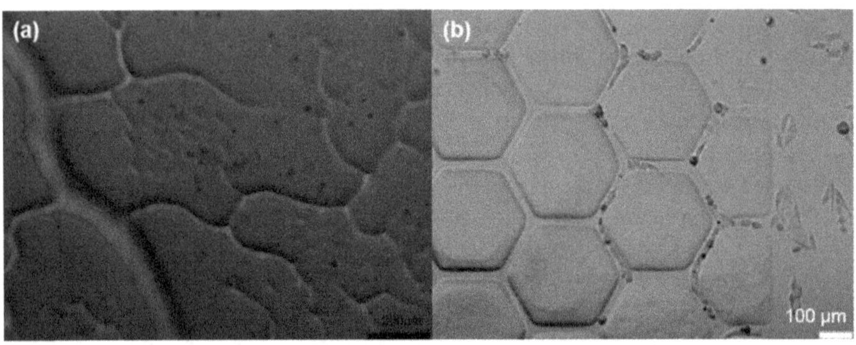


**Figure 6.4** 3D-printed replica of a rat's vascular morphology: (a) microstructure based on a real microCT scan of rat capillaries, and (b) honeycomb structure with seeded HeLa cells. Reproduced from ref. 37 with permission from Springer Nature, Copyright 2013.

channel widths of the capillary structures were controlled by changing the focal plane during continuous exposure. The SL printer successfully produced a precise morphological replica of the rat's vascular structure, which was obtained from a microCT scan of rat capillaries.

### 6.2.2 Cell Sorting

Cell sorting is an essential technique in biological research and medical diagnostics. The majority of microfluidic cell sorting techniques are based on hydrodynamic methods, for example, pinched flow fractionation,[38,39] deterministic lateral displacement,[40,41] hydrodynamic filtration,[42,43] inertial separation,[44–48] and disc microfluidics.[49,50] These techniques are limited by long sample processing times, clogging, high cost, low recovery, and the loss of cell viability.[51–53] However, density-based cell sorting can offer multiple advantages, such as low cell-to-cell variability in a cell type, minimal *g*-force induced cell damage, and low instrumentation costs. Recently, Grover and co-workers performed density-based cell sorting using a polyacrylate composite SL-printed microfluidic chip, as shown in Figure 6.5.[54] The chip was used to sort blood cells based on their type (white and red) and polyethylene and polystyrene microbeads based on a small (2%) density difference. It resulted in an enrichment factor of 1000× for white blood cells and was able to separate microbeads with an efficiency of up to 99%.

Further, in the same year, Syed *et al.*[55] reported the development of a PolyJet-printed miniaturised hydrocyclone for high-throughput cell sorting (Figure 6.6).[55] Four hydrocyclones were printed with a polyacrylate-based material with different overflow diameters, underflow diameters, and number of inlets. Larger underflow diameters were found to give a lower particle cut size, and larger overflow diameters resulted in lower separation efficiencies. These hydrocyclones were used for harvesting marine microalgae giving enrichment factors of up to 7.3 in 11 minutes.

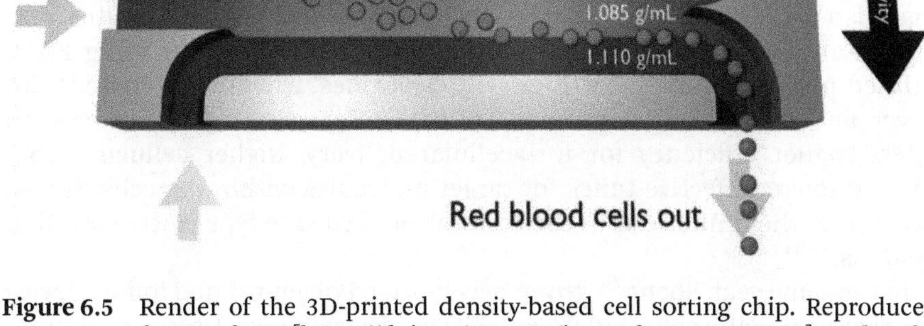


**Figure 6.5** Render of the 3D-printed density-based cell sorting chip. Reproduced from ref. 54 [https://doi.org/10.1371/journal.pone.0180520] under the terms of a CC By 4.0 license [https://creativecommons.org/licenses/by/4.0/].

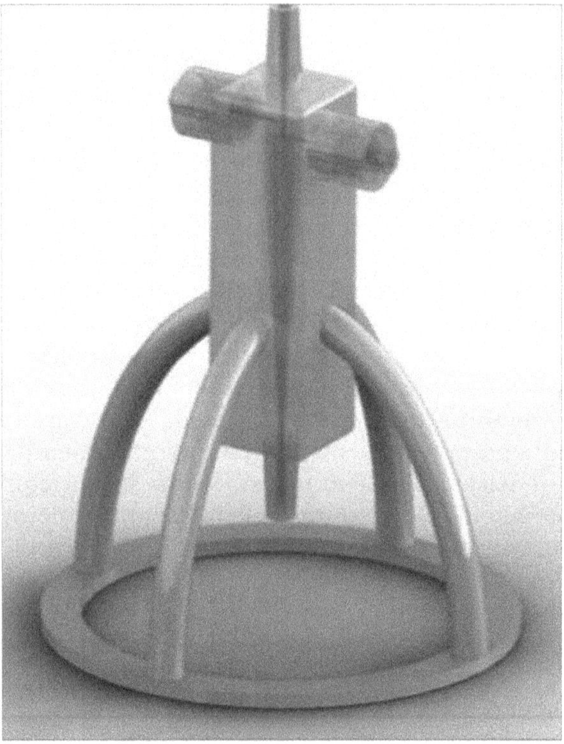

**Figure 6.6** Render of the 3D-printed miniaturised hydrocyclone. Reproduced from ref. 55 with permission from The Royal Society of Chemistry.

### 6.2.3 Electroporation

Electroporation is often used in microbiology to introduce foreign chemicals, drugs, or genetic materials into cells. It is usually performed by applying a high voltage across cells suspended in an electroporation cuvette.[56,57] However, electroporation cuvettes may undergo metal ion dissolution,[56,57] generate excess heat,[57,58] vary local pH,[56,57] distort the electric field,[57] and may also result in low electroporation efficiency and low cell viability.[59] In 2015, Kang *et al.*[59] developed a micro-electroporation system using FDM-printed poly(lactic acid) interdigitated electrodes, which were coated with silver and aluminium (Figure 6.7).[59] The micro-electroporation system offers higher efficiency for intracellular delivery, higher cellular viability, and longer effective times for target molecules within the cells. It also overcomes the limitations of the conventional cuvette type electroporation systems.[56–58]

In the same year, Spence's group developed a PolyJet-printed fluidic device for electrical cell lysing (Figure 6.8).[60] The device allowed easy integration of electrodes through threaded ports, as shown in Figure 6.8(C). Cell lysing of endothelial cells was performed with a low-voltage (500 V max) power

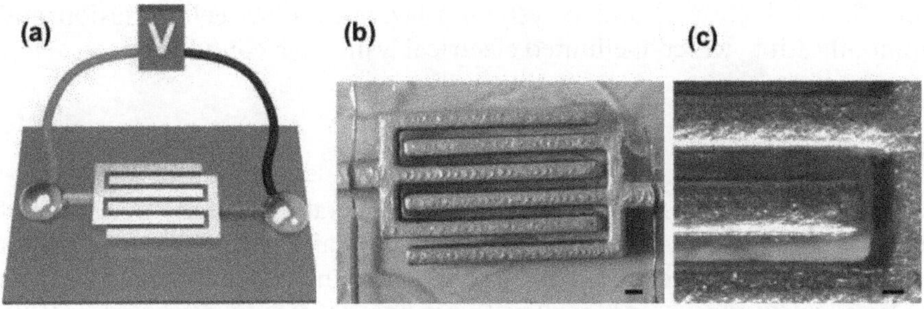


**Figure 6.7** 3D-printed micro-electroporation system: (a) schematic, (b) photograph (scale bar: 1 mm), (c) photograph (scale bar: 500 μm). Reproduced from ref. 59 [https://doi.org/10.1038/srep15835] under the terms of a CC BY 4.0 license [https://creativecommons.org/licenses/by/4.0/].

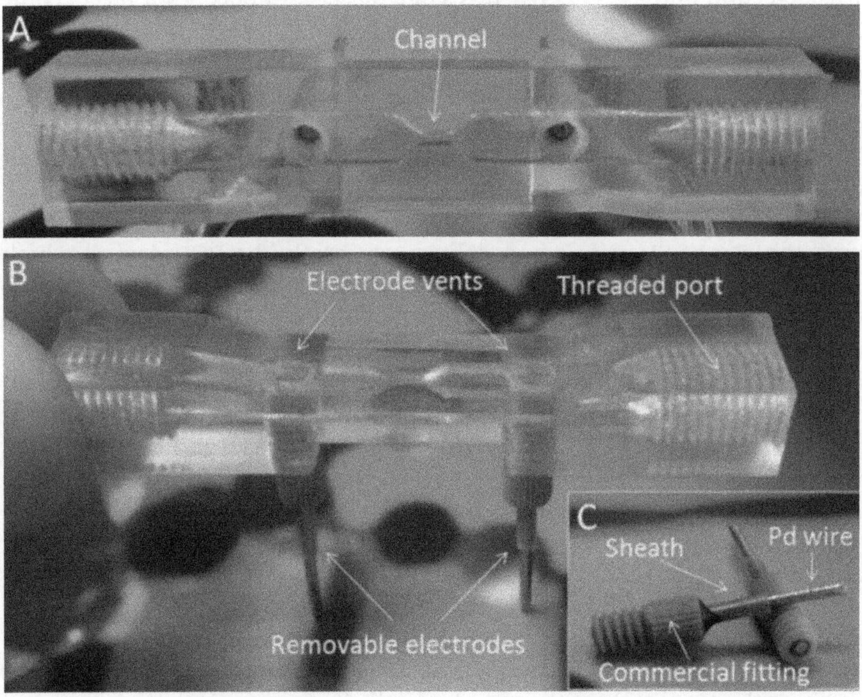


**Figure 6.8** 3D-printed fluidic device for electrical cell lysing: (A) top view, (B) side view, and (C) electrodes with threaded fittings. Reproduced from ref. 60 with permission from American Chemical Society, Copyright 2015.

source within a small section (1 mm wide, 0.8 mm high, and 2 mm long) of the channel. The device was printed in a polyacrylate-based material, whose surface chemistry was not completely understood, due to the proprietary nature of the print material; hence, the fluidic channels were coated with poly(dimethylsiloxane) and polystyrene layers to allow cell adhesion and immobilisation, which facilitated electrical lysing in a confined space.

## 6.3 Molecular Biology

Rapid disease diagnosis is an essential tool to enable the early treatment of patients and the fight to reduce the progression and spread of disease. This is particularly the case for the most destructive infectious diseases, such as malaria, tuberculosis, and acquired immune deficiency syndrome (AIDS). The availability of reliable and low-cost diagnostic tests in local clinics, particularly in regions of limited medical resources and quarantine facilities is of significant importance.[61]

The polymerase chain reaction (PCR) is widely used for the amplification and ultra-sensitive and selective detection of DNA. Its high sensitivity allows virus detection in the very early stages of a disease;[62] hence, PCR has become a common technique for rapid disease diagnosis.[63,64] The Centers for Disease Control and Prevention (CDC) and the World Health Organization (WHO) highly recommend the use of PCR for diseases such as malaria and AIDS.[65,66]

Microfluidic PCR devices have recently gained attention because of their low cost, low reagent and energy consumption, high surface-to-volume ratio, and short analysis time.[62,67–69] Accordingly, Park *et al.*[70] reported upon the use of SL printing to deliver a microfluidic PCR device, which consisted of a 1.47 m long serpentine channel (0.25 mm in depth and 0.45 mm in width) within a footprint of 30 mm × 40 mm. The channel was printed in an open configuration to allow complete removal of the uncured resin. The channel walls had a surface roughness of 0.2 μm and a maximum bump height of 1.7 μm. The device was printed using a polyacrylate-based material and integrated with heaters for PCR. The long channel length allowed 27 thermal cycles along three temperature zones (Figure 6.9): denaturation (90 °C to 95 °C), annealing (55 °C to 65 °C), and extension (70 °C to 77 °C). The reported cost of printing the device was less than $2.

Recently, in 2017, Mulberry *et al.*[61] reported upon the development of a portable and low-cost 3D-printed quantitative PCR (qPCR) device (Figure 6.10).[61] qPCR allows the quantification of DNA during its amplification. It monitors the amplitude of the signal (usually fluorescence) that is generated from DNA-intercalating probes.[66] The device consists of thirteen customised FDM-printed components, including the casing, heating element holders, fan and cooling systems, and the fluorescence detection unit. The device dimensions were 12 cm by 7 cm by 6 cm, and it weighed only 214 grams. It was used for the quantification of lentivirus, which is similar to the human immunodeficiency virus (HIV).

Priye *et al.*[71] 3D-printed snap-on centrifuge rotors to facilitate drone-based centrifugation to prepare samples for on-drone PCR (Figure 6.11). The rotors

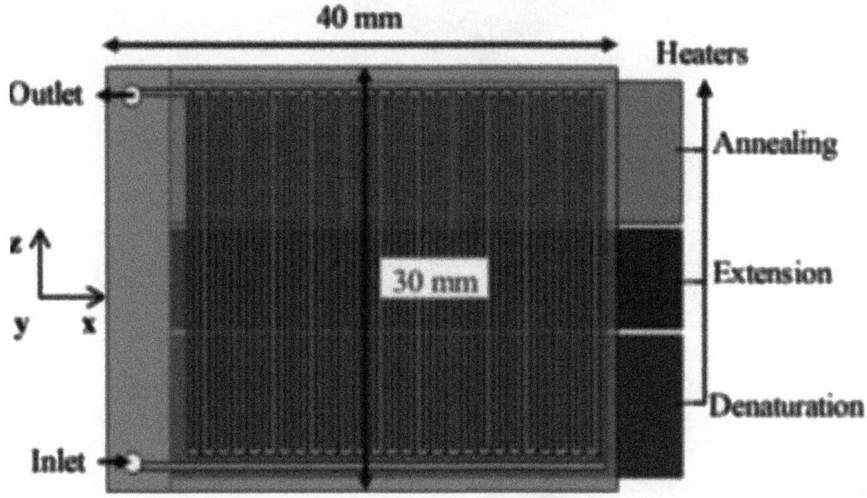


**Figure 6.9** 3D-printed polyacrylate-based PCR device with a 1.47 m long channel running 27 times through three temperature zones. Reproduced from ref. 70 with permission from Elsevier, Copyright 2017.

were compatible with standard spin column based extraction/purification kits and were FDM-printed with a low infill to minimise their weight. The performance of these rotors was comparable to a benchtop centrifuge for pre-concentrating nanosized colloidal materials from suspensions.

3D-printed components have also been used for the development of low-cost rapid diagnostic tests for malaria. Dirkzwager *et al.*[72] used polyacrylate-based SL-printed components to develop simple paper-based syringe tests and magnetic bead-based well tests for malaria (Figure 6.12). The tests were based on the aptamer-tethered enzyme capture (APTEC) assay for *Plasmodium falciparum* lactate dehydrogenase (*Pf*LDH). Both tests were able to detect recombinant *Pf*LDH at ng $mL^{-1}$ concentration level in 20 μL samples.

A range of novel three-dimensionally complex hydrogel biosensors have been developed by Mandon *et al.*, as shown in Figure 6.13.[73] These biosensors were SL-printed in a poly(ethylene glycol) diacrylate hydrogel. Different biosensor shapes with mono-, di-, and tri-component configurations were created, such as a fanciful ball, 3D pixel, propeller, and twin propeller. The fanciful ball shape was an airy structure with thin edifices (Figure 6.13(a)), the 3D pixel shape represented a dense structure (Figure 6.13(b)), and the propeller shape represented a large structure with thin walls (Figure 6.13(c)). Moreover, the propeller structure could provide increased sensitivity by stirring the assay solution by rotating around its axis. These three shapes, along with others, were printed with a hydrogel containing a combination of glucose oxidase and peroxidase enzymes, and these sensors were used for a sequential chemiluminescence assay of glucose and luminol. This

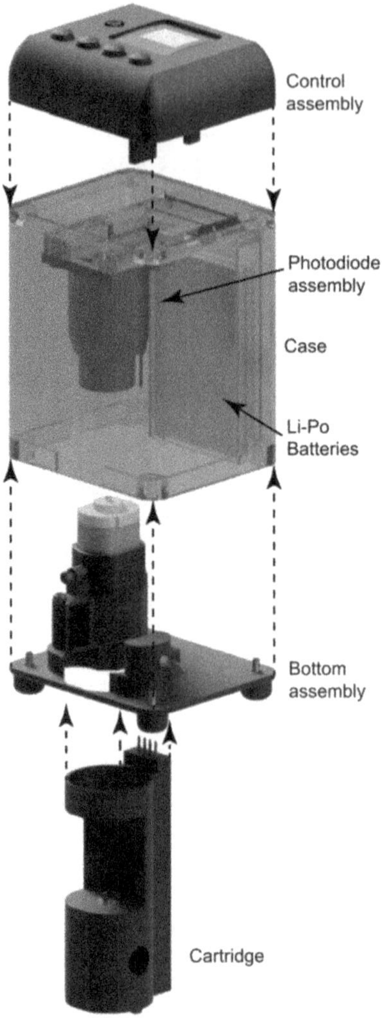


**Figure 6.10** Render of the 3D-printed miniaturised qPCR device. Reproduced from ref. 61 [https://doi.org/10.1371/journal.pone.0179133] under the terms of a CC BY 4.0 license [https://creativecommons.org/licenses/by/4.0/].

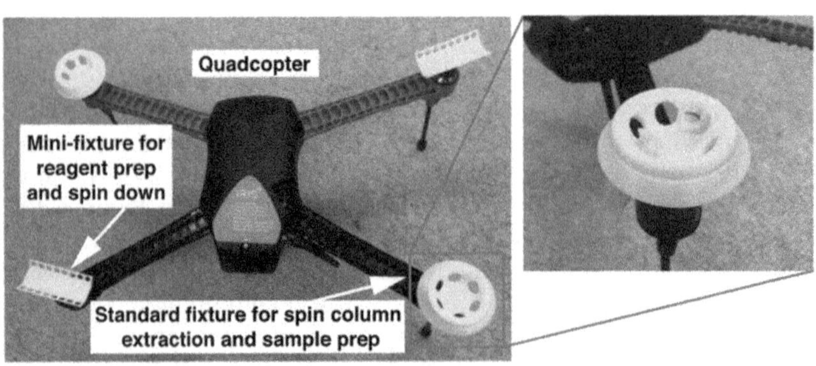


**Figure 6.11** 3D-printed snap-on rotors for an on-drone PCR. Reproduced from ref. 71 with permission from American Chemical Society, Copyright 2016.

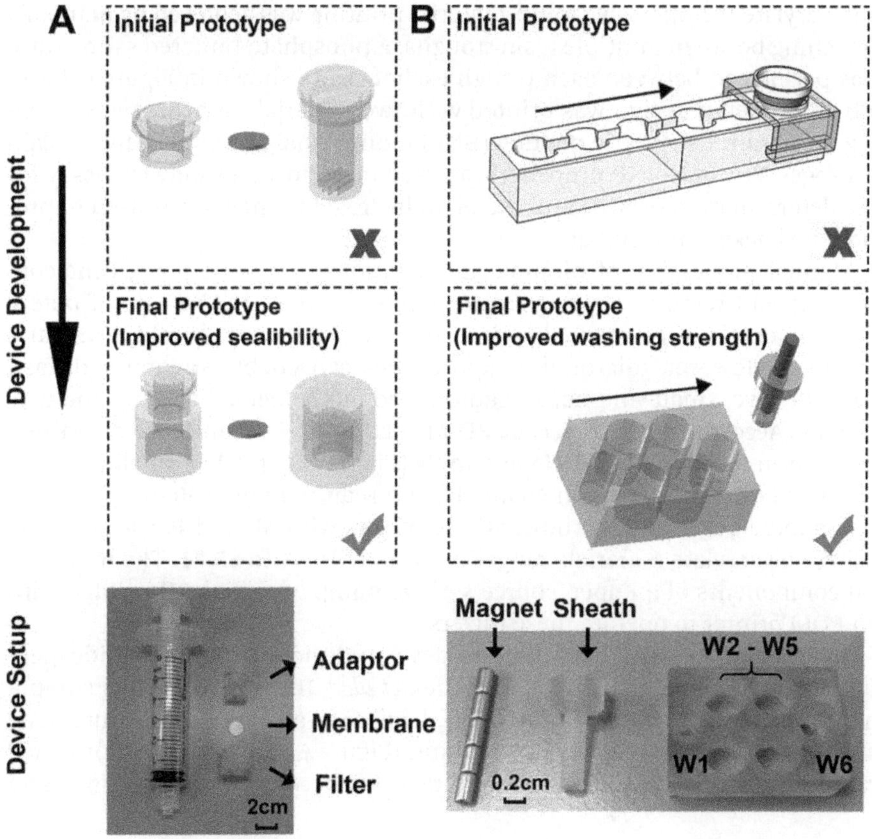


**Figure 6.12** Development of syringe and well tests using 3D-printed components: (a) paper-based syringe tests and (b) magnetic bead-based well tests. Reproduced from ref. 72 with permission from American Chemical Society, Copyright 2016.

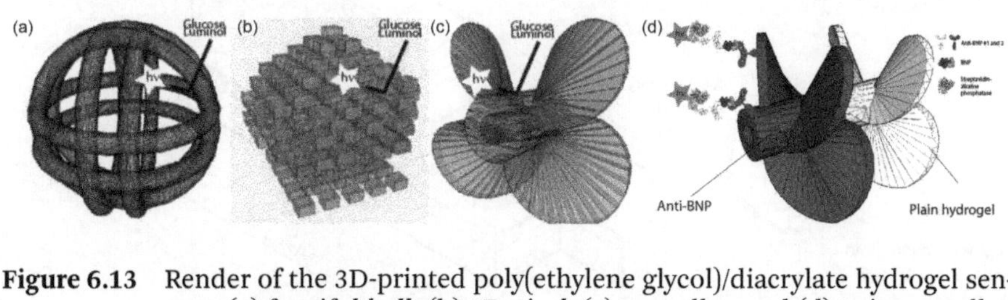


**Figure 6.13** Render of the 3D-printed poly(ethylene glycol)/diacrylate hydrogel sensors: (a) fanciful ball, (b) 3D pixel, (c) propeller, and (d) twin propeller. Reproduced from ref. 73 with permission from American Chemical Society, Copyright 2016.

demonstrated the potential of 3D printing to develop biosensors with previously unimaginable designs.

The same group also successfully SL-printed multi-component structures using multiple resins, specifically hydrogel with enzymes, plain hydrogel, and acrylate.[73] Impressive multi-material printing was achieved by manually switching between multiple resin troughs. A phosphate-buffered saline rinse was performed between each trough exchange. As shown in Figure 6.13(d), a twin propeller design was printed with two materials, a hydrogel containing anti-brain natriuretic peptide (BNP) monoclonal antibodies and a plain hydrogel. The anti-BNP propeller was used to perform a sandwich assay for the determination of BNP, and the plain hydrogel propeller was used to provide the background signal.

Manual processing of biomolecule solutions can be tedious, time-consuming, and require relatively large sample volumes. In the case of potential biohazards, such manual approaches are particularly cumbersome and complex. However, microfluidic approaches can enable streamlined, fast, cost-effective, small-volume, and automated biomolecule handling and processing. Accordingly, Patrick *et al.* 3D-printed micro- and milli-fluidic mixers to perform 'Golden Gate' DNA assembly (Figure 6.14).[74] The smallest mixer consisted of narrow 220 μm channels, and required only 490 nL of sample. The quoted costs of 3D printing these mixers with SL and PolyJet printers using polyacrylate materials ranged from only $0.61 to $5.71. The mechanical components of an open source syringe pump were also 3D-printed with an FDM printer to operate these mixers.

3D printing has also been used to develop devices to study peptide–peptide interactions. For example, Michalek *et al.*[75] 3D-printed various components of a flow-through electrochemical device in an acrylonitrile butadiene styrene copolymer using an FDM printer (Figure 6.15). The device provided rapid automated analysis of the interactions between hemagglutinin and

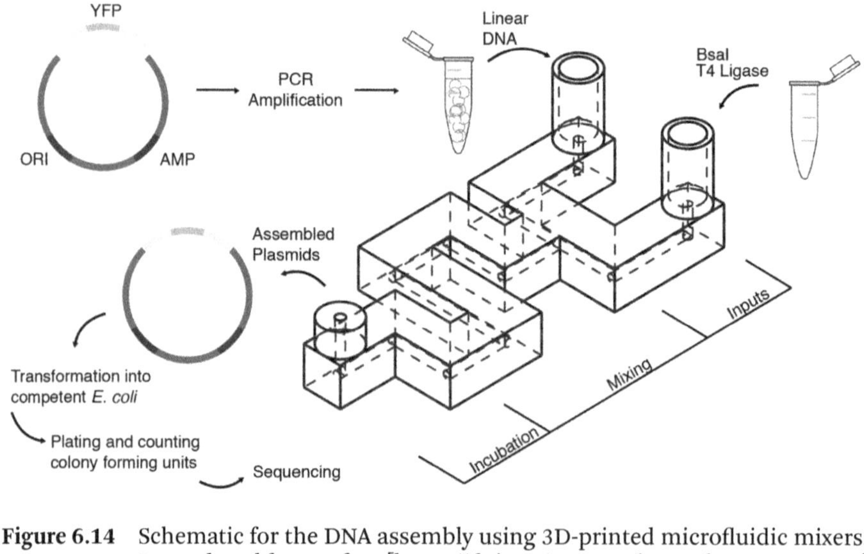


**Figure 6.14** Schematic for the DNA assembly using 3D-printed microfluidic mixers. Reproduced from ref. 74 [https://doi.org/10.1371/journal.pone.0143636] under the terms of a CC BY 4.0 license [https://creativecommons.org/licenses/by/4.0/].

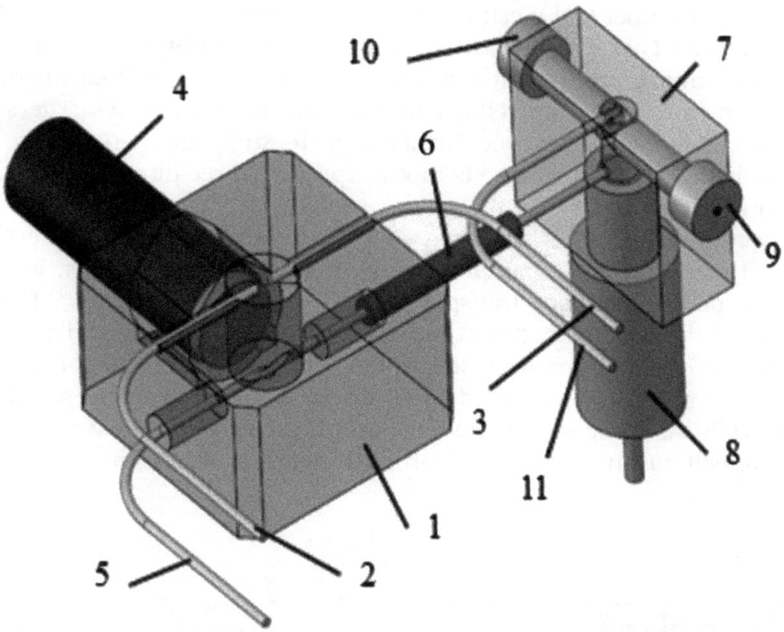


**Figure 6.15** Schematic of the flow-through electrochemical device: (1) 3D-printed cell, (2) sample inlet, (3) washing solution inlet, (4) electromagnet, (5) waste output, (6) connector to the electrochemical detector, (7) electrochemical detector, (8) AgCl reference electrode, (9 and 10) carbon electrodes, and (11) detector output. Reproduced from ref. 75 with permission from John Wiley and Sons, Copyright © 2015 WILEY-VCH Verlag GmbH & Co. KGaA, Weinheim.

metallothionein derived peptides based upon electrochemical signals generated from PbS and CuS quantum dots.

## 6.4 *In vitro* Drug Screening

Development of new drugs is a lengthy and costly process, which typically has a very low final success rate. Hence, *in vitro* drug screening platforms are extremely important within the drug development process for screening drug efficacy before animal testing.[76–78] Current drug screening platforms are being developed based upon three-dimensionally biomimetic complex microstructures and microfluidic networks, to produce improved environments for cell cultures and 'organs-on-a-chip' based investigations. As discussed above, the advent of 3D printing has greatly accelerated research in this area.

Spence and co-workers have reported the use of the 3D-printed fluidic device mentioned above (Figure 6.3) to study dynamic *in vitro* pharmacokinetic profiles of drugs.[79] The use of Transwell inserts allowed free back and forth diffusion of model drugs, such as fluorescein and levofloxacin, between the fluidic channel and a cell culture for studying both loading and clearance profiles. The 3D-printed fluidic device offered automated and parallel pharmacokinetic profiling. It also allowed access to the cellular micro-environment for measuring cellular pharmacodynamics profiles. The same 3D-printed fluidic device was further used to study the penetration and metabolism of combination chemotherapies in three-dimensional colon cancer cell cultures.[80]

However, cell cultures on Transwell inserts often provide a questionable estimation of the absorption of actively transported drugs, as they poorly express membrane transporters such as P-glycoprotein.[81,82] Hence, Shen *et al.*[83] 3D-printed special inserts to produce hanging cell cultures (Figure 6.16). The inserts were 3D-printed with polycarbonate using a PolyJet printer, and they were used to study the pharmacokinetic profile of Caco-2 cells. As compared to the Caco-2 cell cultures on Transwell inserts, hanging cell cultures on the 3D-printed inserts resulted in 30 to 100%

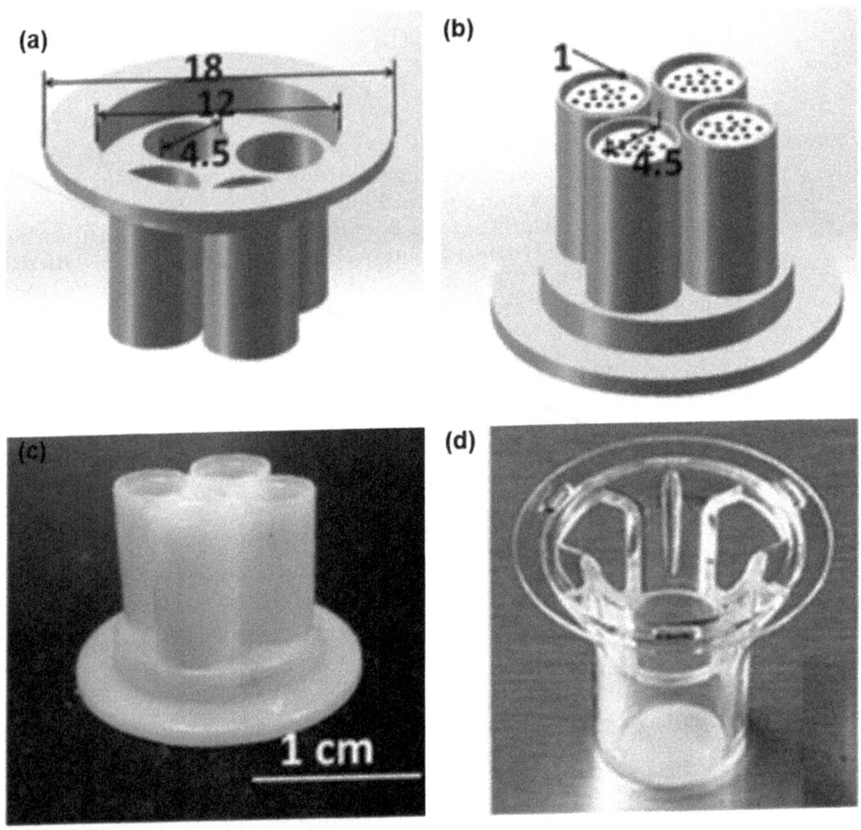


**Figure 6.16** 3D-printed insert for a hanging cell culture: (a and b) renders, (c) photograph, and (d) Transwell insert. Reproduced from ref. 83 with permission, Copyright © IOP Publishing.

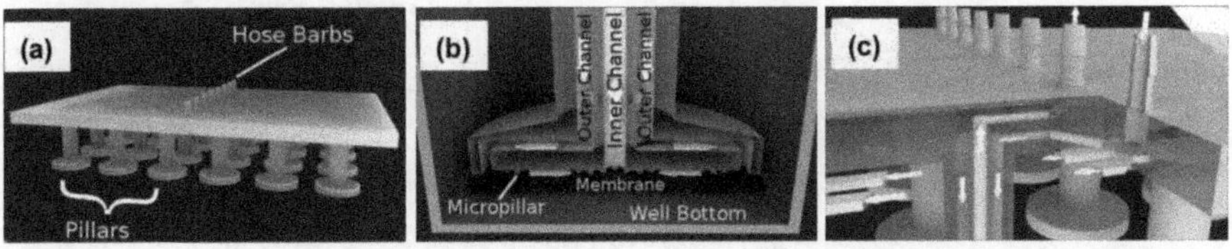


**Figure 6.17** Render of the oxygen control cell culture insert: (a) 3D-printed insert, (b) assembly of the gas permeable poly(dimethylsiloxane) membrane, and (c) cross-section (white arrows: flow of oxygen). Reproduced from ref. 84 [https://doi.org/10.1371/journal.pone.0137631] under the terms of a CC0 1.0 license [https://creativecommons.org/publicdomain/zero/1.0/].

higher brush border enzyme activity, 2- to 7-fold higher P-glycoprotein activity, and better prediction for human oral absorption of eight membrane transporter substrates. The small sample size of the printed inserts should allow their use for high-throughput drug screening by placing multiple inserts in a 24-well plate. Moreover, the inserts can be easily customised to suit specific requirements.

3D-printed polyacrylate-based inserts limit oxygen supply to cell cultures due to their impermeability. Hence, Brennan *et al.*[84] integrated gas permeable poly(dimethylsiloxane) membranes with SL-printed polyacrylate inserts to allow controlled oxygen flow as shown in Figure 6.17. The inserts successfully maintained variable oxygen levels to control gene expression of vascular endothelial growth factor A in A549 cells.

Multiscale, bioinspired, and customisable organ-on-a-chip models are required to perform future *in vitro* drug screening. Such models are already being 3D-printed. For example, Johnson *et al.*[85] developed a nervous system model using the direct ink writing of silicone, polycaprolactone, and cell suspension. This technique has been briefly described in Chapter 4. The 3D-printed model consisted of microchannels for axonal alignment and compartmented chambers for cell isolation (Figure 6.18). It maintained the critical functions of glial cell–axon interfaces, and it was used to study viral infection.

Similarly, Ferraz *et al.*[86] have developed an oviduct-on-a-chip model using SL-printed scaffolds (Figure 6.19). The scaffolds were able to sustain bovine oviduct epithelial cell polarisation during long-term culture (more than six weeks). The combination of 3D printing and microfluidics technology is likely to facillitate rapid and advanced development of other similar organs-on-a-chip in the near future.

## 6.5 Biocompatibility of 3D Print Materials

The composition of most commonly used SL and PolyJet print resins is proprietary, which has severely limited the understanding of their biological and environmental hazards. However, based on their approximate composition and recent reports, it is very clear that the biocompatibility of 3D print materials deserves increased attention. Below, a brief discussion on

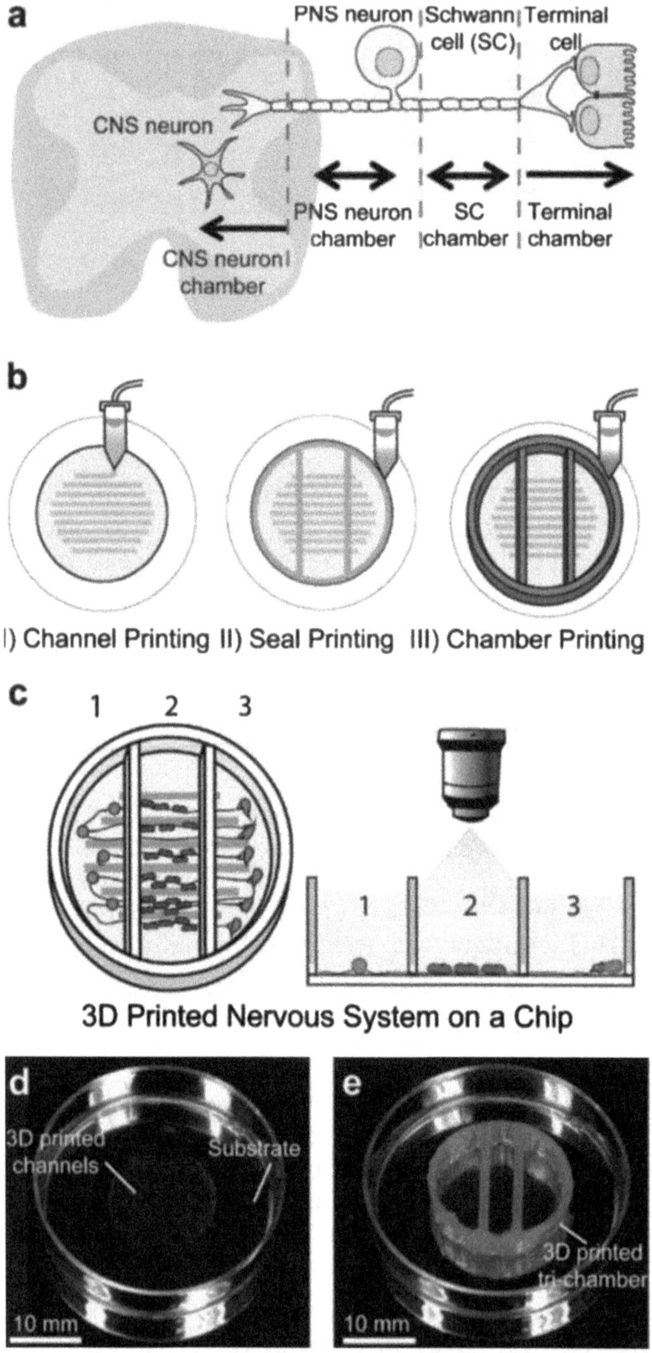


**Figure 6.18** Nervous system model: (a) schematic of the system, (b) direct ink writing of the model nervous system on a chip, (c) schematic of a representative 3D nervous system on a chip: (1) PNS neurons in chamber 1, (2) Schwann cells in chamber 2, and (3) terminal cell junctions in chamber 3, (d) 3D-printed silicon microchannels, and (e) 3D-printed tri-chamber components. Reproduced from ref. 85 with permission from The Royal Society of Chemistry.

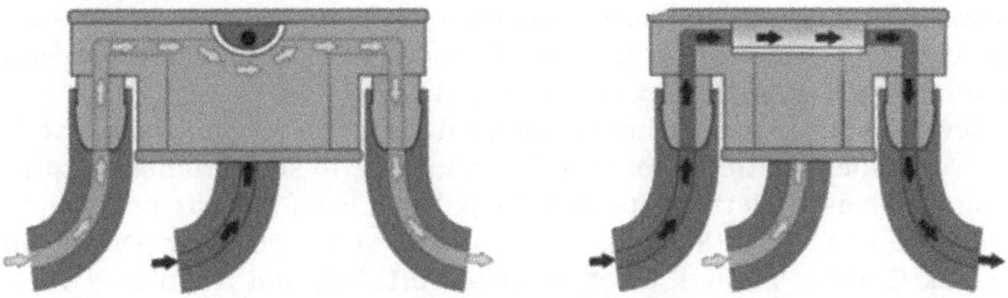

**Figure 6.19** Schematic of the 3D-printed oviduct-on-a-chip model perfused with different types of media/cell (white arrows and black arrows). Reproduced from ref. 86 with permission from The Royal Society of Chemistry.

biocompatibility and the ways to minimise the toxicity of 3D print materials is presented. Readers can further find a detailed discussion of the topic in a recent comprehensive review by Carve *et al.*[87]

Most SL and PolyJet print resins are composed of approximately 90% w/w acrylate- and methacrylate-based monomers, 5% w/w phosphine-oxide based systems, and the rest are auxiliary compounds, such as 2,6-Di-tert-butyl-4-methylphenol, hydroquinone, 2-methyl-1-(4-methylthiophenol)-2-morpholino-propan-1-one, and pentamethyl-piperidyl sebacate, *etc.*[87] The approximate composition of a widely used PolyJet print resin (VeroClear™) is isobornyl acrylate (15%–30%), urethane acrylate (10%–30%), other acrylic monomers (15%–30%), acrylate oligomers (10%–15%), and photoinitiators (1%–2%).[60] Each of these components has been previously reported as potentially toxic to some degree; hence, it not surprising that in their combined formulation as 3D print resins biotoxicity has been demonstrated.

Photopolymers that are regularly used in SL and PolyJet print resins are acrylate and methacrylate monomers and oligomers, urethane acrylate oligomers, tripropylene glycol diacrylate, hydroxyethyl methacrylate, 1,6-bis(2,3-epoxypropoxy) hexane, 3,4-epoxycyclohexylmethyl 3,4-epoxycyclohexanecarboxylate, bisphenol A-glycidyl methacrylate, and tetraacrylate.[87] All of these compounds have previously shown toxicity toward aquatic organisms.[87] Moreover, acrylate and methacrylate monomers and oligomers have also shown toxicity to animals and mammalian cell cultures, and they have identified potential mutagens and reproductive and developmental toxicants.[88–90]

The most commonly used photoinitiators for SL and PolyJet printing are phosphine oxide and benzophenone compounds, hydroxyacetophenone, 1-hydroxy-cyclohexyl-phenyl-ketone, camphorquinone, and triarylsulfonium salts.[87] Once again, these photoinitiators have been shown to exhibit toxicity to both mammals and aquatic organisms.[87,91,92] The toxicity of photoinitiators is primarily linked to the generation of reactive oxygen species, such as peroxides and peroxy radicles as metabolites.[93,94] These species are known to cause damage to DNA and proteins through oxidative stress mechanisms.[95] Moreover, photoinitiators, such as Irgacure 369 and Irgacure 2959, which are usually considered to be biocompatible, have also demonstrated some toxicity in recent studies.[96]

Several aquatic species have been used as biological models to investigate the biocompatibility of 3D print materials. In such studies, 3D print resins such as VisiJet® Crystal and VisiJet® SL Clear (3D Systems), Somos® Watershed 11122XC (Somos), Dreve Fototec SLA 7150 Clear (Dreve Otoplastik GmbH), Form 1 Clear resin (FormLabs), and ABSplus™ P-430 (Solutions 2) have been found to exhibit toxicity.[97,98] Leachates from these polymers have shown significant growth inhibition of freshwater micro-algae (*Pseudokirchneriella subcapitata*), freshwater crustacean (*Ceriodaphnia dubia*), rotifer (*Brachionus calyciflorus*), and zebra fish.[98] The toxicity of these materials to various aquatic life forms is summarised in Table 6.1.

SL and PolyJet printed materials have repeatedly been found to be more toxic than those for FDM, as the former contain a complex mixture of toxic monomers and photoinitiators, whereas, the latter are usually made up of simple thermoplastic-based materials.[98,99] Photopolymers used in SL and PolyJet printing have resulted in complete paralysis of zebrafish larvae after just five minutes of exposure; whereas leachates from acrylonitrile butadiene styrene and poly(lactic acid) did not result in any significant behavioural abnormalities (Figure 6.20).[98] PolyJet print resins, MED610, MED620, and VisiJet® Crystal have been classified as biocompatible materials as per USP

**Table 6.1** Toxicity of commercially available SL and PolyJet print resins. The toxicity is reported after 24 hours of exposure unless otherwise stated. Data from ref. 81.

| 3D print material | Aquatic life-form | Toxicity report |
|---|---|---|
| Watershed 11122XC and Fototec 7150 Clear | Algae | >90% growth inhibition[98] |
| | Rotifer | 100% mortality[98] |
| | Flea | 100% mortality[98] |
| Form Clear | Algae | 60% growth inhibition[98] |
| | Rotifer | 100% mortality[98] |
| | Flea | 100% mortality[98] |
| | Zebrafish | After 72 hours, high rate of mortality and malformations[99] |
| VisiJet® Crystal | Algae | 70% growth inhibition[98] |
| | Flea | 100% mortality[98] |
| | Rotifer | 100% mortality[98] |
| | Zebrafish | After 48 hours, >90% mortality[100] |
| | | After 72 hours, 100% mortality[97,100] |
| VisiJet® Clear | Zebrafish | After 48 hours, >90% mortality of embryos[100] |
| | Algae | >90% growth inhibition[98] |
| | Rotifer | 100% mortality[98] |
| | Flea | 100% mortality[98] |
| MED610/620 | Zebrafish | >50% lethality[100] |

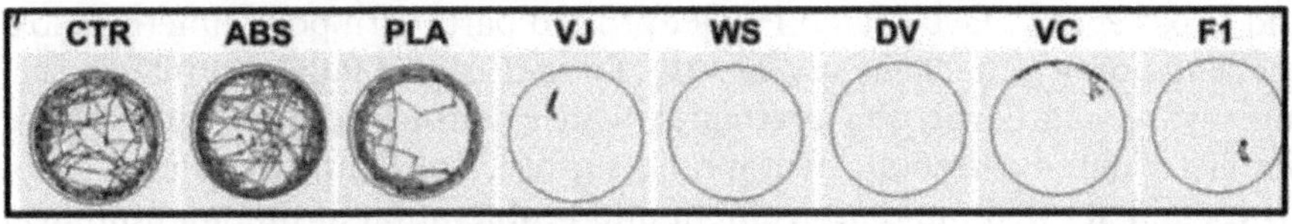


**Figure 6.20** Trajectories of zebrafish larvae after 5 minutes of exposure to different 3D print resin and polymer leachates: Control (CTR), acrylonitrile butadiene styrene (ABS), poly(lactic acid) (PLA), VisiJet® Crystal (VJ), Somos® Watershed 11122XC (WS), Dreve Fototec 7150 Clear (DV), VisiJet® SL Clear (VC), and Form 1 Clear resin (F1). Reproduced from ref. 98 with permission from AIP Publishing.

class IV certification. However, these three materials have also been found to be toxic to zebrafish larvae, as shown in Table 6.1.

The toxicity of 3D-printed objects is primarily linked to the release of unpolymerised monomers, photoinitiators, and free radicals. These compounds can be released either due to the degradation or erosion of the printed surface or due to the extraction of these compounds in the liquid medium.[101] Efforts have been made to minimise the release of these toxic substances to increase the biocompatibility of 3D-printed objects.

One of the primary causes for the excessive release of these substances is incomplete polymerisation during SL and PolyJet printing.[102] It is observed that even the most efficient printing systems can only achieve *ca.* 55–60% polymerisation.[103] Accordingly, SL- and PolyJet-printed materials have been post-cured with heat[104] or UV[99] treatment. Heat treatment adversely affects the appearance of the 3D-printed parts, whereas UV treatment results in minimal changes. However, the effectiveness of UV post-curing is dependent on the device geometry, because three-dimensionally complex designs often produce shadowing effects, which reduces the performance of the post-curing step.[87]

In addition to post-curing, often thorough post-cleaning is also required to achieve the required biocompatibility. In 2004, Popov *et al.* suggested post-cleaning of SL-printed implants with supercritical carbon dioxide to increase their biocompatibility.[105] Recently, Macdonald *et al.* also observed an increase in the biocompatibility of the Dreve Fototec resin printed parts after cleaning with 99% ethanol, however, this approach did not work for the parts printed with VisiJet® Crystal resin.[97] Moreover, the parts printed with VisiJet® Clear resin required multiple cycles of soaking and scrubbing with isopropanol followed by a post-curing step to meet the USP Class IV certification for biocompatibility.[106]

Erosion and leaching of toxic substances are a surface phenomenon. Hence, several groups have modified the surface of 3D-printed objects to minimise their toxicity. For example, Macdonald *et al.*[97] and van den Driesche *et al.*[107] have coated SL-printed parts with wax and parylene-C, respectively, and Gross *et al.*[60] have coated PolyJet-printed parts with poly(dimethylsiloxane) and polystyrene. Recently, in 2016, Lu *et al.* coated PolyJet-printed structures with waterborne poly(urethane), which not only reduced cytotoxicity, but also offered a smooth hydrophilic surface, excellent mechanical properties, and showed rapid attachment and proliferation of human umbilical vein endothelial cells.[108]

Owing to the adverse effects of the 3D print resins on aquatic organisms, the disposal of 3D printing waste has also become an important issue. In the past, direct disposal of acrylic monomers to municipal sewerage systems was considered acceptable.[109] However, an exponential increase in the use of 3D printers with acrylic resins in homes, research laboratories, and industries demands improved procedures and greater education on the safe use and disposal of 3D printing materials.[99]

## 6.6 Outlook

As can be seen from the above examples, 3D printing is increasingly being used to deliver new biochemical assays, in areas such as cellular analysis, rapid diagnostic tests, and pharmacokinetic and pharmacodynamic profiling of drugs. 3D printing has also provided the tools for the development of three-dimensionally complex microfluidic platforms and devices, 3D cell cultures, and organs-on-a-chip. The ability to fabricate bespoke biosensors with entrapped enzymes and antibodies has opened up new fields of investigation in the biosensor community.[73] Table 6.2 summarises the 3D printing techniques, printers, and print materials that have been used for all of the exciting new developments described in this chapter.

3D printing offers extensive design freedom for future modification and improvement of the above-mentioned 3D-printed components. For example, the two-dimensional cell sorting chip design can be further extended to a three-dimensional sorting cube.[54] Cell sorting cubes can contain 1000-fold more channels than their counterpart chips to perform high throughput and high purity cell sorting, and they can be modified for different biochemical applications, such as blood sample preparation, point-of-care diagnostics, and vaccine preparation.[51]

The more widespread use of SL and PolyJet 3D printing techniques in biochemistry is primarily limited by the biotoxicity of the 3D print resins. As discussed, several measures have been proposed to minimise the biotoxicity of the 3D-printed parts, such as UV treatment, heat treatment, and surface modification. An increasing number of biocompatible 3D print resins, ceramics, and metals are now being reported that would significantly further the use of 3D printing in biochemistry. Higher resolution 3D printing would also reduce the roughness of the microfluidic channels and minimise the adsorption of biomolecules.

**Table 6.2** The 3D printing techniques, printers, and print materials that are used for fabricating different biochemical components.

| | Printer | | | |
|---|---|---|---|---|
| Technique | Manufacturer | Model | Print material used | 3D-printed components[a] |
| FDM | Zmorph | Zmorph2.0SX | Acrylonitrile butadiene styrene | PCR device[61] |
| | 3Dfactories | Profi3Dmaker | | Flow-through electrochemical device[75] |
| | MakerGear | M2 | — | Components for a lab-on-a-drone[71] |
| | Rokit | 3DISON | Poly(lactic acid) | Microelectrodes for a micro-electroporation system[59] |
| PolyJet printing | 3D Systems | ProJet 3500 HDPlus | VisiJet® M3 Crystal | Mini-hydrocyclone[55] |
| | Stratasys | Objet350 Connex | VeroClear™ | Fluidic device for Transwell membrane inserts[35,36,79] |
| | | | | Fluidic device for electrical lysis[60] |
| | | | VeroWhitePlus™ | Fluidic device for Transwell membrane inserts[34] |
| | | Objet24 | Polycarbonate | Insert for hanging culture[83] |
| SL | Asiga | Freeform Pico | PlasCLEAR | Biochip for malaria diagnosis[72] |
| | B9Creation | | Poly(ethylene glycol)diacrylate | Hydrogel structures with encapsulated enzymes[73] |
| | EnvisionTEC | Perfactory 3 Mini | PIC 100 | Oviduct-on-a-chip model scaffold[86] |
| | Formlabs | Form 1+ | Polyacrylate composite | Density sorter chip[54] |
| | | | | Microfluidic chip for DNA assembly[74] |
| | Miicraft | — | | PCR device[70] |
| | — | — | Poly(ethyleneglycol)diacrylate | Biomimetic microstructures for cancer cell migration[37] |
| | | | Somos® Watershed XC | Oxygen control insert for a 24-well plate[84] |
| | | | — | Co-extrusion device[20] |

[a]The components have been named in this table as per the names proposed in their respective reports. In the case of the absence of any clearly proposed name, generic names based on their applications have been used.

# References


1. C. M. B. Ho, S. H. Ng, K. H. H. Li and Y.-J. Yoon, *Lab Chip*, 2015, **15**, 3627–3637.
2. I. T. Ozbolat, *3D Bioprinting: Fundamentals, Principles and Applications*, Academic Press, 2016.
3. C. K. Chua and W. Y. Yeong, *Bioprinting: Principles and Applications*, World Scientific Publishing Co Inc, 2014.
4. M. G. Mitchell, *Bioprinting: Techniques and Risks for Regenerative Medicine*, Academic Press, 2017.
5. B. C. Gross, J. L. Erkal, S. Y. Lockwood, C. Chen and D. M. Spence, *Anal. Chem.*, 2014, **86**, 3240–3253.
6. S. V. Murphy and A. Atala, *Nat. Biotechnol.*, 2014, **32**, 773.
7. V. Mironov, T. Boland, T. Trusk, G. Forgacs and R. R. Markwald, *Trends Biotechnol.*, 2003, **21**, 157–161.
8. W. Yang, H. Yu, G. Li, F. Wei, Y. Wang and L. Liu, *Lab Chip*, 2017, **17**, 4243–4252.
9. S. Takayama, E. Ostuni, P. LeDuc, K. Naruse, D. E. Ingber and G. M. Whitesides, *Chem. Biol.*, 2003, **10**, 123–130.
10. R. S. Kane, S. Takayama, E. Ostuni, D. E. Ingber and G. M. Whitesides, in *The Biomaterials: Silver Jubilee Compendium*, Elsevier, 2006, pp. 161–174.
11. L. I. Genes, N. V. Tolan, M. K. Hulvey, R. S. Martin and D. M. Spence, *Lab Chip*, 2007, **7**, 1256–1259.
12. C.-J. Ku, T. D'Amico Oblak and D. M. Spence, *Anal. Chem.*, 2008, **80**, 7543–7548.
13. D. Huh, H. J. Kim, J. P. Fraser, D. E. Shea, M. Khan, A. Bahinski, G. A. Hamilton and D. E. Ingber, *Nat. Protoc.*, 2013, **8**, 2135.
14. J. B. Kim, R. Stein and M. J. O'hare, *Breast Cancer Res. Treat.*, 2004, **85**, 281–291.
15. J. S. Miller, K. R. Stevens, M. T. Yang, B. M. Baker, D.-H. T. Nguyen, D. M. Cohen, E. Toro, A. A. Chen, P. A. Galie and X. Yu, *Nat. Mater.*, 2012, **11**, 768.
16. C. Fischbach, R. Chen, T. Matsumoto, T. Schmelzle, J. S. Brugge, P. J. Polverini and D. J. Mooney, *Nat. Methods*, 2007, **4**, 855.
17. H. A. Svahn and A. v. d. Berg, *Lab Chip*, 2007, **7**, 544–546.
18. S. Utech, R. Prodanovic, A. S. Mao, R. Ostafe, D. J. Mooney and D. A. Weitz, *Adv. Healthcare Mater.*, 2015, **4**, 1628–1633.
19. C. Kim, S. Chung, Y. E. Kim, K. S. Lee, S. H. Lee, K. W. Oh and J. Y. Kang, *Lab Chip*, 2011, **11**, 246–252.
20. K. Alessandri, M. Feyeux, B. Gurchenkov, C. Delgado, A. Trushko, K.-H. Krause, D. Vignjević, P. Nassoy and A. Roux, *Lab Chip*, 2016, **16**, 1593–1604.
21. K. Alessandri, B. R. Sarangi, V. V. Gurchenkov, B. Sinha, T. R. Kießling, L. Fetler, F. Rico, S. Scheuring, C. Lamaze and A. Simon, *PNAS*, 2013, **110**, 14843–14848.

22. J. M. Karp, J. Yeh, G. Eng, J. Fukuda, J. Blumling, K.-Y. Suh, J. Cheng, A. Mahdavi, J. Borenstein and R. Langer, *Lab Chip*, 2007, 7, 786–794.
23. H.-C. Moeller, M. K. Mian, S. Shrivastava, B. G. Chung and A. Khademhosseini, *Biomaterials*, 2008, **29**, 752–763.
24. J. C. Mohr, J. J. de Pablo and S. P. Palecek, *Biomaterials*, 2006, **27**, 6032–6042.
25. M. Andreasson-Ochsner, G. Romano, M. Hakanson, M. L. Smith, D. E. Leckband, M. Textor and E. Reimhult, *Lab Chip*, 2011, **11**, 2876–2883.
26. F. Khan, R. Zhang, A. Unciti-Broceta, J. J. Díaz-Mochón and M. Bradley, *Adv. Mater.*, 2007, **19**, 3524–3528.
27. J. R. Rettig and A. Folch, *Anal. Chem.*, 2005, 77, 5628–5634.
28. E. Afrimzon, G. Botchkina, N. Zurgil, Y. Shafran, M. Sobolev, S. Moshkov, O. Ravid-Hermesh, I. Ojima and M. Deutsch, *Lab Chip*, 2016, **16**, 1047–1062.
29. E. Ostuni, C. S. Chen, D. E. Ingber and G. M. Whitesides, *Langmuir*, 2001, **17**, 2828–2834.
30. D. Falconnet, G. Csucs, H. M. Grandin and M. Textor, *Biomaterials*, 2006, **27**, 3044–3063.
31. A. R. Liberski, J. T. Delaney Jr and U. S. Schubert, *ACS Comb. Sci.*, 2010, **13**, 190–195.
32. B. Ringeisen, P. Wu, H. Kim, A. Pique, R. Auyeung, H. Young, D. Chrisey and D. Krizman, *Biotechnol. Prog.*, 2002, **18**, 1126–1129.
33. Š. Selimović, F. Piraino, H. Bae, M. Rasponi, A. Redaelli and A. Khademhosseini, *Lab Chip*, 2011, **11**, 2325–2332.
34. K. B. Anderson, S. Y. Lockwood, R. S. Martin and D. M. Spence, *Anal. Chem.*, 2013, **85**, 5622–5626.
35. C. Chen, Y. Wang, S. Y. Lockwood and D. M. Spence, *Analyst*, 2014, **139**, 3219–3226.
36. Y. Liu, C. Chen, S. Summers, W. Medawala and D. M. Spence, *Integr. Biol.*, 2015, **7**, 534–543.
37. T. Q. Huang, X. Qu, J. Liu and S. Chen, *Biomed. Microdevices*, 2014, **16**, 127–132.
38. J. Takagi, M. Yamada, M. Yasuda and M. Seki, *Lab Chip*, 2005, **5**, 778–784.
39. J. Shi, H. Huang, Z. Stratton, Y. Huang and T. J. Huang, *Lab Chip*, 2009, **9**, 3354–3359.
40. R. Huang, T. A. Barber, M. A. Schmidt, R. G. Tompkins, M. Toner, D. W. Bianchi, R. Kapur and W. L. Flejter, *Prenatal Diagn.*, 2008, **28**, 892–899.
41. L. R. Huang, E. C. Cox, R. H. Austin and J. C. Sturm, *Science*, 2004, **304**, 987–990.
42. M. Kersaudy-Kerhoas, R. Dhariwal, M. P. Y. Desmulliez and L. Jouvet, *Microfluid. Nanofluid.*, 2009, **8**, 105.
43. M. Yamada and M. Seki, *Lab Chip*, 2005, **5**, 1233–1239.
44. D. Di Carlo, *Lab Chip*, 2009, **9**, 3038–3046.
45. L. Wu, G. Guan, H. W. Hou, A. A. S. Bhagat and J. Han, *Anal. Chem.*, 2012, **84**, 9324–9331.

46. J. Zhou, P. V. Giridhar, S. Kasper and I. Papautsky, *Lab Chip*, 2013, **13**, 1919–1929.
47. J. M. Martel and M. Toner, *Annu. Rev. Biomed. Eng.*, 2014, **16**, 371–396.
48. M. E. Warkiani, A. K. P. Tay, G. Guan and J. Han, *Sci. Rep.*, 2015, **5**, 11018.
49. R. Martinez-Duarte, R. A. Gorkin Iii, K. Abi-Samra and M. J. Madou, *Lab Chip*, 2010, **10**, 1030–1043.
50. Z. Jinlong, G. Qiuquan, L. Mei and Y. Jun, *J. Micromech. Microeng.*, 2008, **18**, 125025.
51. M. Shakeel Syed, M. Rafeie, R. Henderson, D. Vandamme, M. Asadnia and M. Ebrahimi Warkiani, *Lab Chip*, 2017, **17**, 2459–2469.
52. S. Nagrath, L. V. Sequist, S. Maheswaran, D. W. Bell, D. Irimia, L. Ulkus, M. R. Smith, E. L. Kwak, S. Digumarthy and A. Muzikansky, *Nature*, 2007, **450**, 1235.
53. I. Cima, C. Wen Yee, F. S. Iliescu, W. Min Phyo, K. Hon Lim, C. Iliescu and M. Han Tan, *Biomicrofluidics*, 2013, **7**, 011810.
54. N. Norouzi, H. C. Bhakta and W. H. Grover, *PLoS One*, 2017, **12**, e0180520.
55. M. S. Syed, M. Rafeie, R. Henderson, D. Vandamme, M. Asadnia and M. E. Warkiani, *Lab Chip*, 2017, **17**, 2459–2469.
56. J. A. Kim, K. Cho, M. S. Shin, W. G. Lee, N. Jung, C. Chung and J. K. Chang, *Biosens. Bioelectron.*, 2008, **23**, 1353–1360.
57. W. G. Lee, U. Demirci and A. Khademhosseini, *Integr. Biol.*, 2009, **1**, 242–251.
58. T. Geng and C. Lu, *Lab Chip*, 2013, **13**, 3803–3821.
59. S. Kang, K.-H. Kim and Y.-C. Kim, *Sci. Rep.*, 2015, **5**, 15835.
60. B. C. Gross, K. B. Anderson, J. E. Meisel, M. I. McNitt and D. M. Spence, *Anal. Chem.*, 2015, **87**, 6335–6341.
61. G. Mulberry, K. A. White, M. Vaidya, K. Sugaya and B. N. Kim, *PLoS One*, 2017, **12**, e0179133.
62. J. Wu, R. Kodzius, W. Cao and W. Wen, *Microchim. Acta*, 2014, **181**, 1611–1631.
63. Y. Liu, J.-X. Han, H.-Y. Huang and B. Zhu, *Exp. Biol. Med.*, 2005, **230**, 587–591.
64. K. Sen and D. M. Asher, *Transfusion*, 2001, **41**, 1356–1364.
65. B. M. Branson, H. H. Handsfield, M. A. Lampe, R. S. Janssen, A. W. Taylor, S. B. Lyss and J. E. Clark, *MMWR Recomm. Rep.*, 2006, **55**, 1–17.
66. P. B. Bloland and W. H. Organization, *Drug Resistance in Malaria*, 2001.
67. X. Jiang, N. Shao, W. Jing, S. Tao, S. Liu and G. Sui, *Talanta*, 2014, **122**, 246–250.
68. Y. Zhu, Y.-X. Zhang, W.-W. Liu, Y. Ma, Q. Fang and B. Yao, *Sci. Rep.*, 2015, **5**, 9551.
69. A. Zahra, R. Scipinotti, D. Caputo, A. Nascetti and G. de Cesare, *Sens. Actuators, A*, 2015, **236**, 206–213.
70. J. Park and H. Park, *Sens. Actuators, A*, 2017, **268**, 183–187.
71. A. Priye, S. Wong, Y. Bi, M. Carpio, J. Chang, M. Coen, D. Cope, J. Harris, J. Johnson and A. Keller, *Anal. Chem.*, 2016, **88**, 4651–4660.
72. R. M. Dirkzwager, S. Liang and J. A. Tanner, *ACS Sens.*, 2016, **1**, 420–426.

73. C. A. Mandon, L. J. Blum and C. A. Marquette, *Anal. Chem.*, 2016, **88**, 10767–10772.
74. W. G. Patrick, A. A. Nielsen, S. J. Keating, T. J. Levy, C.-W. Wang, J. J. Rivera, O. Mondragón-Palomino, P. A. Carr, C. A. Voigt and N. Oxman, *PLoS One*, 2015, **10**, e0143636.
75. P. Michalek, L. Richtera, L. Krejcova, L. Nejdl, R. Kensova, J. Zitka, P. Kopel, Z. Heger, V. Adam and R. Kizek, *Electrophoresis*, 2016, **37**, 444–454.
76. A. I. Minchinton and I. F. Tannock, *Nat. Rev. Cancer*, 2006, **6**, 583.
77. M. Ginai, R. Elsby, C. J. Hewitt, D. Surry, K. Fenner and K. Coopman, *Drug Discovery Today*, 2013, **18**, 922–935.
78. E. Lengyel, J. E. Burdette, H. A. Kenny, D. Matei, J. Pilrose, P. Haluska, K. P. Nephew, D. B. Hales and M. S. Stack, *Oncogene*, 2013, **33**, 3619.
79. S. Y. Lockwood, J. E. Meisel, F. J. Monsma and D. M. Spence, *Anal. Chem.*, 2016, **88**, 1864–1870.
80. G. J. LaBonia, K. R. Ludwig, C. B. Mousseau and A. B. Hummon, *Anal. Chem.*, 2018, **90**, 1423–1430.
81. Y. Sambuy, I. De Angelis, G. Ranaldi, M. Scarino, A. Stammati and F. Zucco, *Cell Biol. Toxicol.*, 2005, **21**, 1–26.
82. S. Chong, S. A. Dando, K. M. Soucek and R. A. Morrison, *Pharm. Res.*, 1996, **13**, 120–123.
83. C. Shen, Q. Meng and G. Zhang, *Biofabrication*, 2014, **7**, 015003.
84. M. D. Brennan, M. L. Rexius-Hall and D. T. Eddington, *PLoS One*, 2015, **10**, e0137631.
85. B. N. Johnson, K. Z. Lancaster, I. B. Hogue, F. Meng, Y. L. Kong, L. W. Enquist and M. C. McAlpine, *Lab Chip*, 2016, **16**, 1393–1400.
86. M. A. M. M. Ferraz, H. H. W. Henning, P. F. Costa, J. Malda, F. P. Melchels, R. Wubbolts, T. A. E. Stout, P. L. A. M. Vos and B. M. Gadella, *Lab Chip*, 2017, **17**, 905–916.
87. M. Carve and D. Wlodkowic, *Micromachines*, 2018, **9**, 91.
88. formlabs, SAFETY DATA SHEET: Dental LT Clear, https://formlabs.com/media/upload/DentalLTClear-SDS-EN.pdf, accessed 21st November, 2018.
89. F. Rice, *Concise International Chemical Assessment Document*, World Health Organization, 2000.
90. BASF, Safety Data Sheet: METHYL ACRYLATE, https://worldaccount.basf.com/wa/NAFTA/Catalog/ChemicalsNAFTA/doc4/BASF/PRD/30041968/.pdf?asset_type=msds/pdf&language=EN&validArea=US&urn=urn:documentum:ProductBase_EU:09007af8800907ee.pdf, accessed 20th November, 2018.
91. formlabs, SAFETY DATA SHEET: Clear, https://formlabs.com/media/upload/Clear__Resin_SDS_EU.pdf, accessed 20th November, 2018.
92. SIGMA-ALDRICH, SAFETY DATA SHEET: Diphenyl(2,4,6-trimethylbenzoyl)phosphine oxide, https://www.sigmaaldrich.com/MSDS/MSDS/DisplayMSDSPage.do?country=AU&language=en&productNumber=415952&brand=ALDRICH&PageToGoToURL=https%3A%2F%2Fwww.sigmaaldrich.com%2Fcatalog%2Fproduct%2Faldrich%2F415952%3Flang%3Den, accessed 20th November, 2018.

93. M. Sangermano, *Pure Appl. Chem.*, 2012, **84**, 2089–2101.
94. M. A. Lago, A. Rodriguez-Bernaldo de Quiros, R. Sendón, J. Bustos, M. T. Nieto and P. Paseiro, *Food Addit. Contam.*, 2015, **32**, 779–798.
95. S. J. Stohs, *J. Basic Clin. Physiol. Pharmacol.*, 1995, **6**, 205–228.
96. A. K. Nguyen, S. D. Gittard, A. Koroleva, S. Schlie, A. Gaidukeviciute, B. N. Chichkov and R. J. Narayan, *Regener. Med.*, 2013, **8**, 725–738.
97. N. P. Macdonald, F. Zhu, C. Hall, J. Reboud, P. Crosier, E. Patton, D. Wlodkowic and J. Cooper, *Lab Chip*, 2016, **16**, 291–297.
98. F. Zhu, T. Friedrich, D. Nugegoda, J. Kaslin and D. Wlodkowic, *Biomicrofluidics*, 2015, **9**, 061103.
99. S. M. Oskui, G. Diamante, C. Liao, W. Shi, J. Gan, D. Schlenk and W. H. Grover, *Environ. Sci. Technol. Lett.*, 2015, **3**, 1–6.
100. F. Alifui-Segbaya, S. Varma, G. J. Lieschke and R. George, *3D Print. Addit. Manuf.*, 2017, **4**, 185–191.
101. W. Geurtsen, *Crit. Rev. Oral Biol. Med.*, 2000, **11**, 333–355.
102. F. P. Melchels, J. Feijen and D. W. Grijpma, *Biomaterials*, 2010, **31**, 6121–6130.
103. J. Ferracane and J. Condon, *Dent. Mater.*, 1992, **8**, 290–295.
104. Y. Inoue and K. Ikuta, *Procedia CIRP*, 2013, **5**, 115–118.
105. V. Popov, A. Evseev, A. Ivanov, V. Roginski, A. Volozhin and S. Howdle, *J. Mater. Sci.: Mater. Med.*, 2004, **15**, 123–128.
106. D. S. Inc., *VisiJet SL Clear cleaning procedure for USP Class VI*, L. T. Reports, 3D Systems Inc., Rock Hill, SC, USA, 2013.
107. S. van den Driesche, F. Lucklum, F. Bunge and M. J. Vellekoop, *Micromachines*, 2018, **9**, 71.
108. Z. Lu, X. Jiang, X. Zuo and L. Feng, *RSC Adv.*, 2016, **6**, 102381–102388.
109. J. Autian, *Environ. Health Perspect.*, 1975, **11**, 141.

CHAPTER 7

# 3D Printing in Synthetic and Physical Chemistry

## 7.1 Synthetic Chemistry

3D printing has been recently used to develop reactors for synthetic chemistry, enabling rapid design and development of bespoke reactors for chemical reactions. As opposed to conventional techniques, printed reactor designs are not limited in complexity. Moreover, 3D printing allows the successful combination of synthetic chemistry, molecular modelling, and chemical engineering principles to optimise reactor design and reaction parameters.[1] 3D printing has also lowered the required cost and time for the development of new reactors, which has extended their use in resource-limited locations. The use of 3D printing for the development of modern reactors was first introduced by Cronin's group in 2012.[1] Since then different techniques and procedures have been applied for the development of new batch and flow reactors.

### 7.1.1 Batch Reactors

3D printing allows the development of batch reactors with precisely controlled geometry, surface topology, and material composition. Moreover, techniques such as FDM printing allow the printing of reactors with spatially controlled pre-filling of reagents. These capabilities have allowed the development of bespoke reactors with enhanced capabilities as described below. Kitson *et al.* named such reactors 'reactionware'.[2]

3D Printing in Chemical Sciences: Applications Across Chemistry
By Vipul Gupta, Pavel Nesterenko and Brett Paull

Published by the Royal Society of Chemistry, www.rsc.org

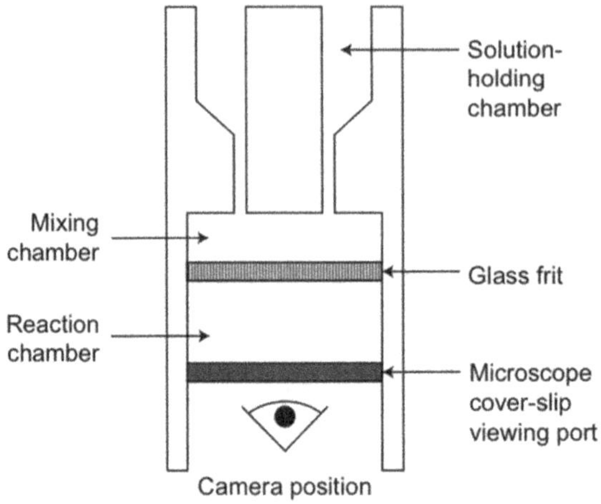


**Figure 7.1** Schematic of a 3D-printed reactionware. Reproduced from ref. 1 with permission from Springer Nature, Copyright 2012.

The first 3D-printed reactionware was developed by Cronin's group in 2012.[1] It was printed with the direct ink writing technique (as briefly described in Chapter 4) using an acetoxysilicone polymer that allowed the preparation of self-healing, reusable, and robust reactors. The reactionware consisted of two solution holding chambers, a mixing chamber, a glass frit, a reaction chamber, and a microscope cover-slip (Figure 7.1). Non-printable components such as the glass frit, cover-slip, and catalysts were inserted during the printing process by pausing and resuming the print cycle (as discussed in Chapter 4). The reactionware was used for the formation and crystallisation of new inorganic nanoclusters and organic heterocyclic compounds in an aqueous medium. Moreover, the outcome of the reactions was successfully modulated by simply altering the reactor configuration, without changing any other reaction conditions.

In 2013, the same research group developed multi-material integrated monolithic reactors to perform multi-step reaction sequences.[3] This time, both direct ink writing and FDM techniques were used to produce the reactor, using acetoxysilicone polymer and polypropylene, respectively. All the required reagents, catalysts, stirrer bars, and purification systems were placed during the printing process. The reactor consisted of three separate reaction chambers (20 mm cube), one for each process. More importantly, the reagents were transferred from one chamber to another by simply rotating the reactor at 90°, as described in Figure 7.2. A 4 mm wide circular passage was included between each chamber to allow easy transfer of the reagents during rotation. The process allowed a three-step reaction sequence to be conducted in a controlled and ordered fashion. The reactor was successfully used for Diels–Alder reactions, giving similar products to reactions in conventional glassware; however, it significantly reduced the requirements for chemical handling and special equipment.

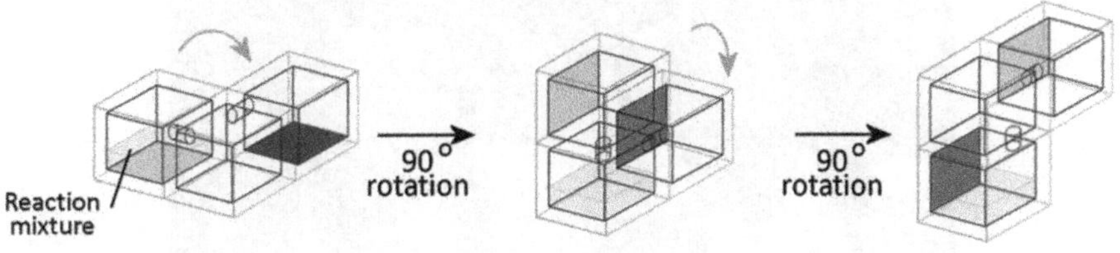


**Figure 7.2** Schematic of the multi-step reaction sequence in monolithic reactionwares. Reproduced from ref. 3 with permission from The Royal Society of Chemistry.

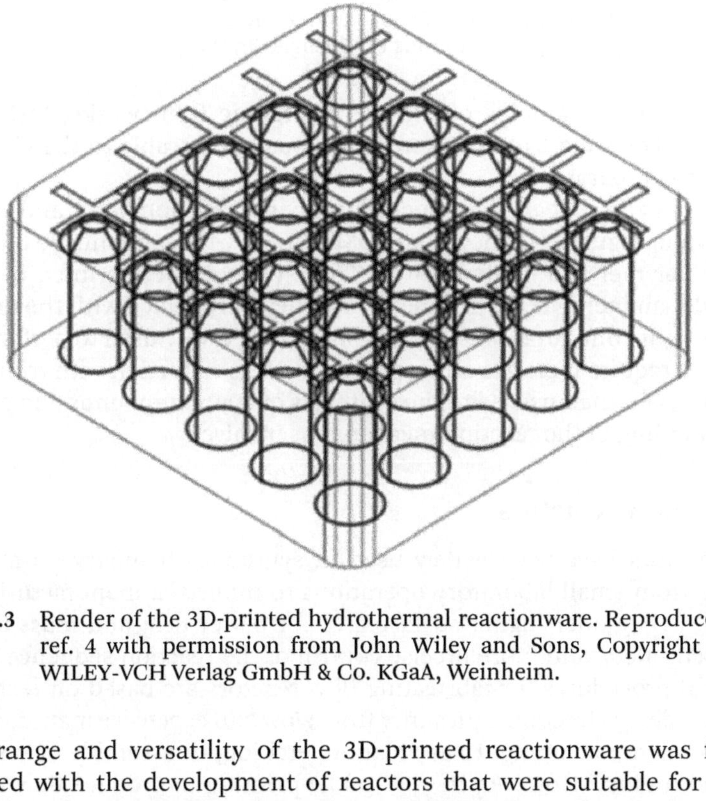

**Figure 7.3** Render of the 3D-printed hydrothermal reactionware. Reproduced from ref. 4 with permission from John Wiley and Sons, Copyright © 2014 WILEY-VCH Verlag GmbH & Co. KGaA, Weinheim.

The range and versatility of the 3D-printed reactionware was further increased with the development of reactors that were suitable for hydrothermal synthesis. In 2014, Cronin's group FDM-printed disposable monolithic hydrothermal reactors with polypropylene, which were successfully used to perform reactions at high temperatures and pressures.[4] The reactors consisted of 5 × 5 arrays of 1 mL reaction chambers (Figure 7.3). They were printed with a 4 mm polypropylene wall, which successfully resisted the elevated pressure resulting from the heating of aqueous and aqueous/DMF systems up to 140 °C for more than 72 hours. Polypropylene starts to soften at *ca.* 150 °C, hence the reactor would not resist high pressures at or above 150 °C. The reactors were used for the high-throughput screening of 25 simultaneous reactions to optimise the synthetic parameters and

**Figure 7.4** 3D-printed acrylonitrile butadiene styrene derived polymer reaction flasks, cuvettes, and NMR tubes. Reproduced from ref. 5 [https://doi.org/10.1002/hlca.201500502] under the terms of a CC BY 4.0 license [https://creativecommons.org/licenses/by/4.0/].

to produce pure samples of the metal–organic frameworks, MIL-96 and HKUST-1. The yield from these reactors was comparable to the yield from traditional apparatus.

In 2016, Lederle *et al.*[5] developed a number of reaction flasks and cuvettes, as shown in Figure 7.4. They were FDM-printed with acrylonitrile butadiene styrene polymer and its derivatives. These reactors were printed in a glove box under an inert gas environment, and they were sealed with the reactants in a gas-tight configuration during printing. They were used to perform reactions that require inert-gas atmospheres, such as reduction with trimethylaluminium, and their use led to the synthesis of new compounds and a greater understanding of the reaction mechanisms involved.

### 7.1.2 Flow Reactors

Flow reactors have been widely used in synthetic chemistry on all scales ranging from small laboratory operations to industrial manufacturing processes.[6–8] They offer higher surface-area for better heat and mass transfer, increased safety, and more precise control of the reaction sequence.[6,8] Conventional procedures for fabricating flow reactors are based on techniques that limit design freedom and suffer from slow and expensive manufacturing, such as micromachining, laser ablation, etching, or assembly of polymeric or metal tubing.[6] However, the use of 3D printing to produce flow reactors can allow virtually full design freedom and complete control over the mixing structures, mixing points, inlets, outlets, flow paths, and residence volumes.[9] Moreover, 3D printing allows rapid production and customisation, facilitating quick design iterations and optimisation.

The first 3D-printed flow reactor was reported by Cronin and co-workers in 2012.[10] They have produced three flow reactors with 0.8 mm I.D. channels in polypropylene using FDM printers (Figure 7.5). The first design, R1, consists of two inlets followed by a reactor. R2 has three inlets, two initial inlets and a third that joins the reactor after a certain volume. R3 is an unconventional reactor design with pre-filled reactant silos. The reactants were inserted into

R3 during the printing process as described above. 3D printing allowed the design, production, and use of these reactors within the same day. These reactors proved to be robust and versatile, and they were successfully used for the synthesis of various organic and inorganic substances. Moreover, reactors R1 and R2 could be recycled for multiple uses.

In the following year, Capel *et al.*[11] reported the development of flow reactors using different 3D printing techniques, specifically FDM, PolyJet, SL, SLS, and SLM. Four different reactor designs were developed (Figures 7.6–7.8).

The reactor 'RD1' (Figure 7.6) was printed with both an FDM and a PolyJet printer using acrylonitrile butadiene styrene, and polyacrylate-based materials, respectively. It consisted of a 510 mm long tube with 3 mm I.D. and 5 mm wall thickness. The use of a high-end FDM printer (Stratasys Dimension) with a soluble support material allowed the printing of this complex design. However, FDM printing failed to provide a liquid-tight seal at the

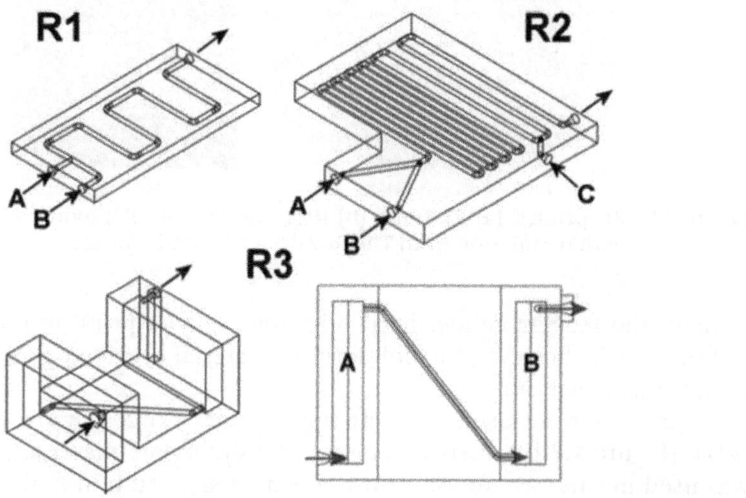


**Figure 7.5** Render of the 3D-printed flow reactors: R1, R2, and R3 as indicated. Reproduced from ref. 10 with permission from The Royal Society of Chemistry.

**Figure 7.6** 3D-printed RD1 flow reactor. Reproduced from ref. 11 with permission from The Royal Society of Chemistry.

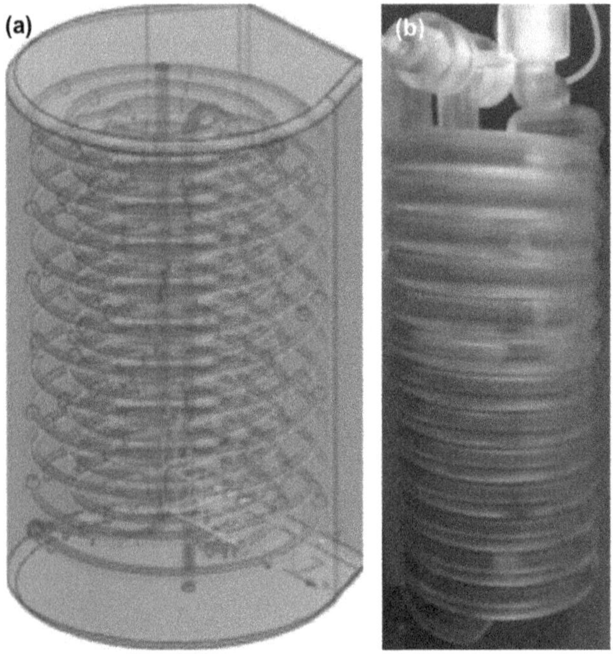


**Figure 7.7** 3D-printed (a) RD2 and (b) RD4 flow reactor. Reproduced from ref. 11 with permission from The Royal Society of Chemistry.

walls of the reactor, rendering it unusable. PolyJet printing provided liquid-tight seals, however, the polyacrylate material provided poor resistance to the organic solvents.

SL printers were used to print the reactors 'RD2' (Figure 7.7(a)) and 'RD4' (Figure 7.7(b)). A special blend of polyacrylate material (Accura 60) was used in this case for SL printing, which showed minimal swelling on immersion for 24 hours in a variety of organic solvents, except tetrahydrofuran and dichloromethane. The RD2 reactor consisted of a 3.3 m long tube with 3 mm I.D., and was found to be leak-proof up to an operating pressure of 3 MPa. The printed reactor was used for the oxidation of an aldehyde to a methyl ester, and resulted in a similar or higher yield than conventional tube reactors. The RD4 reactor was used for the generation of elemental iodine, and its use resulted in up to 17% shorter reaction times compared to its counterpart tube reactor. The RD2 reactor was further printed with an SLS printer using nylon-12. SLS printing resulted in the accumulation of a dense bed of unsintered powder particles in the reactor tube that was difficult to remove, which rendered the SLS-printed reactors unusable.

SLM printing was used to print the RD8 reactor (Figure 7.8) using stainless steel 316L. It was designed with tubes of internal diameters ranging from 1 mm to 2 mm. The SLM-printed reactor was found to be stable at high

**Figure 7.8** 3D-printed RD8 flow reactor. Reproduced from ref. 11 with permission from The Royal Society of Chemistry.

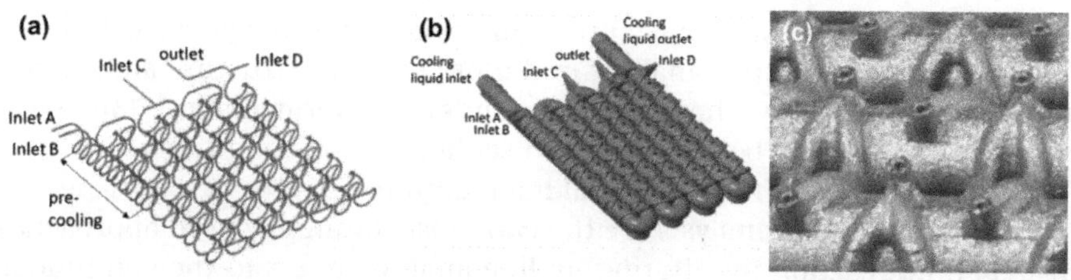


**Figure 7.9** 3D-printed flow reactor: (a) render of the reactor channel, (b) render of the reactor channel attached to a cooling element, and (c) photograph of the shortcut holes. Reproduced from ref. 9 with permission from The Royal Society of Chemistry.

temperatures and also up to an operating pressure of 3 MPa. However, as discussed in Chapter 4, SLM-printed stainless steel and titanium alloy reactors can be stable at operating pressures of more than 20 MPa.

Later in 2017, Gutmann *et al.*[9] used an SLM printer to print much more complex flow reactors (Figure 7.9(a)). The reactor consisted of a 4 m long tube with an I.D. of 0.8 mm and an O.D. of 2.4 mm, and it was wrapped along a cooling tube of 7 mm inner diameter and 9 mm outer diameter (Figure 7.9(b)). However, the reactor channel could not be printed in a continuous configuration, and shortcut holes were incorporated at every turn to assist with the removal of the unmelted powder particles (Figure 7.9(c)). The holes were later manually sealed with laser beam welding. The reactor was operated at minus 65 °C, and it was used to perform a two-step difluoromethylation reaction with gaseous fluoroform. The use of stainless steel build material provided the required chemical, mechanical, and thermal stability for use in organic synthesis.

Recently, Rossi *et al.*[12] 3D-printed flow reactors for the continuous flow synthesis of biologically active chiral 1,2-amino alcohols, such as norephedrine, metaraminol, and methoxamine. The flow reactors were 3D-printed in different shapes, sizes, and materials using an FDM printer. Three different materials were used, namely poly(lactic acid), high-impact polystyrene (HIPS), and nylon. Both poly(lactic acid) and nylon were found to be resistant to organic solvents, hence they were used for the copper-catalysed stereoselective Henry reaction. These reactors were operated at minus 20 °C, and gave higher yields and enantioselectivity compared to their counterpart coiled tube reactors. Moreover, a 3D-printed reactor with 10 mL reactor volume was also developed to scale-up the synthesis.

The above reports have showcased the potential to develop novel 3D-printed micro- and milli-fluidic reactors. The reactors have already proved useful for a wide variety of chemical reactions. Moreover, the design freedom offered by 3D printing paves the way for the development of flow reactors that were considered too geometrically complex in the recent past.

### 7.1.3 In-line and On-line Reaction Monitoring

3D-printed flow reactors can be further interfaced with analytical techniques to allow high-throughput analysis and more efficient structure elucidation and identification of synthesised compounds and intermediates.[13] Moreover, real-time analysis can be used to optimise the reaction conditions through a feedback loop with minimal manual intervention.[14]

In-line and on-line analysis are the two most commonly used approaches for real-time reaction monitoring. In-line analysis refers to the continuous analysis of the product stream, whereas on-line analysis involves the analysis of selected aliquots in a parallel stream.[15] The techniques most commonly used for in-line analysis are spectroscopy techniques, such as UV-Vis, IR, fluorescence, Raman, and NMR spectroscopy;[16] whereas techniques such as chromatography and mass spectrometry are commonly used for on-line analysis.[17,18] 3D-printed reactionware and flow reactors can be easily customised to interface with these analytical techniques.

The foundation for in-line reaction monitoring in 3D-printed reactors was already laid by Cronin's group's first report on printed reactionware.[1] The reactionware included a glass slide coated with a thin, transparent layer of indium tin oxide and 3D-printed electrodes for *in situ* spectroelectrochemistry. The electrodes were printed using direct ink writing of a conductive paste of acetoxysilicone polymer and carbon black.

The same group further interfaced the flow reactors (similar to the reactors shown in Figure 7.5) with a high-resolution electrospray ionisation mass spectrometer (ESI-MS)[16] and an ATR-IR flow-cell.[19] Conventionally, interfacing flow devices with ESI-MS has required expensive and complex microfluidic devices. However, 3D printing allowed easy integration using standard screw fittings. Both ESI-MS and ATR-IR were used for real-time in-line analysis, and the obtained data was used to control the reaction conditions, such as the reactant flow-rates and order. The effect of the reactor configuration on the reaction product was also studied in real-time using these simple interfaces. Moreover, multiple flow reactors were integrated in series using

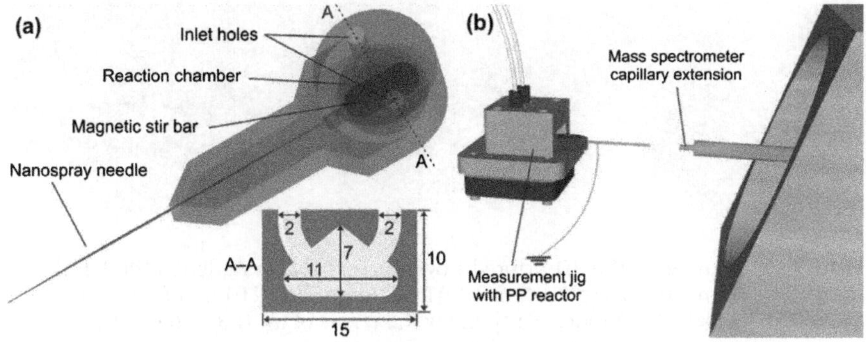


**Figure 7.10** 3D-printed miniaturised reactor with an integrated nano-ESI capillary and magnetic stir bar: (a) render of the device, and (b) render of the set-up with a 3D-printed jig. Dimensions in the cross-section are in mm. Reproduced from ref. 20 with permission from The Royal Society of Chemistry.

standard connections to perform multi-step reactions, such as imine synthesis followed by its reduction.[19]

In 2017, Scotti *et al.* developed a miniaturised 3D-printed reactor with an integrated nano-ESI capillary and a magnetic stir bar (Figure 7.10(a))[20] for direct coupling with mass spectrometry. The reactor was printed with an FDM printer using polypropylene, and the stir bar and nano-ESI capillary were integrated during the printing cycle. A jig to hold the reactor and to incorporate fluidic connections, plus a small fan with magnets to activate the stir bar were also printed (Figure 7.10(b)). The use of a nano-ESI capillary circumvented the need for additional pumps, because the ESI process was able to pull the reaction solution directly from the reaction chamber.

Capel *et al.* developed three SL and SLM printed flow reactors to integrate with commercially available flow analysis systems, liquid chromatography, and spectroscopic equipment (Figure 7.11).[21] The RD1 reactor was designed to fit within the DAD compartment of an Agilent 1100 HPLC, and it consisted of a 1.6 m long, 1.5 mm I.D. channel (Figure 7.11(a)). The reactor was SL-printed with a polyacrylate composite with a material cost of less than $20. It also served as an in-line spectroscopic flow-cell with a path length of 6 mm. The RD2 reactor was designed to fit within the column oven of an Agilent 1100 system (Figure 7.11(b)). It was printed with an SLM printer using Ti-6Al-4V as the build material to allow stability at high temperatures. It consisted of a 3.2 m long, 2 mm I.D. channel within a cuboid of 100 mm × 20 mm × 20 mm, which provided a reaction volume of 10 mL. The RD3 reactor was designed to fit within the DAD compartment, and it was printed with an SLM printer using Ti-6Al-4V to allow better solvent and thermal resistance (Figure 7.11(c)). These bespoke reactors could be used to optimise reactions using spectroscopic and chromatographic analysis while harmoniously sitting within a standard HPLC system.

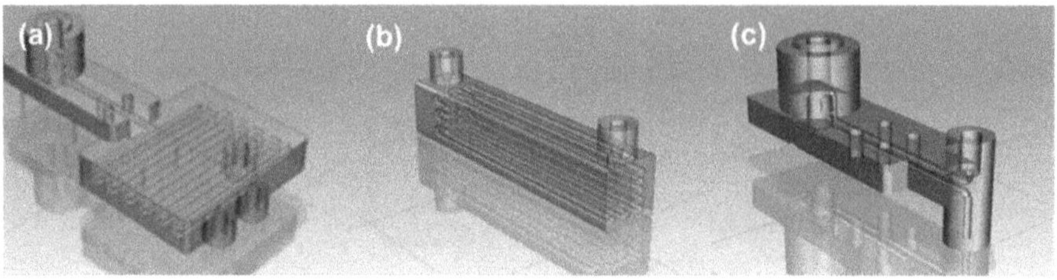


**Figure 7.11** Render of the 3D-printed flow reactors for an Agilent 1100 HPLC system: (a) RD1, (b) RD2, and (c) RD3. Reproduced from ref. 21 [https://doi.org/10.3762/bjoc.13.14] under the terms of a CC By 4.0 license [https://creativecommons.org/licenses/by/4.0/].

## 7.2 Physical Chemistry and Electrochemistry

The use of 3D printing in the field of physical chemistry and electrochemistry has been somewhat limited by the availability of functional print materials. However, recently, many new materials have been developed in-house for specific applications, with notable examples including lithium oxide-based inks to produce interdigitated microbatteries,[22] graphene oxide based inks for lithium-ion batteries,[23] and graphene and phenolic powders for methanol fuel cells.[24] For greater discussion on such novel composite materials for 3D printing, readers are directed to the recent review by Kalsoom *et al.*,[25] and to that of del Campo *et al.*[26] for a brief review of composite materials and their applications in the development of electrochemical flow devices. Herein, some of the recent reports on the use of commercial 3D printers and materials in the field electrochemistry are described below.

Electrolysers and redox flow batteries are becoming an important aspect of renewable energy. Electrolysers can be used for water electrolysis to generate hydrogen renewably, and redox flow batteries can be used to store energy generated from different sources. Both electrolysers and redox flow batteries require the construction of flow plates with complex channel configuration for the distribution of reagents and products in an electrolyser, and distribution of electrolytes in a redox flow battery.

In 2014, Cronin's group reported the preparation of flow plates for a proton exchange membrane (PEM) based electrolyser.[27] These flow plates were 3D-printed with an inexpensive FDM printer using polypropylene. The polypropylene plates were initially coated with silver paint to make them conductive for silver electrodeposition (Figure 7.12). The 3D-printed plates were thoroughly cleaned with detergent in a sonication bath before painting, as a failure to do so resulted in delamination of the coating during the curing process. The silver electrodeposited plates resulted in the desired electrical conductivity. However, the performance of these plates quickly deteriorated during electrolysis. Hence, a thin layer of gold was further sputter-coated. These 3D-printed plates require less than 25% of the materials' cost of

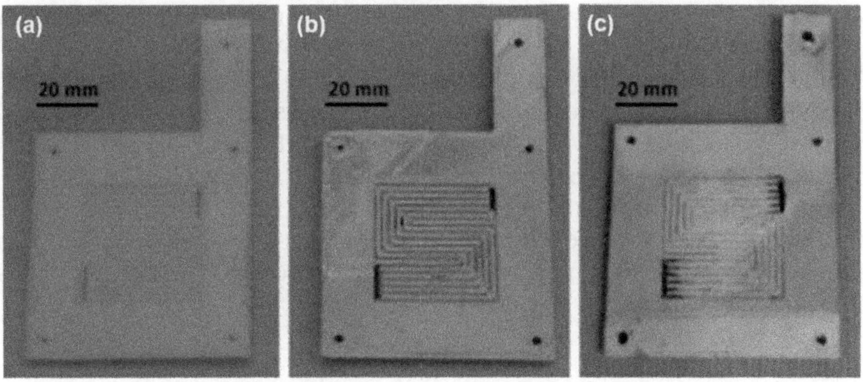


**Figure 7.12** 3D-printed flow plates for a proton exchange membrane (PEM) based electrolyser: (a) uncoated polypropylene, (b) after silver painting, and (c) after electrodeposition of silver. Reproduced from ref. 27 with permission from The Royal Society of Chemistry.

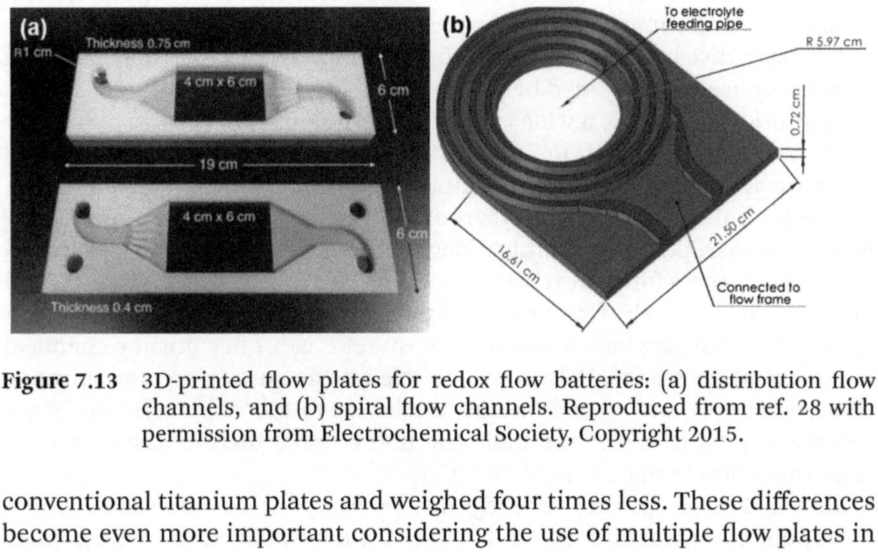


**Figure 7.13** 3D-printed flow plates for redox flow batteries: (a) distribution flow channels, and (b) spiral flow channels. Reproduced from ref. 28 with permission from Electrochemical Society, Copyright 2015.

conventional titanium plates and weighed four times less. These differences become even more important considering the use of multiple flow plates in an electrolyser.

Similarly, Arenas *et al.*[28] reported the development of flow plates for a redox flow battery (Figure 7.13). The longest dimension of these flow plates was approximately 20 cm, a length which usually results in high thermal stress and deformation during FDM printing with a low-cost printer.[29] Hence, a high-end FDM printer (Stratasys Dimension) was used to print these flow plates using acrylonitrile-butadiene-styrene. The thermal stress could be further reduced with the use of a poly(lactic acid) build material. However, poly(lactic acid) does not provide the required chemical resistance to the electrolytes used. The printed flow plates provided a non-porous and impermeable structure for the distribution of electrolytes, and a surface roughness of only 10 μm was observed, which is encouragingly low for FDM printing.

Computational fluid dynamic simulations are being increasingly used to develop new flow bodies for redox flow batteries and electrolysers,[30] and these simulations usually require experimental validation. 3D printing can be used to produce physical prototypes of complex flow models for such validation. For example, Ponce de Leon *et al.*[29] developed a 3D-printed flow-cell to determine mass transfer characteristics, such as the correlation between Sherwood and Reynolds numbers during the reduction of ferricyanide ions at nickel electrodes.

## 7.3 Chemical and Thermal Stability of 3D Print Materials

The stability of 3D print materials at high temperatures and with different solvents is a crucial factor to consider while developing 3D-printed reactors. A wide variety of print materials are available commercially, hence it is impractical to comment on the chemical and thermal stability of each material. A brief discussion on the stability of the material types that are most commonly used is presented below.

FDM printers can use a wide variety of thermoplastics, such as poly(lactic acid), polycarbonate, acrylonitrile-butadiene-styrene, polyetherimide, and polypropylene (as discussed in Chapter 1). Some of these thermoplastics with ester linkages, such as polycarbonate, poly(lactic acid), and polyetherimide are susceptible to chemical degradation. Moreover, the glass transition temperatures for polycarbonate and poly(lactic acid) are only 147 °C and 60–65 °C, respectively. However, materials such as polypropylene, polyphenylsulfone, and acrylonitrile-butadiene-styrene can offer greater chemical and thermal resistance. Polypropylene offers resistance to organic solvents, acids, and bases, and it has a thermostability up to 150 °C.[19] Acrylonitrile-butadiene-styrene is one of the most thermally stable thermoplastics, with a glass transition temperature of 220 °C.

SL and PolyJet printers mostly use polyacrylate-based resins, which offer poor chemical and thermal stability due to weak acrylate and epoxy backbones. Most of these materials suffer significant degradation and swelling in organic solvents. However, there are some exceptions, such as Accura 60. As mentioned above, Accura 60 demonstrated minimal swelling in a wide variety of organic solvents.[11] Polyacrylate-based materials can offer some resistance to dilute acids and bases, but the thermal stability of these materials is generally poor. Accura Bluestone, which is marketed as a high-temperature material has a glass transition temperature of only 83 °C.[11]

As discussed in Chapter 1, SLS printers can use a variety of polymers, such as nylon, polyether ether ketone, polypropylene, and acrylonitrile-butadiene-styrene. Most of these polymers are thermostable, however, they vary greatly regarding their chemical stability. Polymers such as nylon are unstable in strong acids and bases because of their amide linkages, however, they are resistant to organic solvents. Polymers such as polyether ether ketone are highly resistant to a wide variety of chemicals, and they are routinely used in chemical applications.

SLM printers use metal powders such as stainless steel, aluminium, and titanium alloy. These materials generate a highly stable oxide layer, which gives them an extremely high resistance to a wide variety of chemicals. Moreover, these metals generally have a very high melting point. The melting points of aluminium and titanium alloy are 660 °C and 1600 °C, respectively.

## 7.4 Outlook

As compared to conventional glass blowing and microfabrication techniques, 3D printing offers rapid and cost-effective development of bespoke reactors for both synthetic chemistry and electrochemistry. Table 7.1 lists the different techniques, printers, and print materials that have been used for printing the various reactors that are discussed in this chapter.

The seamless integration of analytical techniques with 3D-printed reactors allows the development of automated flow systems with feedback loops for self-optimisation. Coupling 3D printing with digital design technology further allows systematic optimisation of the reactor configuration to provide greater control of chemical reactions.[1] 3D printing also provides an opportunity to experimentally validate computational simulations by producing physical prototypes of the model domain. Moreover, the 3D print files can be easily shared among different research groups and users for modification or direct printing.

The most commonly used printing technique in synthetic chemistry and electrochemistry is FDM printing. The popularity of FDM printers is only partially attributed to the fact that they are inexpensive and easy to operate. The most significant advantage offered by these printers in the development of reactors is the ability to pause and resume a print, while allowing localised deposition of additional reagents. This unique ability provides the potential to significantly advance the field of reactor development. Moreover, FDM printers offer a broad choice of thermoplastic polymers, including chemical, and up to a certain degree temperature resistive polymers, such as polypropylene, polyphenylsulfone, and acrylonitrile-butadiene-styrene. Similarly, SLM printers can be used to develop chemically and thermally stable reactors, as they can print with metals such as stainless steel (316L) and titanium alloy (Ti-6Al-4V), which are typical materials for conventional metal reactors. However, both FDM and SLM printers suffer from poor resolution and difficulty in printing highly complex channels.

SLS printers also offer the advantage of printing with a wide variety of polymers. However, their use in synthetic chemistry and electrochemistry has been limited by the high cost of SLS printers and the porous nature

**Table 7.1** The 3D printing techniques, printers, and print materials that are used for fabricating different components for synthetic or physical chemistry.

| 3DP technique | Printer: Manufacturer | Printer: Model | Print material used | 3D-printed components[a] |
|---|---|---|---|---|
| FDM | 3D Systems | 3D Touch | Polypropylene | Hydrothermal reactionware[4]<br>Flow reactors[3,10,16,19]<br>Electrolyser flow plate[27] |
| | | UP Plus | Acrylonitrile butadiene styrene | Electrochemical flow-cell[29] |
| | Stratasys | Dimension | Acrylonitrile butadiene styrene | Flow reactor[11]<br>Redox flow battery flow plate[28] |
| | — | — | Acrylonitrile butadiene styrene | Reaction flasks[5] |
| | | | High-impact polystyrene | Flow reactor[12] |
| | | | Nylon | Flow reactor[12] |
| | | | Poly(lactic acid) | Flow reactor[12] |
| PolyJet printing | 3D Systems | ProJet 3000 Plus | Polyacrylate composite resin | Flow reactor[11] |
| SL | 3D Systems | Viper | Acura® 60 | Flow reactor[21] |
| | envisionTEC | Ultra® 3SP | Acrylonitrile butadiene styrene | Electrochemical flow-cell[29] |
| | Viper | si2 | Accura® 60 | Flow reactor[11] |
| SLM | EOS | EOSINT M 280 | Stainless steel (316L) | Flow reactor[9] |
| | Realizer | SLM 50 | | Flow reactor[11] |
| | Renishaw | AM250 | | Flow reactors[11,21] |
| | | | Titanium alloy (Ti-6Al-4V) | Flow reactors[21] |
| SLS | — | — | Nylon-12 | Flow reactor[11] |

[a]The components have been named in this table as per the names proposed in their respective reports. In the case of the absence of any clearly proposed name, generic names based on their applications have been used.

(leakage) of the printed reactors. Both SL and PolyJet printers currently are unable to print with chemically and thermally stable materials, which significantly limits their usage in the field.

The list of 3D-printable materials is growing rapidly, and the advent of 3D printing with materials such as glass[31] and graphene[23] will further promote the use of 3D printing for developing high temperature and chemically resistant devices. Research groups are already working on the development of new functional polymers, 3D print materials with embedded catalysts, and surface modification processes for improving the performance of 3D-printed reactors.[4] Moreover, multi-material 3D printing would provide a further means to develop complex reactors and devices in a single-bodied arrangement.

## References


1. M. D. Symes, P. J. Kitson, J. Yan, C. J. Richmond, G. J. Cooper, R. W. Bowman, T. Vilbrandt and L. Cronin, *Nat. Chem.*, 2012, **4**, 349–354.
2. P. J. Kitson, S. Glatzel, W. Chen, C.-G. Lin, Y.-F. Song and L. Cronin, *Nat. Protoc.*, 2016, **11**, 920–936.
3. P. J. Kitson, M. D. Symes, V. Dragone and L. Cronin, *Chem. Sci.*, 2013, **4**, 3099–3103.
4. P. J. Kitson, R. J. Marshall, D. Long, R. S. Forgan and L. Cronin, *Angew. Chem., Int. Ed.*, 2014, **53**, 12723–12728.
5. F. Lederle, C. Kaldun, J. C. Namyslo and E. G. Hübner, *Helv. Chim. Acta*, 2016, **99**, 255–266.
6. M. B. Plutschack, B. Pieber, K. Gilmore and P. H. Seeberger, *Chem. Rev.*, 2017, **117**, 11796–11893.
7. M. Movsisyan, E. Delbeke, J. Berton, C. Battilocchio, S. Ley and C. Stevens, *Chem. Soc. Rev.*, 2016, **45**, 4892–4928.
8. B. Gutmann, D. Cantillo and C. O. Kappe, *Angew. Chem., Int. Ed.*, 2015, **54**, 6688–6728.
9. B. Gutmann, M. Köckinger, G. Glotz, T. Ciaglia, E. Slama, M. Zadravec, S. Pfanner, M. C. Maier, H. Gruber-Wölfler and C. O. Kappe, *React. Chem. Eng.*, 2017, **2**, 919–927.
10. P. J. Kitson, M. H. Rosnes, V. Sans, V. Dragone and L. Cronin, *Lab Chip*, 2012, **12**, 3267–3271.
11. A. J. Capel, S. Edmondson, S. D. Christie, R. D. Goodridge, R. J. Bibb and M. Thurstans, *Lab Chip*, 2013, **13**, 4583–4590.
12. S. Rossi, R. Porta, D. Brenna, A. Puglisi and M. Benaglia, *Angew. Chem., Int. Ed.*, 2017, **56**, 4290–4294.
13. Q. Xue, F. Foret, Y. M. Dunayevskiy, P. M. Zavracky, N. E. McGruer and B. L. Karger, *Anal. Chem.*, 1997, **69**, 426–430.
14. D. Fabris, *Mass Spectrom. Rev.*, 2005, **24**, 30–54.
15. D. L. Browne, S. Wright, B. J. Deadman, S. Dunnage, I. R. Baxendale, R. M. Turner and S. V. Ley, *Rapid Commun. Mass Spectrom.*, 2012, **26**, 1999–2010.
16. J. S. Mathieson, M. H. Rosnes, V. Sans, P. J. Kitson and L. Cronin, *Beilstein J. Nanotechnol.*, 2013, **4**, 285.
17. C. J. Welch, X. Gong, J. Cuff, S. Dolman, J. Nyrop, F. Lin and H. Rogers, *Org. Process Res. Dev.*, 2009, **13**, 1022–1025.
18. E. Garcia-Egido, V. Spikmans, S. Y. Wong and B. H. Warrington, *Lab Chip*, 2003, **3**, 73–76.
19. V. Dragone, V. Sans, M. H. Rosnes, P. J. Kitson and L. Cronin, *Beilstein J. Org. Chem.*, 2013, **9**, 951.
20. G. Scotti, S. M. Nilsson, M. Haapala, P. Pöhö, G. B. af Gennäs, J. Yli-Kauhaluoma and T. Kotiaho, *React. Chem. Eng.*, 2017, **2**, 299–303.
21. A. J. Capel, A. Wright, M. J. Harding, G. W. Weaver, Y. Li, R. A. Harris, S. Edmondson, R. D. Goodridge and S. D. Christie, *Beilstein J. Org. Chem.*, 2017, **13**, 111.

22. K. Sun, T. S. Wei, B. Y. Ahn, J. Y. Seo, S. J. Dillon and J. A. Lewis, *Adv. Mater.*, 2013, **25**, 4539–4543.
23. K. Fu, Y. Wang, C. Yan, Y. Yao, Y. Chen, J. Dai, S. Lacey, Y. Wang, J. Wan and T. Li, *Adv. Mater.*, 2016, **28**, 2587–2594.
24. K. Alayavalli and D. L. Bourell, *Rapid Prototyping J.*, 2010, **16**, 268–274.
25. U. Kalsoom, P. N. Nesterenko and B. Paull, *RSC Adv.*, 2016, **6**, 60355–60371.
26. F. J. del Campo, *Electrochem. Commun.*, 2014, **45**, 91–94.
27. G. Chisholm, P. J. Kitson, N. D. Kirkaldy, L. G. Bloor and L. Cronin, *Energy Environ. Sci.*, 2014, **7**, 3026–3032.
28. L. Arenas, F. Walsh and C. P. de León, *ECS J. Solid State Sci. Technol.*, 2015, **4**, P3080–P3085.
29. C. Ponce de Leon, W. Hussey, F. Frazao, D. Jones, E. Ruggeri, S. Tzortzatos, R. Mckerracher, R. Wills, S. Yang and F. Walsh, *Chem. Eng. Trans.*, 2014, **41**, 1–6.
30. M. R. Cruz-Díaz, E. P. Rivero, F. J. Almazán-Ruiz, Á. Torres-Mendoza and I. González, *Chem. Eng. Process.*, 2014, **85**, 145–154.
31. F. Kotz, K. Arnold, W. Bauer, D. Schild, N. Keller, K. Sachsenheimer, T. M. Nargang, C. Richter, D. Helmer and B. E. Rapp, *Nature*, 2017, **544**, 337.

CHAPTER 8

# 3D Printing in Chemical Education

## 8.1 Introduction

Most chemistry disciplines rely heavily on the understanding of the three-dimensional spatial arrangement of atoms within a molecule, molecular complexes, and crystals. This provides an essential foundation for concepts such as crystal structure, molecular symmetry, stereochemistry, and macromolecular structures, to name a few. Conventionally, the combination of verbal explanation and 2D textbook images is used to describe these concepts. Textbook images have improved significantly in the last few years, and associated technology has seen modern chemistry teaching utilising virtual 3D models to move spatial understanding even further. However, these advances still lack the kinaesthetic learning experience that physical models provide.

The use of three-dimensional physical molecular models has no doubt resulted in greater understanding of complex molecular structures within the classroom.[1] Physical models demand greater attention from students and help them to retain and use concepts more effectively over time. Such physical models for chemical education have been produced using various materials over the years, including paper,[2,3] polystyrene spheres,[4–7] plastic beads,[8] glass beads,[9] and even bottle caps.[10] These models are assembled manually, and they are limited to simple structures. 3D printing allows rapid design, production, and easy customisation of three-dimensionally complex physical models, and hands-on experience for students in design, fabrication, and use. As a complete process, this presents a novel and modern activity for gaining greater student engagement.

3D Printing in Chemical Sciences: Applications Across Chemistry
By Vipul Gupta, Pavel Nesterenko and Brett Paull

Published by the Royal Society of Chemistry, www.rsc.org

To the best of the authors' knowledge, the first 3D-printed chemical education models were reported back in 2006 by Herman *et al.*[11] The models were printed using an SLS printer, and they were used to explain protein structure and function in secondary school and undergraduate classes. They received an overwhelmingly positive response from students, teachers, and researchers. Students found the models interesting and were motivated to learn more about the invisible molecular world. One of the teachers commented; *"I would almost go so far as to say that models are essential to teaching protein structure. A 2-D representation in a textbook, or even the computer models with a 3-D feel, just cannot convey all aspects of structure".*

## 8.2 Resources for 3D Printing Chemistry Models

Many methods and protocols have been reported for the 3D printing of physical models from crystallographic information files (CIF) and protein databank (PDB) files. These protocols allow conversion of CIF and PDB files into STL files or VRML files.

Molecular modelling software such as Mercury, Vesta, Visual Molecular Dynamics (VMD) and Chimera have been used with computer-aided design software such as Blender and Magics to convert CIF files to VRML files. These methods utilise open source, easy-to-use, and widespread software to generate models for both discrete molecules and segments of an infinite structure.[12–17] Blender has also been used with plug-in software such as Python Molecular Viewer[12] and MolPrint3D[18] to customise the model. MolPrint3D allows ball-and-stick models to be selectively split into multiple fragments to enhance their printing performance.[18] The software adds a 'pin' to the bond and a 'hole' in the atom to quickly assemble the printed fragments in a 'Plug-and-Play' fashion. It also provides automated and interactive tools to orientate models on the print bed, add supports and struts, and manipulate space-filling CPK models for multi-colour printing.

Software such as Cif2VRML,[19] WinXMorph,[20,21] and WinTensor[22] have been developed to convert CIF files directly to STL or VRML files.[22] The National Institutes of Health (NIH) has introduced an open, comprehensive, and interactive website, NIH 3D Print Exchange, for searching, browsing, downloading, and sharing 3D-printable models. CIF or PDB files obtained from online databases can be directly converted into STL files by uploading them to the NIH 3D Print Exchange, eliminating the need for extra software. Recently, Jones *et al.*[23] have applied this resource to generate 3D-printable models for a fluorescent tag bound to fatty acid (structure optimised in GAUSSIAN), horseradish peroxidase enzyme (obtained from a PDB database), graphene sheet (built with VMD), silicene (built with Materials Studio), and functionalised silicene (structure optimised by Vienna *Ab initio* Simulation Package) (Figure 8.1). The models created with the 3D Print Exchange are usually stored on the server and can be

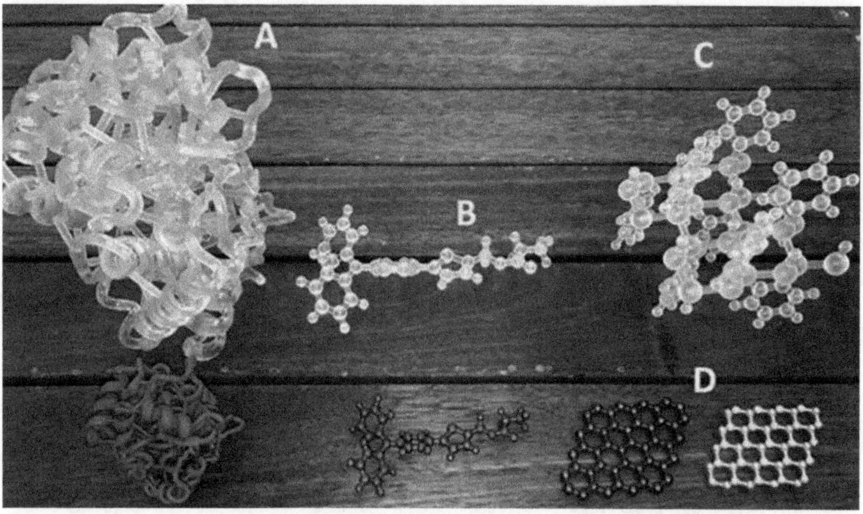


**Figure 8.1** Models 3D-printed using NIH 3D Print Exchange website. (A) horseradish peroxidase printed by SL (top), and Shapeways (bottom), (B) fluorescent molecule printed by FDM (top), and Shapeways (bottom), (C) SL-printed functionalised silicone, and (D) Shapeways printed graphene (left), and silicone (right). Reproduced from ref. 23 with permission from American Chemical Society, Copyright 2018.

shared with others. A similar online resource is the JMOL Crystal Symmetry Explorer web tool;[24] however, it is not as simple to use as the NIH 3D Print Exchange.[23]

## 8.3 Organic Chemistry

Ball and stick models have been extensively used in organic chemistry for decades. However, the molecular structures produced are limited in type and accuracy.[23] Moreover, ball and stick models for structures with more than 15 carbon atoms are fragile, cumbersome, and relatively expensive when multiple models are required.[17] However, 3D printing offers the opportunity to design and produce three-dimensionally complex and more realistic models that can circumvent the above-mentioned limitations, and indeed involve the students in their design and production.

Chemistry model kits are usually limited in size and content, which restricts their use to simple molecules for teaching essential concepts such as chirality and stereochemistry.[10,25,26] Recently, Van Wieren *et al.* 3D-printed multi-colour models for complex chiral compounds such as *cis*- and *trans*-dichloro-bis(2-picolylamine)-iron(II) (Figure 8.2) to aid in stereochemistry discussions.[17] The models were 3D-printed with a binder jetting printer (Zprinter 450), which offers multi-colour printing. The ability to

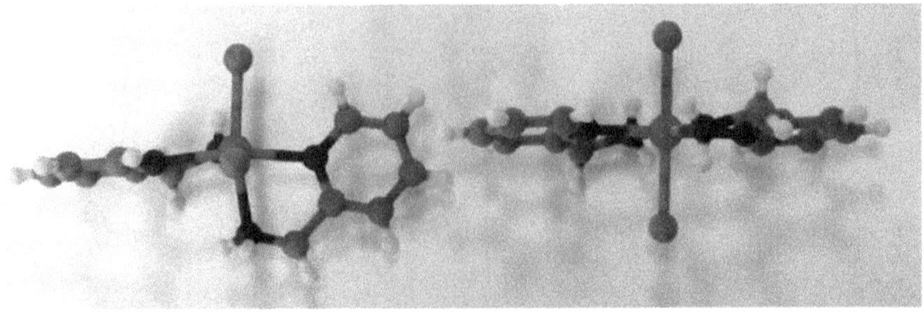

**Figure 8.2** Binder jetting printed multi-colour models. Left to right: *cis*-dichloro-bis-(2-picolylamine)-iron(II) and *trans*-dichloro-bis(2-picolylamine)-iron(II). Reproduced from ref. 17 with permission from American Chemical Society, Copyright 2017.

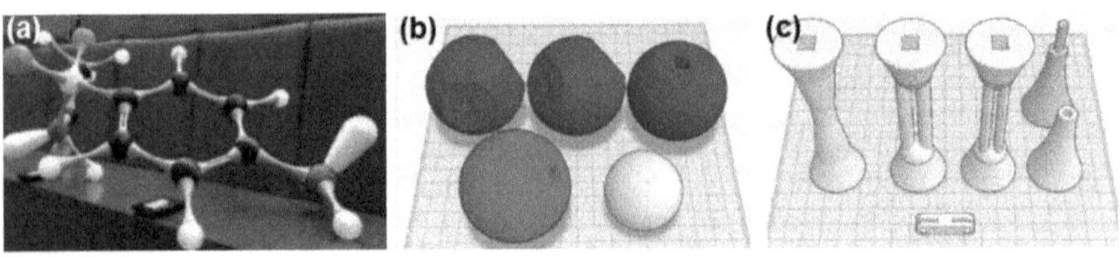


**Figure 8.3** A large molecular model for classroom demonstration. (a) A photograph of the FDM-printed model, (b) renders of the atoms, and (c) renders of the bonds and clips. Reproduced from ref. 27 with permission from American Chemical Society, Copyright 2017.

3D-print with multiple colours enables the production of visually attractive models that can also aid in discussion and explanation of difficult concepts.

As discussed above, conventional molecular model kits are limited in size, and often not ideal for 'front-of-theatre' teaching and demonstrations. Hence, Penny *et al.*[27] attempted to address this with large 3D-printed molecular models for use in such open teaching spaces (Figure 8.3(a)). The models were designed in multiple modular components, such as atoms and bonds, and these components were assembled using a semi-rigid clip (Figure 8.3(b-c)). The carbon atom was 62 mm in diameter, and the bonds were 104 mm in length. Both atoms and bonds were 3D-printed with an FDM printer using poly(lactic acid) and acrylonitrile butadiene styrene, respectively. These 3D-printed models were used to teach topics such as molecular shape, stereochemistry, Newman projections, hybridisation, and aromaticity. The models were presented to a class of more than 180 students. The majority of students (more than 70%) agreed that the 3D-printed models helped them understand molecular structure and bonding and that the models were an improvement over 2D images.

In addition to the 3D-printed molecular models, 3D-printed atomic models have also been used to teach essential concepts, such as Bohr atomic theory, bond polarity, and hybridisation. Smiar *et al.* FDM-printed atomic

**Figure 8.4** FDM-printed Bohr model of boron. Reproduced from ref. 28 with permission from American Chemical Society, Copyright 2016.

models that consisted of a central nucleus with rotating orbitals and clip-on electrons (Figure 8.4).[28] The model could be used to generate atoms from hydrogen to neon, and it was used to teach Bohr atomic theory. The ability of FDM printers to print with different infill percentages was further used to demonstrate the higher electronegativity of oxygen compared to nitrogen. The oxygen atom was printed with a higher infill percentage to make it denser compared to the nitrogen atom. Moreover, different colours were used to differentiate σ- and π- bonds and to demonstrate hybridisation. Further, Carroll *et al.* FDM-printed calculated geometries of stable and highly-strained olefins and conjugated π-systems to add to the discussion.[29]

The above-mentioned FDM-printed atomic models were limited to the 2D representations of the Bohr model. Hence, Gupta *et al.* used PolyJet printing to develop more realistic three-dimensional counterparts (Figure 8.5).[30] These models consisted of three-dimensionally nested spherical geometries representing neutrons, protons, the nucleus, and electron shells. The atomic models of hydrogen, helium, and lithium were 3D-printed to scale to physically demonstrate the trends observed across the groups and periods of the periodic table.

The translation of 2D representations of the valence shell–electron pair repulsion (VSEPR) theory into imaginary 3D structures is often challenging for students.[31] Dean *et al.* constructed 3D models of the basic VSEPR

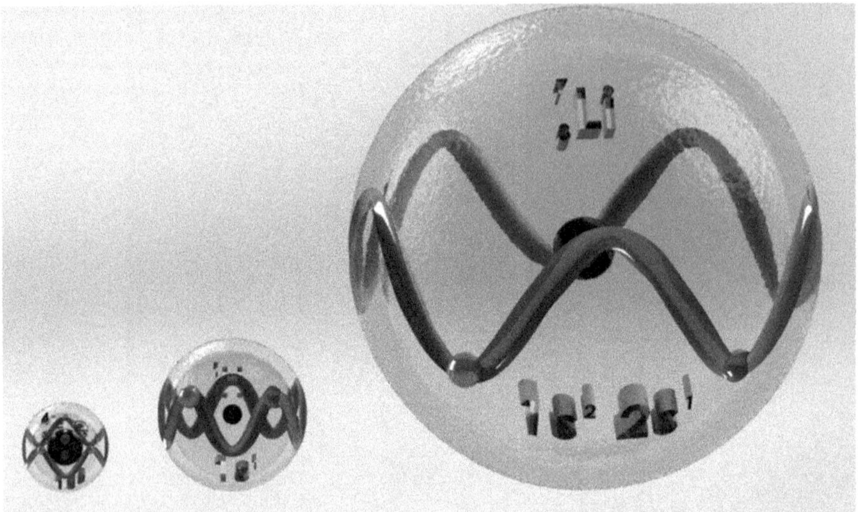

**Figure 8.5** Render of the three-dimensional Bohr atomic models of helium, hydrogen, and lithium (left to right).

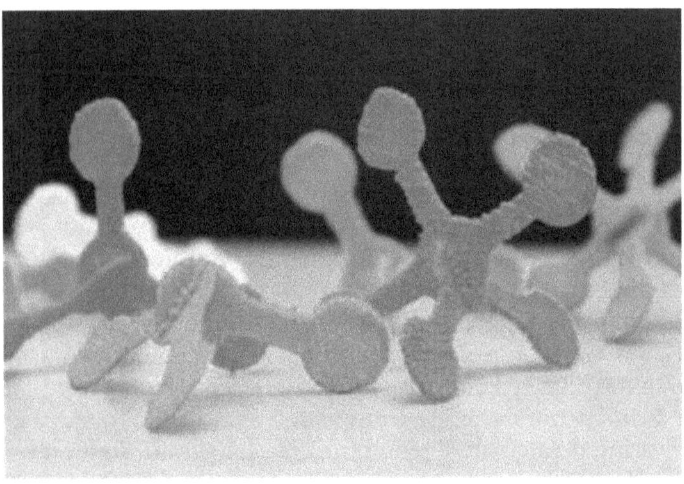

**Figure 8.6** VSEPR models developed using 3D printing pens. Reproduced from ref. 31 with permission from American Chemical Society, Copyright 2016.

shapes, specifically trigonal pyramidal, tetrahedral, seesaw, trigonal bipyramidal, square pyramidal, and octahedral (Figure 8.6).[31] Instead of using standard 3D printers, they successfully used 3D printing pens to develop these models for greater interactive learning. The students were provided with 2D templates of the VSEPR shapes and 3D printing pens, and they were asked to construct 3D models. This exercise presented a novel and highly interactive approach towards 3D printing models, however, a steep learning curve was associated with accurately manipulating the pens, and significant time was required to draw a structure. Moreover, the models produced with the 3D printing pens using acrylonitrile butadiene styrene were not found to be sufficiently robust for easy handling and storage.

Recently, Higman *et al.*[32] were able to develop tangible models of reaction progress surfaces from IR and NMR spectroscopy, and HPLC data (Figure 8.7). Mathematica software was used to convert the raw data from these techniques into STL format for 3D printing. The cost of 3D printing these models in poly(lactic acid) using an FDM printer was less than $1. The models were

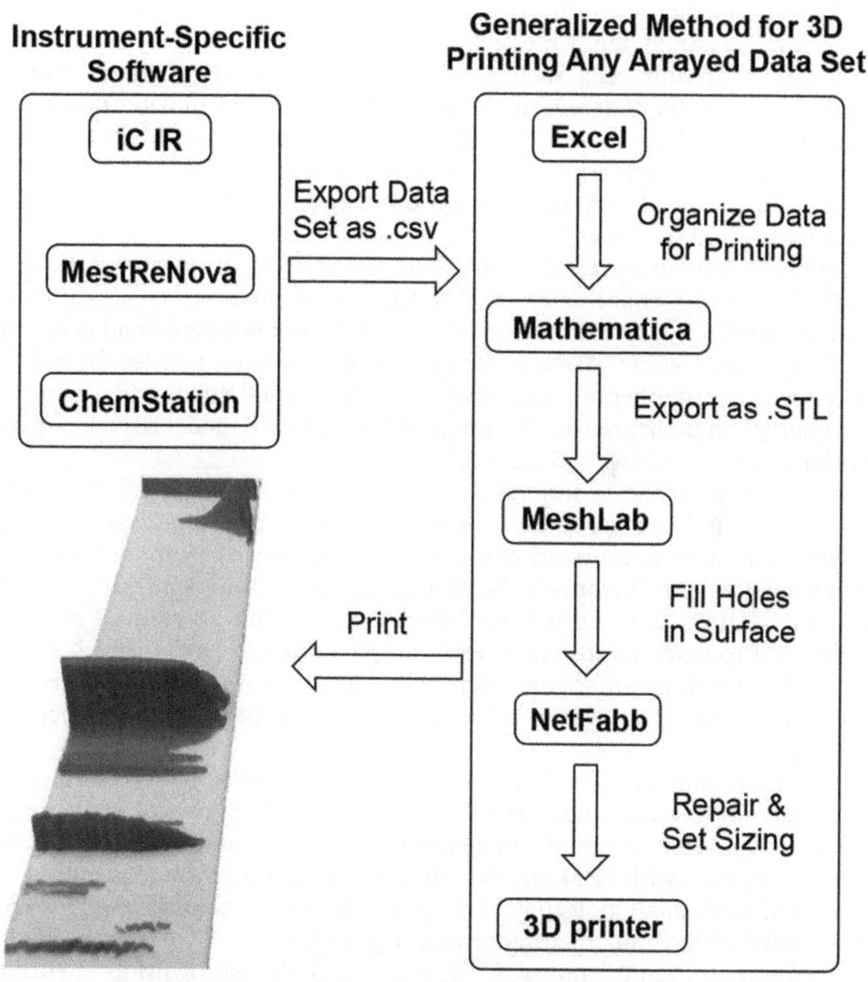


**Figure 8.7** A general flowchart for 3D printing of physical models of arrayed data. Reproduced from ref. 32 with permission from American Chemical Society, Copyright 2017.

used to analyse the trends in the concentrations of starting materials, intermediates, products, and by-products, and it provided a detailed understanding of the reaction progress and pathways.

## 8.4 Inorganic Chemistry

Crystallography models have been around for more than 200 years,[33] and today commercial model kits have become an integral part of every inorganic chemistry class. Commercial kits are designed with general use in mind, and they often fail to aid individual class discussion and the curriculum;[34] whereas 3D printing can be used to design and develop bespoke models to assist in individual lectures. This has provided the motivation to develop 3D-printed crystallography models.

A variety of unit cells such as simple cubic,[34] body-centered cubic,[34] face-centered cubic,[34] perovskite,[15] rock salt,[15] tetragonal spinel,[15] fluorite,[15] and corundum[15] have been 3D-printed. The simple cubic, body-centered cubic, and face-centered cubic unit cells were used in a 300-level general education course to teach crystallography, and they received an overwhelmingly positive student response.[34]

Similarly, Scalfani *et al.*[14] 3D-printed models of $[B_{12}H_{12}]^{2-}$, $[Ge(iPr_2NHC)_3]^{2+}$, $CaF_2$, the Δ and Λ enantiomers of $[Cd(en)_3]^{2+}$, *trans*- and *cis*-$[FeCl_2$(2-(aminomethyl)pyridine$)_2]$, sphalerite and wurtzite ZnS, and staggered and eclipsed $Fe(C_2H_5)$. Casas *et al.*[35] 3D-printed a series of dissection puzzles for cubic, octahedral, trapezohedral, and rhombic dodecahedral geometries to teach symmetry and point groups. The range of FDM-printed models developed by Scalfani *et al.*[14] is shown in Figure 8.8.

In addition to the symmetry and point group models, Scalfani *et al.* also 3D-printed nanostructured morphologies of AB diblock copolymers (spheres on a body-centered cube, hexagonally packed cylinders, gyroid, and lamellae) and linear ABC block copolymer (cylinders at the lamellar interface, and a knitting pattern) (Figure 8.9).[36] The 3D-printed models were used to teach nanostructures in a secondary science outreach activity, and to teach polymer characterisation and crystallography in a graduate course. These models could be further used in the analysis of electron micrographs.

In 2014, Kitson *et al.*[13] and Chen *et al.*[12] used 3D printing to develop monochromatic and multi-colour physical models of X-ray crystal structures, respectively. Monochromatic models of two polyoxommolybdate clusters were 3D-printed with FDM and SL printers (Figure 8.10); whereas multi-colour models of small molecules and metal–organic frameworks (MOF) were 3D-printed with a binder jetting printer (Figure 8.11).

Furthermore, Van Wieren *et al.*[17] also used a binder jetting printer to print multi-colour models of metal complexes, such as ferrocene, $[Th(NO_3)_6]^{4-}$, and bis-terpyridine ruthenium (II) (Figure 8.12). These models were printed using Chimera and Magics software as described above. An issue with the use

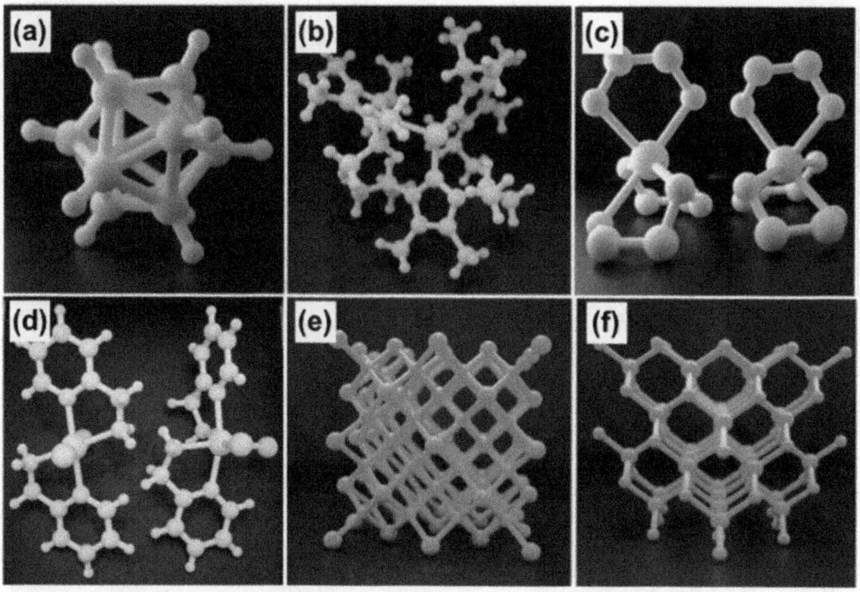


**Figure 8.8** FDM-printed models of (a) $[B_{12}H_{12}]^{2-}$, (b) $[Ge(iPr_2NHC)_3]^{2+}$, (c) Δ and Λ enantiomers of $[Cd(en)_3]^{2+}$, (d) *trans*- and *cis*-$[FeCl_2(2\text{-(aminomethyl)pyridine})_2]$, (e) $CaF_2$, and (f) sphalerite ZnS. Adapted from ref. 14 with permission from American Chemical Society, Copyright 2014.

**Figure 8.9** FDM-printed AB diblock copolymer morphologies. Reproduced from ref. 36 with permission from American Chemical Society, Copyright 2015.

of Chimera software for preparing inorganic models for 3D printing is that it presents ionic bonds as dashed lines, which cannot be directly 3D-printed. Hence, Scalfani *et al.* proposed the replacement of ionic 'dashed' bonds with covalent 'solid' bonds in the software to enable successful printing.[14]

In 2015, Stone-Sundberg *et al.* 3D-printed various models of atomic crystal structures, small molecules, and large molecules to demonstrate their anisotropic properties and morphologies.[33] They developed a new program (Cif2VRML) to obtain 3D-printable STL and VRML files from CIF and PDB files as described above. Figure 8.13 shows their multi-coloured and monochromatic PolyJet-printed models of caffeine and nicotine.

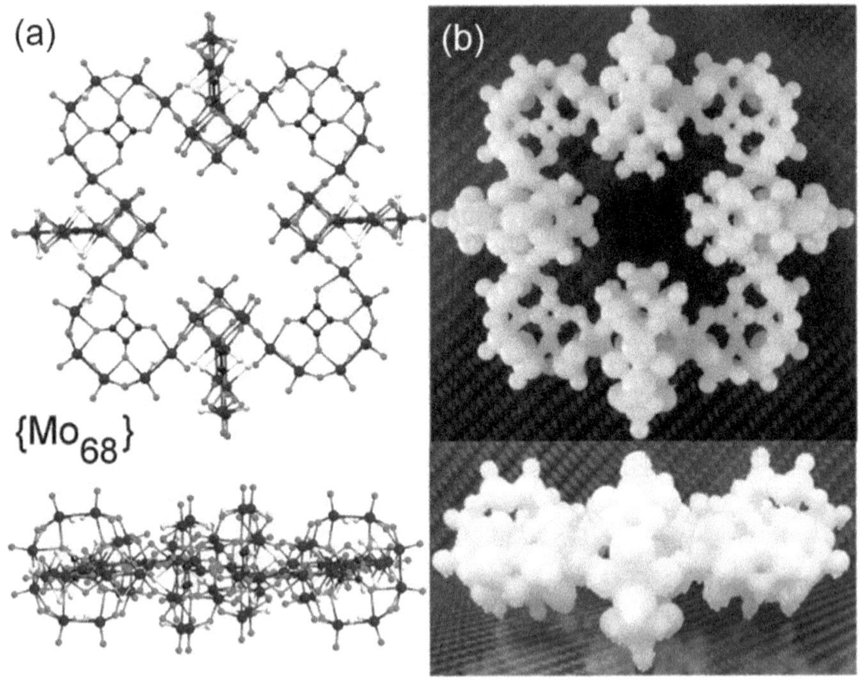


**Figure 8.10** X-ray crystal structure of a polyoxomolybdate cluster: (a) representative ball and stick model, and (b) 3D-printed monochromatic model. Reproduced from ref. 13 with permission from American Chemical Society, Copyright 2014.

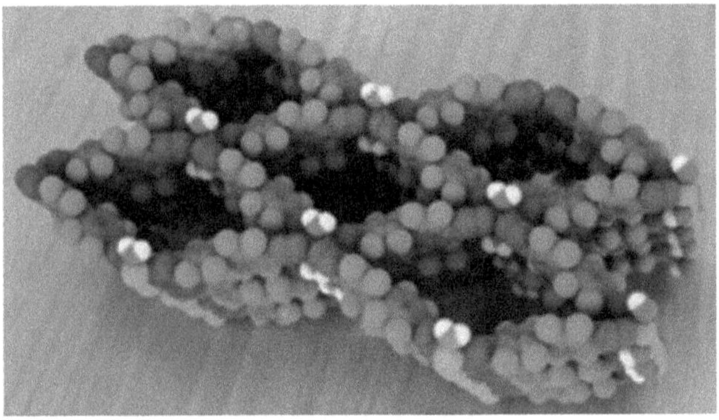

**Figure 8.11** 3D-printed multi-colour model of the MOF structure. Reproduced from ref. 12 with permission from The Royal Society of Chemistry.

## 8.5 Physical Chemistry

Concepts such as potential energy surfaces, reaction trajectories, and atomic and molecular orbitals are the basic foundation of the physical chemistry curriculum. These concepts require students to interpret and

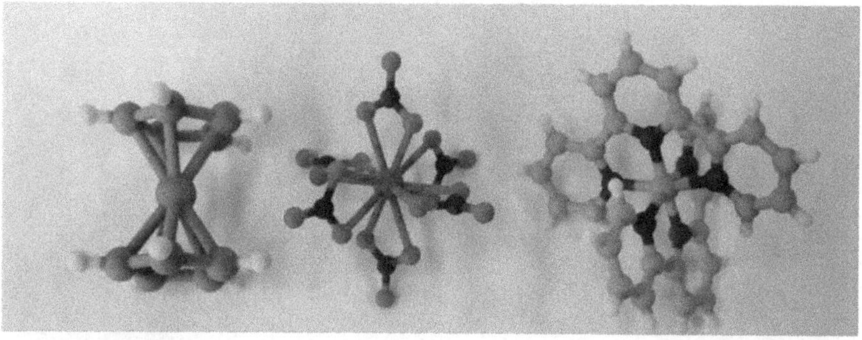

**Figure 8.12** Binder jetting printed multi-colour models. Left to right: ferrocene, $[Th(NO_3)_6]^{4-}$, and bis-terpyridine ruthenium (II). Reproduced from ref. 17 with permission from American Chemical Society, Copyright 2017.

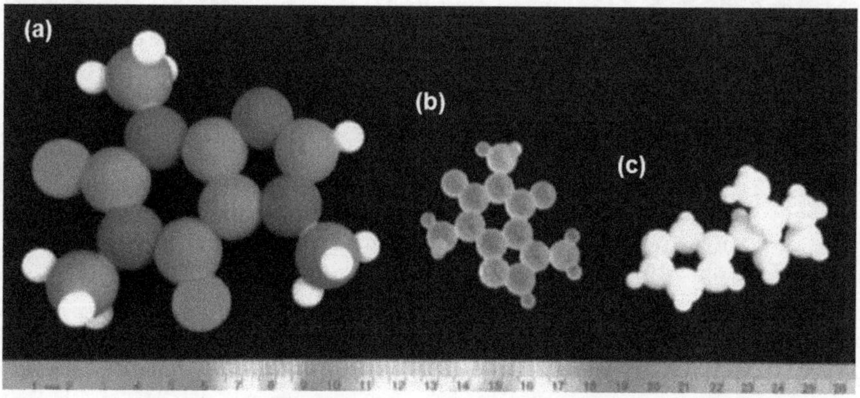


**Figure 8.13** PolyJet-printed models of small molecules: (a) multi-colour model of caffeine, (b) monochromatic model of caffeine, and (c) monochromatic model of nicotine. Reproduced from ref. 33 with permission from John Wiley and Sons, Copyright © 2015 WILEY-VCH Verlag GmbH & Co. KGaA, Weinheim.

visualise multi-dimensional mathematical functions and relationships. Two-dimensional sketches and virtual plots often fail to describe these complex functions and lead to confusion.[37,38] Physical models have been used in the past to explain complex concepts, such as potential energy surfaces.[39–42] However, these models were constructed manually with materials such as plywood and clay, which limited their widespread use. Recently, similar models have been 3D-printed to improve physical and geometric accuracy and ease of production.

In 2014, Lolur *et al.*[43] 3D-printed models for ozone and carbon dioxide potential energy surfaces using Matlab software (Figure 8.14). Individual and combined potential energy surfaces were 3D-printed to demonstrate how multiple surfaces behave and intersect with each other. Further in

**Figure 8.14** 3D-printed models of two different potential energy surfaces of ozone. The 3D-printed balls represent oxygen atoms. Reproduced from ref. 43 with permission from American Chemical Society, Copyright 2014.

2015, Kaliakin *et al.*[44] 3D-printed potential energy surfaces for the hydrogen exchange reaction and the rotation of methyl groups in 1-fluoro-2-methylpropene. They also 3D-printed free energy surfaces for reactions such as protein folding, the dissociation of triatomic molecules, and surface adsorption. Additionally, simple models of double minimum, quadruple minimum, and parabolic surfaces were also 3D-printed. These potential and free energy surfaces were used to teach basic physical chemistry concepts, such as vibrational spectroscopy and reaction kinetics and dynamics. These models were generated using Mathematica software and were FDM-printed (Figure 8.15).

The isoenergy approach allows visualisation of the full dimensionality of the potential energy function.[45,46] Hence, Teplukhin *et al.*[47] 3D-printed physical models for the isoenergy function of a triatomic molecule, ozone, allowing the visualisation of various features of the global energy function, such as covalent wells, transition states, the van der Waals region, and reaction channels (Figure 8.16). Researchers also found these 3D-printed models to be useful for generating numerical calculation grids.

The discussion of hydrogenic orbitals is a fundamental topic, and students encounter it throughout the chemistry curriculum. Models made from

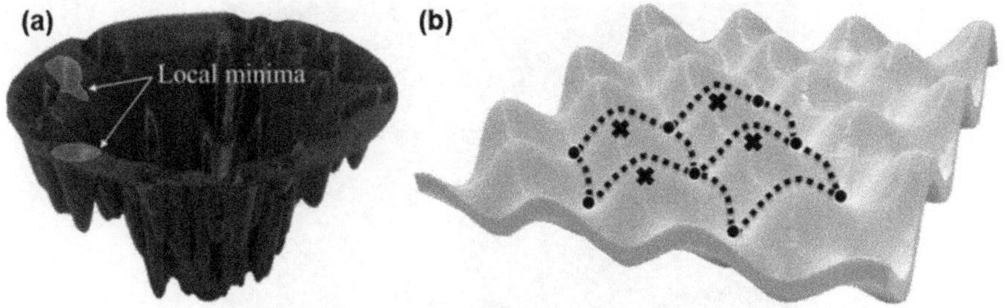


**Figure 8.15** FDM-printed models of (a) a protein folding free energy surface, and (b) an adsorption potential energy surface. Reproduced from ref. 44 with permission from American Chemical Society, Copyright 2015.

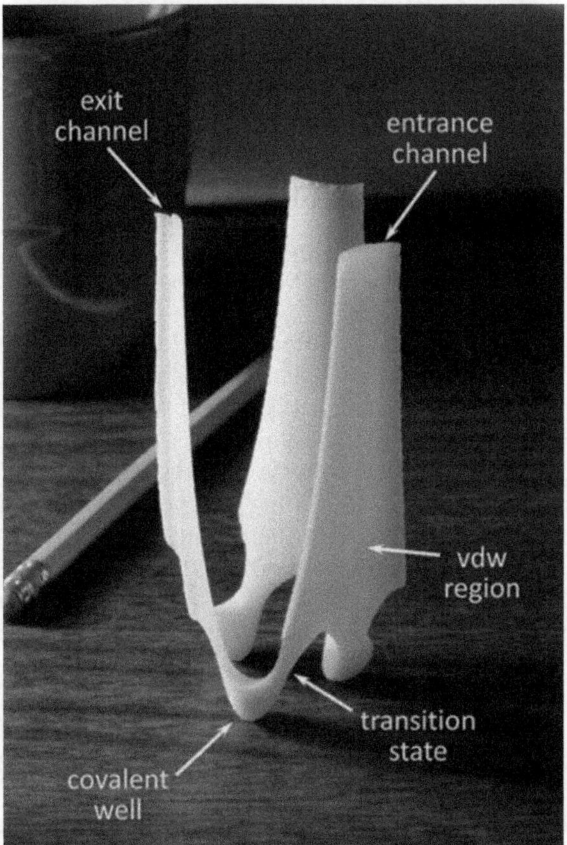


**Figure 8.16** FDM-printed potential energy function of ozone. Reproduced from ref. 47 with permission from American Chemical Society, Copyright 2015.

**Figure 8.17** FDM-printed hydrogenic orbital model. Reproduced from ref. 50 with permission from American Chemical Society, Copyright 2016.

materials such as balloons and Styrofoam are often used to teach these concepts.[48,49] Commercial models are generally artistic renders based on a qualitative interpretation of the isosurfaces.[50] Both approaches are qualitative and indirect, hence they lack the subtle details of the orbital structure. Recently, Griffith *et al.* used mathematical software such as CalcPlot3D (freeware) and Matlab (license required) with 3D printing to generate quantitatively accurate hydrogenic orbital models (Figure 8.17).[50] CalcPlot3D was used to directly save quantum mechanically simulated orbitals in the STL format. The cost of 3D printing a $2p_y$ orbital with an FDM printer was *ca.* \$2.50, whereas the price for a set of three 2p orbital models from Klinger educational is \$140.[50] The FDM-printed models were also found to be sufficiently robust for in-class handling.

The first accurate three-dimensional molecular orbital images were reported by Jorgensen and Salem in 1970.[51] Today, a few commercial models are available for simple demonstrations; however, accurate three-dimensional molecular orbital models are still limited.[52] Hence, Robertson *et al.*[52] 3D-printed molecular orbitals, such as C–H and C–C σ-bonding and antibonding orbitals of ethane, and the highest occupied and lowest unoccupied orbitals of 1,3-butadiene (Figure 8.18). These models were demonstrated for teaching concepts such as introductory quantum

**Figure 8.18** 3D-printed molecular orbitals for (A) LUMO of ethene and the HOMO of 1,3-butadiene, and (B) π lone-pair HOMO of water and σ* LUMO of chloromethane superimposed on 2-chlorotetrahydropyran to illustrate the anomeric effect. Reproduced from ref. 52 with permission from American Chemical Society, Copyright 2015.

chemistry, the anomeric effects in carbohydrate chemistry, Diels–Alder reactions, and the stereochemistry of electrocyclic ring closing and opening reactions.

## 8.6 Biochemistry

Biomacromolecules such as DNA, RNA, and proteins have complex structures, short- and long-range conformational constraints, and specific folding behaviours. A variety of physical biomolecule models are available commercially.[53–55] However, in most instances, these lack the true conformational degrees of freedom, scale, and folded structure of these macromolecules.[56] Hence, visualisation of three-dimensional biomacromolecular structures has been primarily limited to virtual models that do not allow kinaesthetic learning. The use of 3D-printed biomolecular models can overcome these limitations and assist in a better understanding of their complex structure.

The structure-function relationship of biomolecules is one of the most important concepts in biochemistry and chemistry in general. Hence, it was not a surprise that the first 3D-printed chemical education models were used to teach the structure-function relationship of a protein.[57] In 2006, Herman *et al.*[11] 3D-printed models of influenza hemagglutinin protein with an SLS printer. Hemagglutinin binds to the mammalian cell at neutral pH. As the pH decreases due to receptor-mediated endocytosis, the protein undergoes a conformational change that leads to viral fusion with the cell membrane. All three subunits of the hemagglutinin protein were 3D-printed individually and colour-coded to show the different regions (Figure 8.19). The 3D-printed subunits were assembled and disassembled by the students to understand the pH-induced conformational change.

Similarly, in 2015, isolated helices of leucine zippers were 3D-printed and colour-coded to describe their structure-function relationship by Meyer *et*

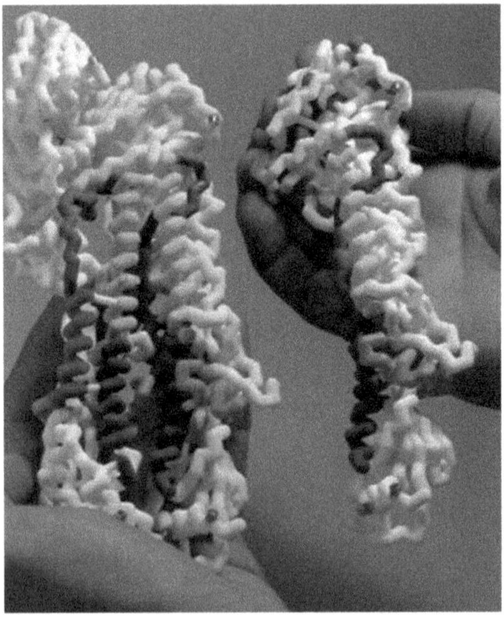

**Figure 8.19** SLS-printed three-subunit hemagglutinin protein. Reproduced from ref. 11 with permission from John Wiley and Sons, Copyright © 2006 International Union of Biochemistry and Molecular Biology, Inc.

**Figure 8.20** FDM-printed and manually painted model of the Fos–Jun leucine zipper. Reproduced from ref. 58 with permission from American Chemical Society, Copyright 2015.

*al.*[58] Isolated helices of the Fos–Jun leucine zipper were 3D-printed with an FDM printer, and they were colour-coded to show their surface amino acid residues (Figure 8.20). Students were able to study the hydrophobic and ionic interactions between the helices by fitting them together and observing the interaction between different amino acid residues.

The assembly of complex structures can be further facilitated by incorporating external accessories such as magnets and actuators into 3D-printed models. Olson *et al.*[59] 3D-printed twelve intermediates of poliovirus and incorporated an oriented magnet to represent its electrostatic complementarity. Just shaking these twelve intermediates in a container resulted in their self-assembly, which presented a unique demonstration of the self-assembly of viral capsids (Figure 8.21).

Further, Chakraborty *et al.*[56] used 3D-printed models and circular magnetic arrays to demonstrate short-range conformational barriers and intrabackbone hydrogen bonding of polypeptide chains (Figure 8.22). The models' flexible backbone chain allowed construction of all the common protein secondary structure elements, which were strengthened due to strong magnetic array connections. They printed an accurate representation of the α-helix length and curvature and parallel β-sheets, helping to provide enhanced understanding of the folding and unfolding of tertiary structures.

**Figure 8.21** Self-assembly of the 12-piece viral capsid. Reproduced from ref. 59 [https://doi.org/10.1073/pnas.0709489104] under the terms of a CC BY 4.0 license [https://creativecommons.org/licenses/by/4.0/].

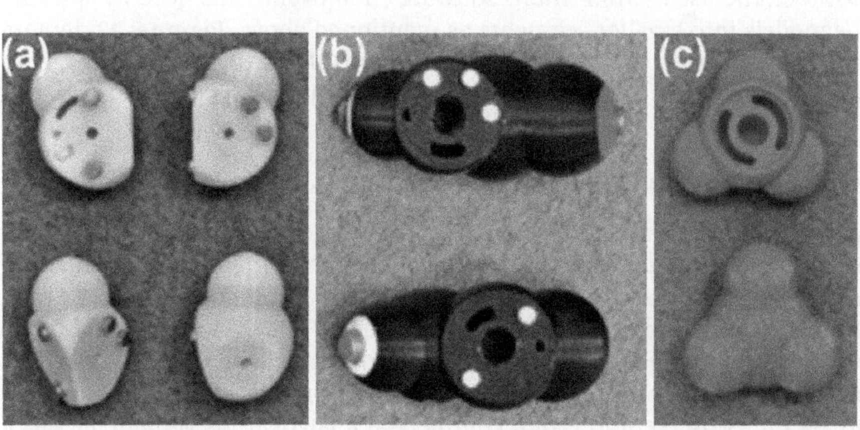


**Figure 8.22** FDM-printed repeating units of polypeptide chains: (a) α-carbon unit of L-amino acid, (b) amide unit, and (c) methyl-group unit for the L-alanine side chain. Reproduced from ref. 56 with permission from PNAS.

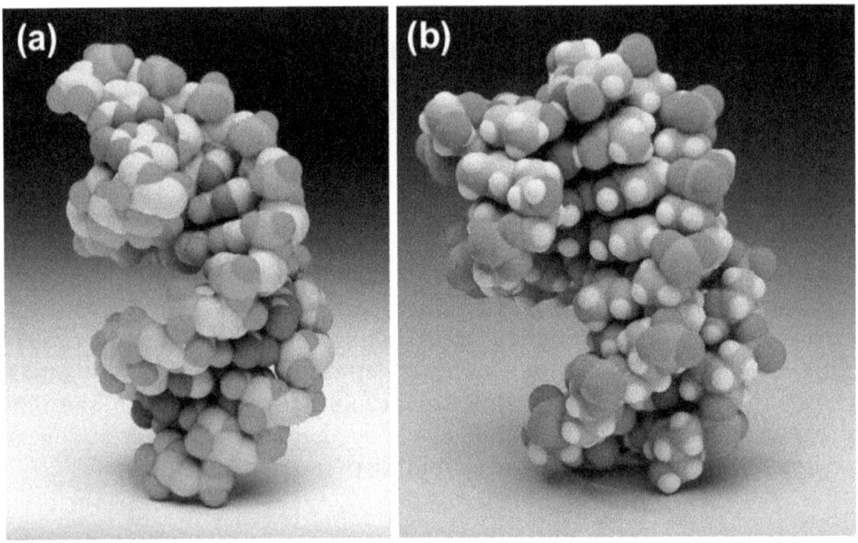


**Figure 8.23** Binder jetting printed model of (a) DNA and (b) RNA. Reproduced from ref. 17 with permission from American Chemical Society, Copyright 2017.

Recently, software such as MolPrint3D and Magics have been used to develop large multi-colour realistic models of one of the most complex biomolecules, including DNA and RNA. On occasion, a large loss of data may occur when handling structures with more than 1000 components (atoms and bonds) using Magics.[17] The loss can be prevented by converting the VRML file into ZPR format.[17] Printing such large models often requires advanced 3D printers and the wastage of a large amount of support material. However, the use of MolPrint3D software circumvents this need by splitting the models into smaller segments as mentioned above. Figure 8.23 demonstrates an example of 3D-printed multi-colour DNA and RNA models. These models were 3D-printed with a binder jetting printer by van Wieren *et al.*[17]

## 8.7 Analytical Chemistry

A comprehensive analytical chemistry education requires extensive training on analytical instruments. However, analytical instruments are usually only introduced in the upper division undergraduate chemistry courses, primarily due to their high cost. These instruments are often few and usually used by groups of students rather than being individually assigned, which can mean that one or two members of the group work on the instrument while others only observe. Moreover, commercial analytical instruments act as a 'black box' for the students and limit their understanding of the underlying mechanisms. Recently, working models of different analytical instruments have been 3D-printed to provide increased availability of the instruments and greater student engagement.

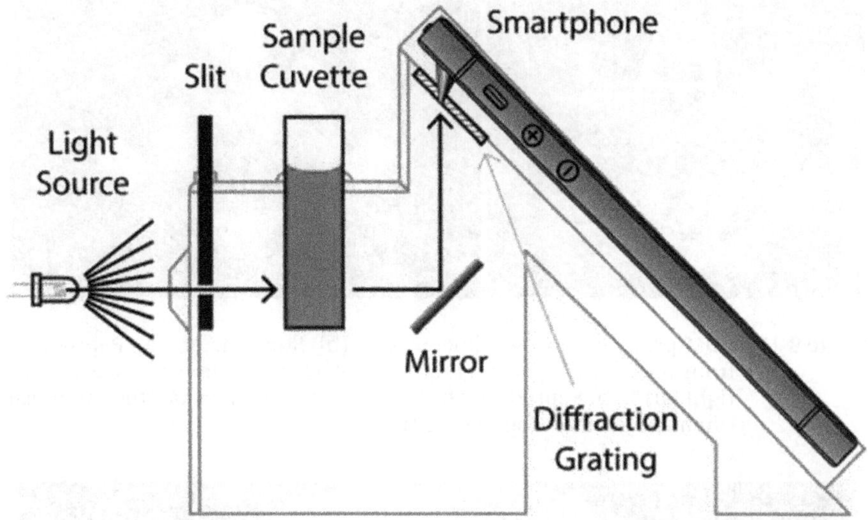


**Figure 8.24** Schematic of the SpecPhone. Reproduced from ref. 60 with permission from American Chemical Society, Copyright 2016.

Absorbance spectroscopy is one of the most widely used techniques in chemical, biochemical, and medical laboratories. It can be used to perform both quantitative and qualitative analysis. Quantitative analysis is primarily based on a simple principle of the Beer–Lambert law. Hence, Grasse *et al.*[60] 3D-printed a smartphone-based spectrophotometer, SpecPhone, to teach the Beer–Lambert law through demonstration. Its optical assembly consisted of a light source, a slit, a cuvette holder, a mirror, and a diffraction grating (Figure 8.24). The housing of the SpecPhone was 3D-printed to allow easy and intuitive assembly of all the components.

Similarly, Porter *et al.* provided students with 3D-printed pieces, optical circuit components, and a digital multimeter to build a colourimeter (Figure 8.25(a)).[61] The cost of the entire colourimeter kit was less than $25. The same set-up was also modified to construct a filter fluorometer by aligning the light source and the detector in an orthogonal arrangement as opposed to a linear arrangement (Figure 8.25(b)).[62] The fluorometer successfully delivered quantitative analysis of luminescent solutes even at ppb levels. These models offer a cheaper alternative to conventional spectrophotometers and fluorometers and result in an interactive learning experience by breaking the 'black box' perception of analytical instruments.

Diffraction devices are commonly found in analytical instruments, especially optical spectrometers.[63] Hence, Piunno 3D-printed a demonstration kit to explain the concepts of diffraction and interference.[63] The demonstration kit consisted of 3D-printed sinusoidal waves of different wavelengths, a 3D-printed platform to arrange the waves at different angles, and a dispersive element (digital versatile disc section) (Figure 8.26). Multiple sinusoidal waves were arranged on the platform at different angles to demonstrate constructive and destructive interference between them. Moreover, interferences

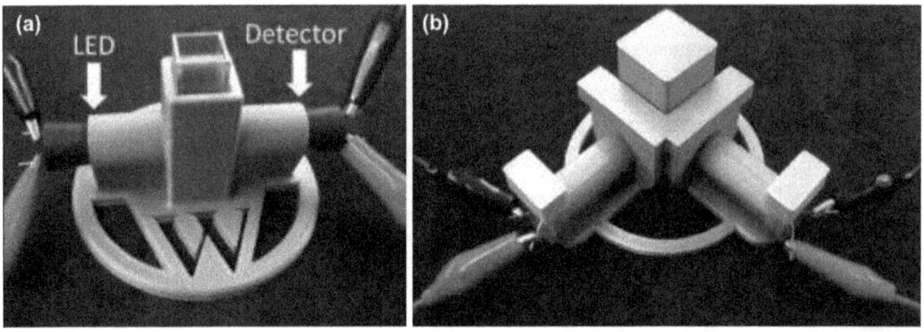


**Figure 8.25** FDM-printed (a) colourimeter and (b) fluorometer. (a) Reproduced from ref. 61 with permission from American Chemical Society, Copyright 2016. (b) Reproduced from ref. 62 with permission from American Chemical Society, Copyright 2017.

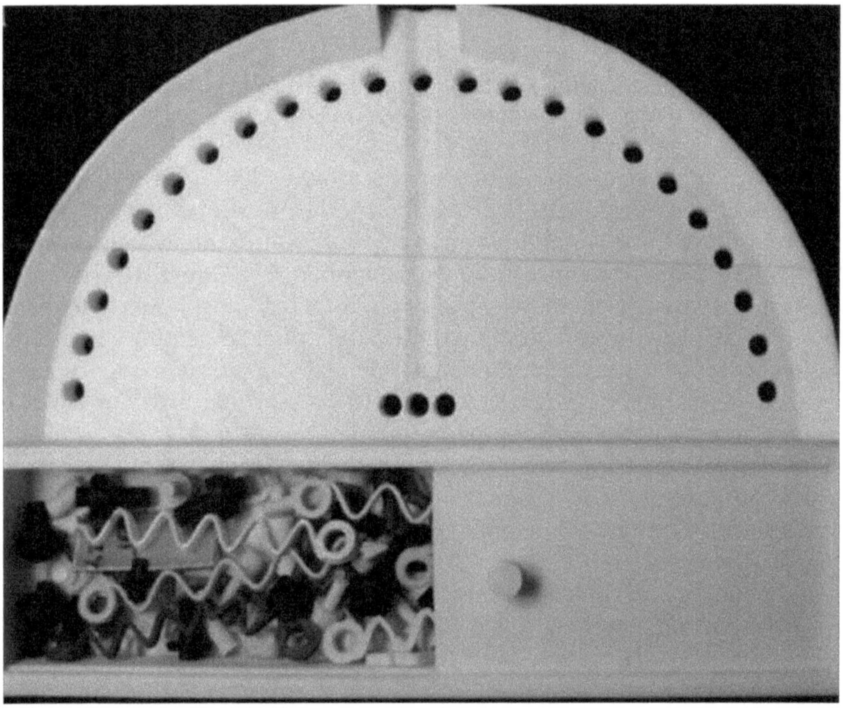

**Figure 8.26** 3D-printed demonstration kit for diffraction and interference. Reproduced from ref. 63 with permission from American Chemical Society, Copyright 2017.

were also observed experimentally using a laser pointer and the dispersive element available in the kit.

Kosenkov *et al.* 3D-printed a temperature-controlled sample holder to perform transient-absorption spectroscopy of the *cis–trans* isomerisation of *N*,*N*-dimethyl-4,4′-azodianiline.[64] The sample holder was developed by

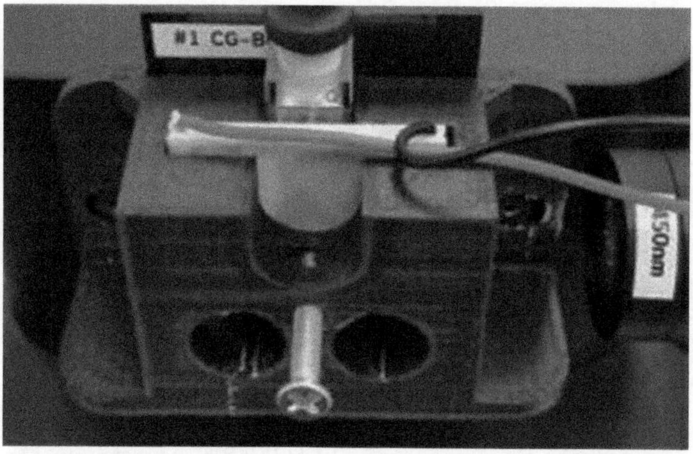


**Figure 8.27** FDM-printed temperature-controlled sample holder to perform transient-absorption spectroscopy. Reproduced from ref. 64 with permission from American Chemical Society, Copyright 2016.

incorporating a Peltier thermoelectric element and a temperature sensor in the FDM-printed acrylonitrile butadiene styrene holder body (Figure 8.27). The students were asked to build an experimental set-up for the transient-absorption spectroscopy and use it to obtain and analyse the kinetic data to calculate the activation energy, enthalpy, entropy, and Gibbs free energy for the transition.

Laboratory demonstrations of liquid phase photocatalysis can be performed through many relatively simple experiments.[65–67] However, the demonstration of gas phase photocatalysis (for example, in a discussion of air pollution) often requires elaborate analytical instruments for gas measurements.[68] In 2014, Stefanov *et al.* developed a 3D-printed gas-phase photocatalytic reactor to demonstrate gas-phase photocatalysis, gas sensing, air pollution monitoring, and photodegradation of air pollutants.[68] The reactor consisted of a photocatalytic filter, an air quality sensor, and an analogue-to-digital converter (Figure 8.28). All the structural components of the reactor were 3D-printed with an FDM printer. Recently, in 2017, Lu *et al.*[69] also developed a two- and four-point convertible probe system using FDM-printed structural components to allow students to perform conductivity measurements in the lab. These in-house developed systems not only offer low-cost instrumental resources for lab demonstrations, but they also provide students with an invaluable experience to develop and customise elaborate analytical instruments.

Epifluorescence microscopy is a powerful tool that can be used to visualise cells and their internal compartments, monitor chemical environments, and study production and distribution of proteins.[70] Although compound light microscopes are readily available in secondary education and undergraduate laboratories, epifluorescence microscopes are usually limited in number. Hence, Stewart *et al.*[71] 3D-printed an adapter to convert a conventional

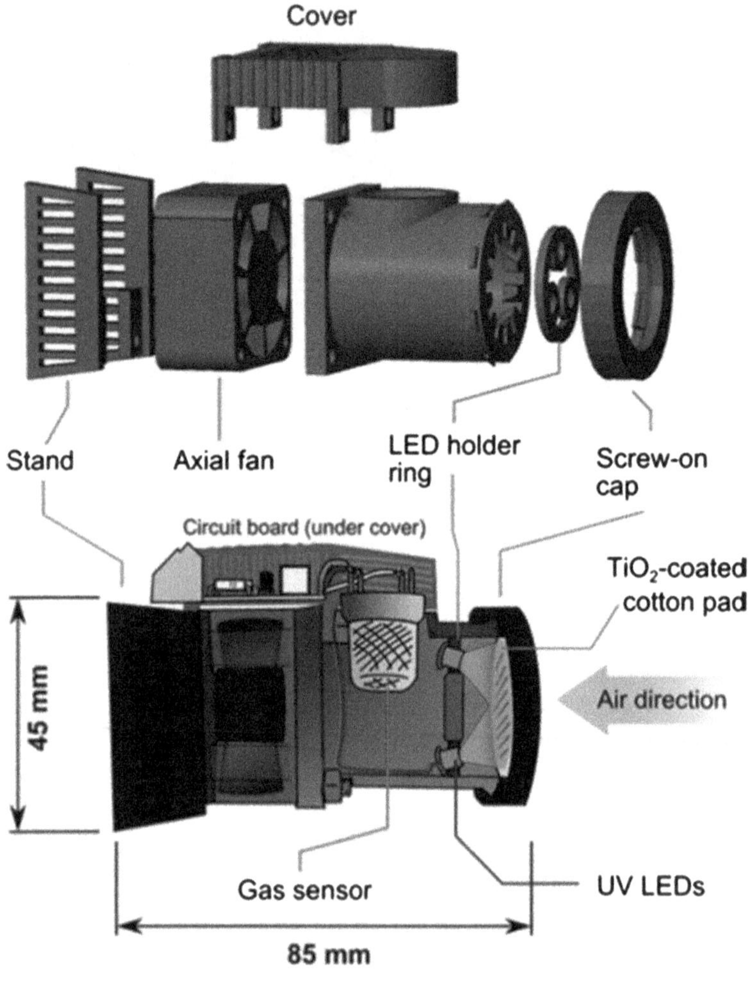


**Figure 8.28** Render of the FDM-printed photocatalytic gas phase reactor. Reproduced from ref. 68 with permission from American Chemical Society, Copyright 2015.

compound light microscope into an epifluorescence microscope (Figure 8.29). The adapter consisted of five 3D-printed parts, specifically a drawer and its handle, an excitation filter holder, a dichroic mirror holder, and an emission filter holder. The microscope can be used to teach the principles of biochemistry, photochemistry, histochemistry, and cell and molecular biology.

## 8.8 Outlook

3D printers are rapidly becoming ubiquitous in schools and universities. As per a 2016 American Library Association report, more than 425 public libraries had 3D printing facilities.[72] Several commonly used chemical and mathematical software packages have started supporting STL and VRML

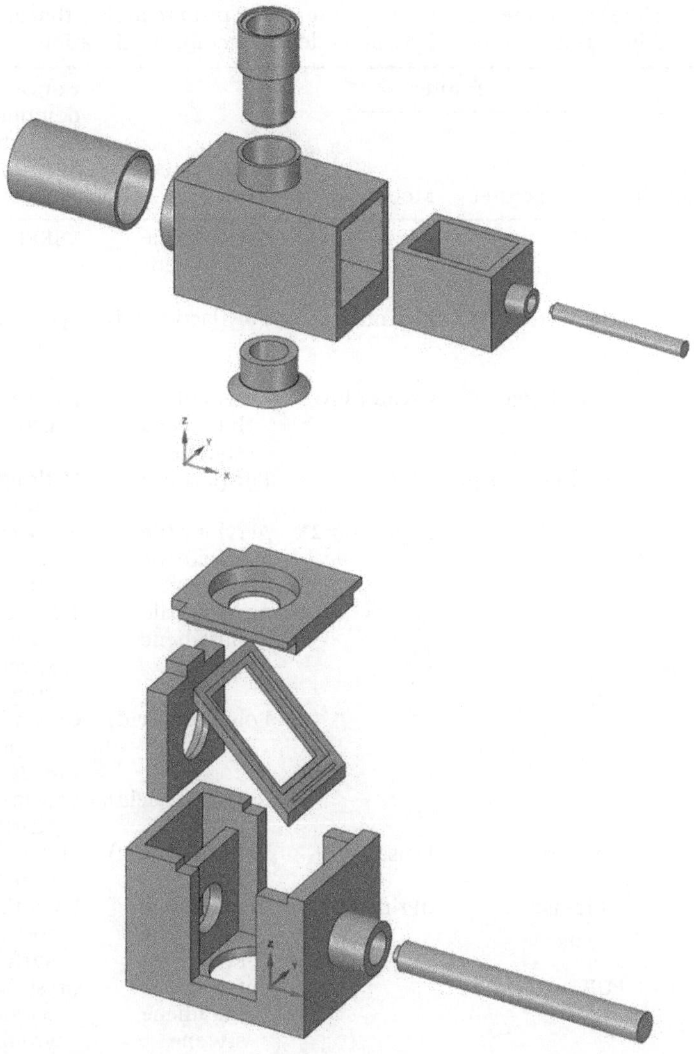

**Figure 8.29** Render of the 3D-printed adapter to convert a conventional compound light microscope into an epifluorescence microscope. Reproduced from ref. 71 with permission from American Chemical Society, Copyright 2016.

file formats to enable the direct 3D printing of simulated models and structures. 3D printing has allowed the development of bespoke, robust, and inexpensive models that can be comfortably used in a classroom setting. Table 8.1 lists the different 3D printing techniques, printers, and print materials that are used to develop physical models for teaching various concepts and principles of chemistry, as highlighted in this chapter.

Unlike conventional models, 3D-printed models are not restricted by the size and complexity of the desired geometry. Moreover, the models are cheap

**Table 8.1** The 3D printing techniques, printers, and print materials that are used for fabricating different physical models for chemical education.

| 3DP technique | Printer: Manufacturer | Printer: Model | Print material used | Concepts demonstrated through 3D-printed models |
|---|---|---|---|---|
| FDM | 3D Printing Pen | | Acrylonitrile butadiene styrene | VSEPR theory[31] |
| | 3D Systems | 3D Touch | Poly(lactic acid) | Symmetry and point groups[14] |
| | FlashForge | Creator Pro | Acrylonitrile butadiene styrene | Molecular orbitals[29] |
| | Fusion3 Design | F306 | Poly(lactic acid) | Hydrogenic orbitals[50] |
| | MakerBot | Replicator 2X | Acrylonitrile butadiene styrene | Block copolymer nanostructure[36] |
| | | | Acrylonitrile butadiene styrene | Transient-absorption spectroscopy[64] |
| | | — | Poly(lactic acid) | Crystallographic unit cells[15] |
| | | | Poly(lactic acid) | Leucine zipper structure[58] |
| | Makerfarm | Prusa i3 | Poly(lactic acid) | Molecular structure[18] |
| | Stratasys | uPrint SE Plus | Acrylonitrile butadiene styrene | Potential and free energy surfaces[44] |
| | UP | | Acrylonitrile butadiene styrene | Crystal symmetry and point groups[35] |
| | Velleman | K8200 | Poly(lactic acid) | Gas-phase photocatalysis reactor[68] |
| | — | — | Acrylonitrile butadiene styrene | Polypeptide chain structure[56] |
| | | | Acrylonitrile butadiene styrene and poly(lactic acid) | Molecular and orbital structure[27] |
| PolyJet printing | Stratasys | Objet Eden 260VS | VeroClear™ | Bohr atomic model[30] |
| | — | — | VisiJet® M3 Crystal | Molecular structure[33] |

**Table 8.1** (*continued*)

| | Printer | | | Concepts demonstrated through 3D-printed models |
|---|---|---|---|---|
| 3DP technique | Manufacturer | Model | Print material used | |
| Binder Jetting | Zcorporation | Zprinter 450 | — | Molecular structure[17] |
| SL | 3D Systems | Projet 7000HD | Polyacrylate composite | Molecular structure[23] |
| | EOS | Formiga P100 | Polyamide | Crystal symmetry and point groups[35] |
| | Formlabs | Form 1 | Polyacrylate composite | X-ray crystal structure[13] |
| SLS | Zcorporation | — | Nylon | Protein structure[57] |

and easy to produce, increasing availability and hence student interaction. 3D printing also facilitates the development of accurate models and their precise reproduction. 3D-printed instruments break the 'black box' notion of commercial instruments to allow a better understanding of their underlying principles.

The advent of low-cost, multi-colour, and multi-material 3D printers would further assist the development of novel chemical education models. Multi-colour 3D printers would allow the development of more realistic, attractive, and descriptive physical models. Lowering the cost of 3D printing would obviate the need for sharing and would even allow students to take models home for further study. Incorporation of magnets, sensors, actuators, and microprocessors in the 3D-printed models would further produce highly interactive models.[56] Molecular image recognition systems such as Microsoft Kinect would facilitate better correspondence between physical and virtual models.[13] A shared library of 3D-printed molecular structures, amino acids, crystallography, and molecular dynamics models would allow better integration of physical models in the chemistry curriculum.

Currently, most libraries and teaching exercises rely on FDM 3D printers because of their low cost and ease of use. Several reports have been published that demonstrate the emission of harmful ultrafine particles and volatile organic compounds during FDM printing.[73–79] The safe use of FDM and other 3D printers by students would require the establishment of standard operating procedures and safety guidelines.[80] The two most commonly used materials with FDM printers are poly(lactic acid) and acrylonitrile butadiene styrene. Generally, FDM printing with poly(lactic acid) is a safer choice than acrylonitrile butadiene styrene, because the former requires lower print temperatures and results in lower levels of emission compared to the latter.[80] It is recommended that exhaust ventilation should be mandated with the use of

FDM printers and the number of FDM printers and their hours of operation in an enclosed space should be limited.[80]

3D printing provides interactive and hands-on learning experiences. Improvements in the cost, performance, and safety of 3D printers would probably make them an integral part of STEM education. In the words of Herman *et al.*;[11] "*The stories that we would like to share with our students have become complex, compelling, and three-dimensional. The tools we use to communicate these stories to students should possess the same qualities.*"

## References

1. Y. J. Dori and M. Barak, *J. Educ. Technol. Soc.*, 2001, **4**, 61–74.
2. F.-c. He, L.-b. Liu and X.-y. Li, *J. Chem. Educ.*, 1990, **67**, 556.
3. F.-c. He, L.-b. Liu and X.-y. Li, *J. Chem. Educ.*, 1990, **67**, 650.
4. T. H. Bindel, *J. Chem. Educ.*, 2002, **79**, 468.
5. T. R. Gibb Jr and H. Bassow, *J. Chem. Educ.*, 1957, **34**, 99.
6. J. P. Birk and J. Foster, *J. Chem. Educ.*, 1989, **66**, 1015.
7. A. Mann, *J. Chem. Educ.*, 1973, **50**, 652.
8. C. Chuang, B.-Y. Jin, C.-C. Tsoo, N. Y.-W. Tang, P. S. M. Cheung and L. A. Cuccia, *J. Chem. Educ.*, 2012, **89**, 414–416.
9. A. Scattergood, *J. Chem. Educ.*, 1937, **14**, 140.
10. D. Siodłak, *J. Chem. Educ.*, 2013, **90**, 1247–1249.
11. T. Herman, J. Morris, S. Colton, A. Batiza, M. Patrick, M. Franzen and D. S. Goodsell, *Biochem. Mol. Biol. Educ.*, 2006, **34**, 247–254.
12. T.-H. Chen, S. Lee, A. H. Flood and O. Š. Miljanić, *CrystEngComm*, 2014, **16**, 5488–5493.
13. P. J. Kitson, A. Macdonell, S. Tsuda, H. Zang, D.-L. Long and L. Cronin, *Cryst. Growth Des.*, 2014, **14**, 2720–2724.
14. V. F. Scalfani and T. P. Vaid, *J. Chem. Educ.*, 2014, **91**, 1174–1180.
15. P. P. Rodenbough, W. B. Vanti and S.-W. Chan, *J. Chem. Educ.*, 2015, **92**, 1960–1962.
16. S. Rossi, M. Benaglia, D. Brenna, R. Porta and M. Orlandi, *J. Chem. Educ.*, 2015, **92**, 1398–1401.
17. K. Van Wieren, H. N. Tailor, V. F. Scalfani and N. Merbouh, *J. Chem. Educ.*, 2017, **94**, 964–969.
18. P. J. Paukstelis, *J. Chem. Educ.*, 2018, **95**, 169–172.
19. W. Kaminsky, T. Snyder, J. Stone-Sundberg and P. Moeck, *Powder Diffr.*, 2014, **29**, S42–S47.
20. W. Kaminsky, *J. Appl. Crystallogr.*, 2005, **38**, 566–567.
21. W. Kaminsky, *J. Appl. Crystallogr.*, 2007, **40**, 382–385.
22. W. Kaminsky, T. Snyder, J. Stone-Sundberg and P. Moeck, *Z. Kristallogr. - Cryst. Mater.*, 2015, **230**, 651–656.
23. O. A. Jones and M. J. Spencer, *J. Chem. Educ.*, 2017, **95**, 88–96.
24. V. F. Scalfani, A. J. Williams, V. Tkachenko, K. Karapetyan, A. Pshenichnov, R. M. Hanson, J. M. Liddie and J. E. Bara, *J. Cheminf.*, 2016, **8**, 66.

25. B. Mohamed-Salah and D. Alain, *Chem. Educ. Res. Pract.*, 2016, **17**, 862–877.
26. E. B. Flint, *J. Chem. Educ.*, 2011, **88**, 907–909.
27. M. R. Penny, Z. J. Cao, B. Patel, B. Sil dos Santos, C. R. Asquith, B. R. Szulc, Z. X. Rao, Z. Muwaffak, J. P. Malkinson and S. T. Hilton, *J. Chem. Educ.*, 2017, **94**, 1265–1271.
28. K. Smiar and J. Mendez, *J. Chem. Educ.*, 2016, **93**, 1591–1594.
29. F. A. Carroll and D. N. Blauch, *J. Chem. Educ.*, 2017, **94**, 886–891.
30. R. A. Vipul Gupta, N. P. Macdonald, P. N. Nesterenko and B. Paull, unpublished work.
31. N. L. Dean, C. Ewan and J. S. McIndoe, *J. Chem. Educ.*, 2016, **93**, 1660–1662.
32. C. S. Higman, H. Situ, P. Blacklin and J. E. Hein, *J. Chem. Educ.*, 2017, **94**, 1367–1371.
33. J. Stone-Sundberg, W. Kaminsky, T. Snyder and P. Moeck, *Cryst. Res. Technol.*, 2015, **50**, 432–441.
34. P. Moeck, J. Stone-Sundberg, T. J. Snyder and W. Kaminsky, *J. Mater. Educ.*, 2014, **36**, 77–96.
35. L. Casas and E. Estop, *J. Chem. Educ.*, 2015, **92**, 1338–1343.
36. V. F. Scalfani, C. H. Turner, P. A. Rupar, A. H. Jenkins and J. E. Bara, *J. Chem. Educ.*, 2015, **92**, 1866–1870.
37. G. Cakmakci, *J. Chem. Educ.*, 2010, **87**, 449–455.
38. S. Bauer and C. Wilcox Jr, *J. Chem. Educ.*, 1995, **72**, 13–16.
39. J. L. Dye, *J. Chem. Educ.*, 1957, **34**, 215–217.
40. J. E. Hulse, R. A. Jackson and J. S. Wright, *J. Chem. Educ.*, 1974, **51**, 78.
41. S. Moss and C. Coady, *J. Chem. Educ.*, 1983, **60**, 455.
42. K. H. Tonge, *J. Chem. Educ.*, 1988, **65**, 65.
43. P. Lolur and R. Dawes, *J. Chem. Educ.*, 2014, **91**, 1181–1184.
44. D. S. Kaliakin, R. R. Zaari and S. A. Varganov, *J. Chem. Educ.*, 2015, **92**, 2106–2112.
45. M. Ayouz and D. Babikov, *J. Chem. Phys.*, 2013, **138**, 164311.
46. D. Babikov, P. Zhang and K. Morokuma, *J. Chem. Phys.*, 2004, **121**, 6743–6749.
47. A. Teplukhin and D. Babikov, *J. Chem. Educ.*, 2014, **92**, 305–309.
48. G. Martins, *J. Chem. Educ.*, 1964, **41**, 658.
49. B. E. Hoogenboom, *J. Chem. Educ.*, 1962, **39**, 40.
50. K. M. Griffith, R. de Cataldo and K. H. Fogarty, *J. Chem. Educ.*, 2016, **93**, 1586–1590.
51. R. B. Woodward and R. Hoffmann, *Angew. Chem.*, 1969, **8**, 781–853.
52. M. J. Robertson and W. L. Jorgensen, *J. Chem. Educ.*, 2015, **92**, 2113–2116.
53. G. Anderson, *US Pat.*, 2006/0099877 A1, 2006.
54. P. H. Buist and A. A. Raffler, *US Pat.*, 5030103, 1991.
55. L. W. Fullerton, M. Roberts and J. L. Richards, *US Pat.*, 7750781 B2, 2010.
56. P. Chakraborty and R. N. Zuckermann, *PNAS*, 2013, **110**, 13368–13373.
57. T. Herman, J. Morris, S. Colton, A. Batiza, M. Patrick, M. Franzen and D. S. Goodsell, *Biochem. Mol. Biol. Educ.*, 2006, **34**, 247–254.
58. S. C. Meyer, *J. Chem. Educ.*, 2015, **92**, 2120–2125.
59. A. J. Olson, Y. H. Hu and E. Keinan, *PNAS*, 2007, **104**, 20731–20736.

60. E. K. Grasse, M. H. Torcasio and A. W. Smith, *J. Chem. Educ.*, 2015, **93**, 146–151.
61. L. A. Porter Jr, B. M. Washer, M. H. Hakim and R. F. Dallinger, *J. Chem. Educ.*, 2016, **93**, 1305–1309.
62. L. A. Porter Jr, C. A. Chapman and J. A. Alaniz, *J. Chem. Educ.*, 2017, **94**, 105–111.
63. P. A. Piunno, *J. Chem. Educ.*, 2017, **94**, 615–620.
64. D. Kosenkov, J. Shaw, J. Zuczek and Y. Kholod, *J. Chem. Educ.*, 2016, **93**, 1299–1304.
65. M. W. Pitcher, S. M. Emin and M. Valant, *J. Chem. Educ.*, 2012, **89**, 1439–1441.
66. X. Chen, S. M. Halasz, E. C. Giles, J. V. Mankus, J. C. Johnson and C. Burda, *J. Chem. Educ.*, 2006, **83**, 265.
67. M. Tatarko, J. Tricker, K. Andrzejewski, J. A. Bumpus and H. Rhoads, *J. Chem. Educ.*, 1999, **76**, 1680.
68. B. I. Stefanov, D. Lebrun, A. Mattsson, C. G. Granqvist and L. Österlund, *J. Chem. Educ.*, 2014, **92**, 678–682.
69. Y. Lu, L. M. Santino, S. Acharya, H. Anandarajah and J. M. D'Arcy, *J. Chem. Educ.*, 2017, **94**, 950–955.
70. D. M. Freifelder, *Physical Biochemistry: Application to Biochemistry and Molecular Biology (Life Sciences/Biochemistry)*, WH Freeman, New York, 1982.
71. C. Stewart and J. Giannini, *J. Chem. Educ.*, 2016, **93**, 1310–1315.
72. Z. Lichaa and C. Wapner, *Progress in the Making: Librarians' Practical 3D Printing Questions Answered*, American Library Association, 2016.
73. Y. Kim, C. Yoon, S. Ham, J. Park, S. Kim, O. Kwon and P.-J. Tsai, *Environ. Sci. Technol.*, 2015, **49**, 12044–12053.
74. P. Azimi, D. Zhao, C. Pouzet, N. E. Crain and B. Stephens, *Environ. Sci. Technol.*, 2016, **50**, 1260–1268.
75. B. Stephens, P. Azimi, Z. El Orch and T. Ramos, *Atmos. Environ.*, 2013, **79**, 334–339.
76. Y. Zhou, X. Kong, A. Chen and S. Cao, *Procedia Eng.*, 2015, **121**, 506–512.
77. Y. Deng, S.-J. Cao, A. Chen and Y. Guo, *Build. Environ.*, 2016, **104**, 311–319.
78. L. S. Hughes, G. R. Cass, J. Gone, M. Ames and I. Olmez, *Environ. Sci. Technol.*, 1998, **32**, 1153–1161.
79. P. Steinle, *J. Occup. Environ. Hyg.*, 2016, **13**, 121–132.
80. N. Bharti and S. Singh, *J. Chem. Educ.*, 2017, **94**, 879–885.

# *Subject Index*